STUDY OF AGRICULTURAL SYSTEMS

STUDY OF AGRICULTURAL SYSTEMS

Edited by

G. E. DALTON

Department of Agriculture and Horticulture, University of Reading, England

APPLIED SCIENCE PUBLISHERS LTD
LONDON

APPLIED SCIENCE PUBLISHERS LTD
RIPPLE ROAD, BARKING, ESSEX, ENGLAND

ISBN: 0 85334 640 2

WITH 49 TABLES AND 72 ILLUSTRATIONS

© APPLIED SCIENCE PUBLISHERS LTD 1975

All rights reserved. No part of this publication may be reproduced, stored in a retrieval system, or transmitted in any form or by any means, electronic, mechanical, photocopying, recording, or otherwise, without the prior written permission of the publishers, Applied Science Publishers Ltd, Ripple Road, Barking, Essex, England

Printed in Great Britain by Galliard (Printers) Ltd Great Yarmouth

Foreword

C. R. W. SPEDDING

Grassland Research Institute, Hurley, Berkshire, and Department of Agriculture and Horticulture, University of Reading, England

Agriculture is practised in the form of production systems, enterprises or farming systems, and the purposes of agricultural practice generally include economic considerations. Agricultural systems thus have to be looked at from an economic point of view, although they cannot be completely understood at this level alone. Since successful manipulation of such systems has to be based on an understanding of constituent processes, it is also necessary to take full account of their underlying biology and the impact that management can have upon them.

The understanding of whole agricultural systems thus requires a synthesis of several biological disciplines, management and economics. The risks of oversimplification and superficial treatment are obvious. The most common solution, of studying the constituent parts separately, leaves the essential synthesis to be undertaken by those engaged at the level of enterprise studies. Because the components interact in a complicated manner, it seems likely that teams representing all the relevant disciplines are required to integrate the results of both research and practice into useful models. Experience in fields other than agriculture suggests that a 'systems approach' could provide the necessary framework on which such teams would build.

The main objectives of the symposium were to explore the possible ways of studying whole agricultural systems, with special reference to systems methodology but not excluding any other approaches. It was also intended to define the role of bio-economic models in both agricultural research and practice and to chart the areas in which biologists and economists can most usefully collaborate.

Editor's Preface

This book reports the proceedings of an international symposium held in the Department of Agriculture and Horticulture, University of Reading. The papers are published in the order in which they were presented at the symposium under the various session titles.

The symposium organising committee, which consisted of Professor C. R. W. Spedding, Mr. J. A. C. Gibb, o.b.e., Dr. J. Pearce and myself, are most grateful for the willing co-operation and enthusiasm of all those who made the symposium such a success. Specifically our thanks are due to the contributors and their organisations, the discussion leaders, the participants, Professor E. H. Roberts (Head of the Department of Agriculture and Horticulture), Applied Science Publishers Ltd. and to the Royal Society for its financial support.

My job as both the secretary of the symposium organising committee and proceedings editor was made much easier by the willing assistance (much of it unpaid) of all concerned in the Department of Agriculture and Horticulture. I am especially thankful for the efforts of the following people: Mary Bather, Valerie Craig, Brenda Howson, John Jones, Bill Mayon-White, Tony Molyneux, Bill Rawlins, Sheila Smith, Jean Walsingham and Beth Winwood.

G. E. DALTON
Reading

List of Contributors

Z. ABRAMSKY
 Natural Resource Ecology Laboratory, Colorado State University, Fort Collins, Colorado 80523, USA.

G. R. ALLEN
 Department of Agriculture, School of Agriculture Building, Aberdeen, Scotland.

G. W. ARNOLD
 CSIRO, Division of Land Resources Management, Private Bag, PO Wembley, WA 6014, Australia.

D. BENNETT
 CSIRO, Division of Land Resources Management, Private Bag, PO Wembley, WA 6014, Australia.

N. R. BROCKINGTON
 Grassland Research Institute, Hurley, Maidenhead, Berks. SL6 5LR, England.

P. J. CHARLTON
 Department of Agriculture and Horticulture, University of Reading, Reading, Berks. RG6 2AT, England.

J. CLARK
 Economics Division, West of Scotland Agricultural College, Auchincruive, Ayr, Scotland.

List of Contributors

G. R. CONWAY
: *Environmental Resource Management Research Unit, Department of Zoology and Applied Entomology, Imperial College of Science and Technology, Silwood Park, Ascot, Berks., England.*

R. N. CURNOW
: *Department of Applied Statistics, University of Reading, Reading, Berks. RG6 2AT, England.*

J. B. DENT
: *Department of Farm Management, Lincoln College, University of Canterbury, Canterbury, New Zealand.*

G. F. DONALDSON
: *Agriculture and Rural Development Department, International Bank for Reconstruction and Development, 1818 H Street NW, Washington, DC 20433, USA.*

A. N. DUCKHAM, C.B.E.
: *Department of Agriculture and Horticulture, University of Reading, Reading, Berks. RG6 2AT, England.*

J. EADIE
: *Hill Farming Research Organisation, Bush Estate, Penicuik, Midlothian, Scotland.*

A. C. EGBERT
: *Agriculture and Rural Development Department, International Bank for Reconstruction and Development, 1818 H Street NW, Washington, DC 20433, USA.*

F. ESTÁCIO
: *Fundação Calouste Gulbenkian, Centro de Estudos de Economia Agraria, Oeiras, Lisbon, Portugal.*

G. S. INNIS
: *Faculty of Wildlife Science, Utah State University, Logan, Utah 84321, USA.*

J. N. R. JEFFERS
: *Institute of Terrestrial Ecology, Merlewood Research Station, Grange-over-Sands, Cumbria, England.*

List of Contributors

A. B. S. KING
  Centre for Overseas Pest Research, College House, Wrights Lane, London W8, England.

J. P. MCINERNEY
  Department of Agricultural Economics, University of Manchester, Manchester M13 9PL, England.

T. J. MAXWELL
  Hill Farming Research Organisation, Bush Estate, Penicuik, Midlothian, Scotland.

G. A. NORTON
  Environmental Resource Management Research Unit, Department of Zoology and Applied Entomology, Imperial College of Science and Technology, Silwood Park, Ascot, Berks., England.

J. PHILLIPSON
  Animal Ecology Research Group, Department of Zoology, University of Oxford, Parks Road, Oxford OX1 3PS, England.

N. J. SMALL
  Environmental Resource Management Research Unit, Department of Zoology and Applied Entomology, Imperial College of Science and Technology, Silwood Park, Ascot, Berks., England.

C. R. W. SPEDDING
  Department of Agriculture and Horticulture, University of Reading, Reading, Berks. RG6 2AT, England, and Grassland Research Institute, Hurley, Maidenhead, Berks. SL6 5LR, England.

P. R. STREET
  Department of Agriculture and Horticulture, University of Nottingham, Sutton Bonington, Loughborough, Leics., England.

S. C. THOMPSON
  CANFARM Data Systems, Canada Department of Agriculture, PO Box 1024, Guelph, Ontario, Canada.

K. J. THOMSON
: Department of Agricultural Economics, University of Newcastle upon Tyne, Newcastle NE1 7RU, England.

M. UPTON
: Department of Agricultural Economics and Management, University of Reading, Reading RG6 2AT, Berks., England.

G. M. VAN DYNE
: Natural Resource Ecology Laboratory, Colorado State University, Fort Collins, Colorado 80523, USA.

Contents

Foreword v
C. R. W. SPEDDING

Editor's Preface vii

List of Contributors ix

Introductory Paper

1 The Study of Agricultural Systems 3
C. R. W. SPEDDING

Part 1: Systems Methodology

2 Agricultural Systems Models and Modelling: An Overview . 23
G. M. VAN DYNE and Z. ABRAMSKY

3 The Application of Systems Theory in Agriculture . . 107
J. B. DENT

4 The Problem of Finding an Optimum Solution . . . 129
G. W. ARNOLD and D. BENNETT

5 Constraints and Limitations of Data Sources for Systems
  Models 175
J. N. R. JEFFERS

Discussion Report   R. N. CURNOW and M. UPTON . . 187

Part 2: Application of a Systems Approach in Practice

6 A Systems Approach to the Control of the Sugar Cane Froghopper 193
G. R. CONWAY, G. A. NORTON, N. J. SMALL and A. B. S. KING

Discussion Report   J. PHILLIPSON 231

7 The Practical Application of Bio-economic Models . . 235
P. J. CHARLTON and P. R. STREET

8 The Study of Agricultural Systems: Application to Farm Operations 267
G. F. DONALDSON

9 Economic Analysis of Farms 307
S. C. THOMPSON

10 Regional Agricultural Planning 317
A. C. EGBERT and F. ESTÁCIO

Discussion Report   J. CLARK and J. P. MCINERNEY   361

Part 3: Applications of a Systems Approach to Research

11 The Use of a Systems Approach in Biological Research . 369
G. S. INNIS

Discussion Report   N. R. BROCKINGTON   393

12 Systems Research in Hill Sheep Farming 395
J. EADIE and T. J. MAXWELL

13 Systems Analysis in Relation to Agricultural Policy and Marketing 415
G. R. ALLEN

Discussion Report   A. N. DUCKHAM and K. J. THOMSON . 431

Index 435

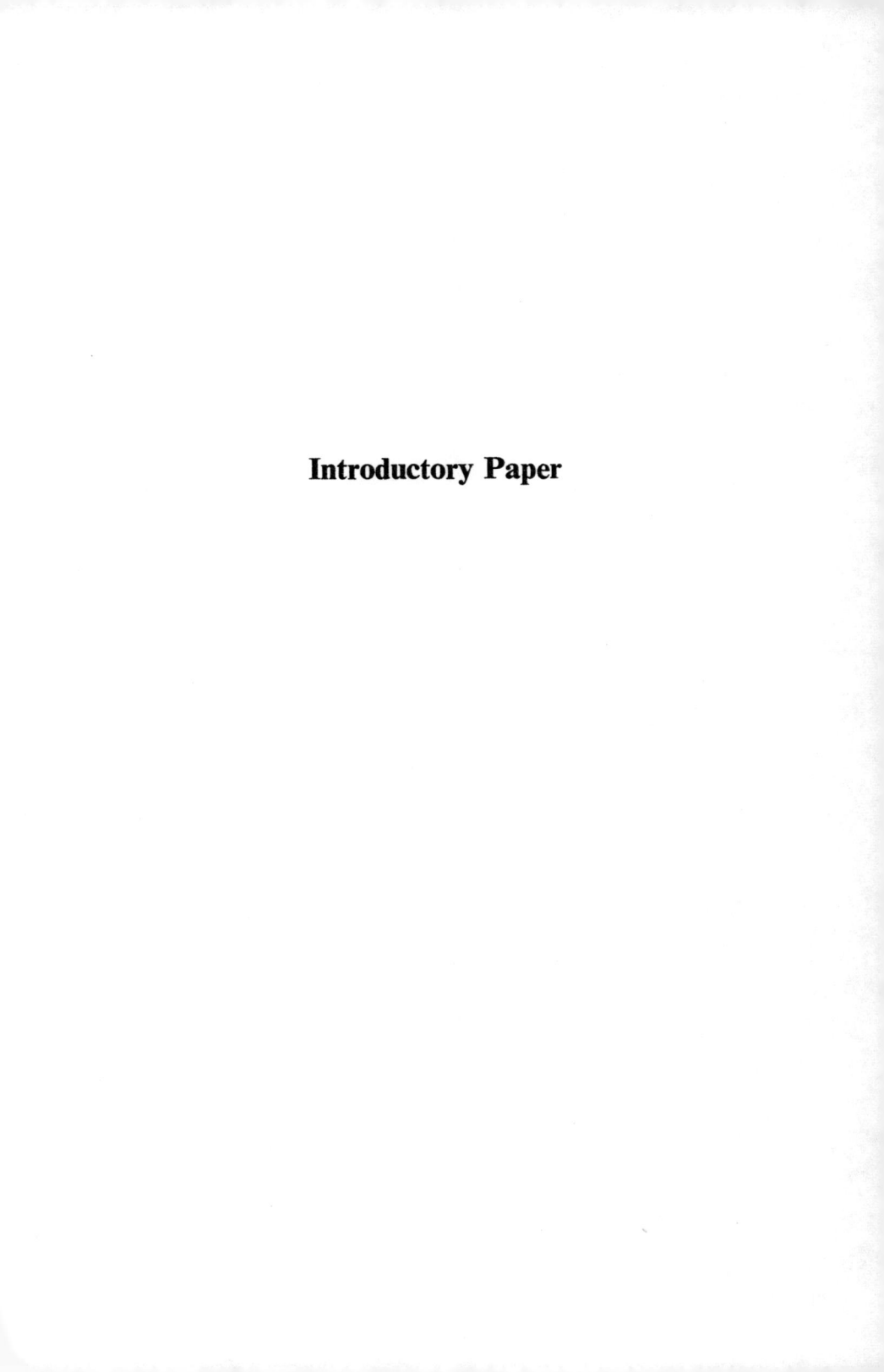

Introductory Paper

1

The Study of Agricultural Systems

C. R. W. SPEDDING

Grassland Research Institute, Hurley, Berkshire, and Department of Agriculture and Horticulture, University of Reading, England

1.1. INTRODUCTION

A symposium on this subject presupposes that such systems should be studied and that the latter is not a straightforward matter.

Since agricultural production is based on the operation of agricultural systems, the importance of understanding them may seem obvious.

It is perhaps useful to question this assumption at the outset, on the grounds that much of all our lives is based on the sufficiently effective use of systems (from cars to television sets) that we would not claim to understand.

This should lead us to probe further what we mean by 'understand', in this context, since clearly we understand these systems sufficiently for our purposes.

The important point, therefore, is to distinguish the various levels of understanding that are required for our various purposes.

The latter may be listed as follows:

(i) to operate systems;
(ii) to repair them;
(iii) to improve them;
(iv) to construct new ones.

It is worth noting that, for most kinds of systems, different people with different skills, knowledge and understanding, are normally involved for these four purposes. For mechanical systems (such as a bicycle), it is quite possible to visualise people who fulfil one or more of these purposes but who are quite incapable of fulfilling others. If agricultural systems are to be studied, therefore, the method of study should be related to the level of understanding to be achieved and this should be related to the purpose for which such understanding is to be harnessed.

What, then, are the problems and difficulties of studying agricultural systems? The importance of observation should not be underestimated, in gaining familiarity with the content and operation of agricultural systems and in gaining knowledge that cannot easily be acquired in other ways.

The problems of observation are the lack of measurement, lack of records and lack of definition of what is measured or recorded. The last is particularly associated with management: grazing management is often not describable because no consistent rules are applied, the same term is often used to describe different methods and the effects of the same methods may vary with the year and environment in which they are applied.

Controlled experimentation, on the other hand, is often impracticable, especially in terms of attempts to study the consequences of varying one factor at a time (Morley & Spedding[1]; Spedding & Brockington[2]). This is partly because agricultural systems are physically too large and too costly and partly because they are too complex.

Complexity, however, is often a property of the observer rather than the system (Spedding[3]): any simple object is complex if viewed in molecular terms, for example.

But agricultural systems are complex even at one level, without recourse to microscopes. First, they are multidisciplinary in content: this means that there are not only many different components but that many of these components differ from each other quite fundamentally. Secondly, the number of theoretical combinations and permutations of values of these components is enormous.

If we consider our attitudes to other complex systems, such as individual animals, it is clear that the number of permutations may be a major factor.

An individual may be regarded as just as complex as an agricultural system, when each is viewed at an appropriate level of detail, and yet the individual may frequently be studied by direct experimentation. If it is objected that the scale of experimentation is not necessarily less for individual animals, because we are not generally interested in an individual except as a sample of a class and many individuals are therefore required, it should also be noted that the same argument applies to systems of production.

However, direct experimentation on individual animals is undertaken on the animals as they exist. Theoretically, there are many other ways of combining the constituent parts of an animal but, in the course of evolution, these have been found to be largely non-viable.

In the case of agricultural systems, it is still conceivable that many very different combinations of their components would be possible and might even be better: certainly they appear to be worth exploring but the cost and difficulty of doing so are considerable.

One solution, therefore, is to limit direct experimentation to those systems that have proved themselves viable in some relevant environmental or economic circumstances. This approach, of course, is only relevant to the modification of existing systems, for repairs or improvement.

Even for existing systems, however, useful study depends upon the capacity to identify particular systems.

1.2. THE IDENTIFICATION OF AGRICULTURAL SYSTEMS

It is essential to be able to generalise to some extent from the results of an experiment.

Experimentation on a particular cow or group of cows, for example, is only of value in so far as an important 'class' of cows is represented. If the results of research only applied to the particular animals studied, it would be of very limited value. But it is obvious that such results do not apply to all animals or even, in the case cited, to all cows: it is necessary to specify such attributes as age, size, breed, physiological state, plane of nutrition, disease status and management. In the case of cows, no one would deny that there are very many different combinations of these attributes and that the results of experiments on one combination cannot necessarily be applied to another. That is one reason why continuing programmes of research, even on one breed, are required over long periods of time.

When an experiment *is* done, therefore, it is essential to identify the animals used as belonging to an identifiable class (or group of attributes) to which the results will apply. It is usually necessary, in addition, to demonstrate that this class is of sufficient importance (economically, or by virtue of its numbers or genetic influence) to justify the cost of experimentation.

Exactly the same argument applies to the study of agricultural systems: it is essential to be able to identify each one as a member of a class and to have a means of assessing the relative importance of the class.

Identification is always important, even in the narrowest sense that it is possible to recognise another example of the same kind. But this sort of identification is not quite enough. For instance, an ornithologist can identify a woodpecker by its characteristic flight but, in fact, this information is very slight; it just so happens that it is diagnostic in a given set of circumstances. In general, even when trivial information is sufficient for identification, it is extracted as required from much greater and more detailed knowledge.

It is also necessary to recognise *all* the features that, in combination, are diagnostic of the class. So the need for identification in the study of agricultural systems leads to the need for an adequate classification.

1.3 THE CLASSIFICATION OF AGRICULTURAL SYSTEMS

Clearly only identifiable objects can be classified and, for practical purposes, it is also necessary to name them.

Even so, there are a great many different ways of classifying anything and the 'rightness' or 'appropriateness' of any one way must depend entirely on the purpose for which classification is undertaken.

It is easily argued that progress in the study of agricultural systems depends upon classification but it is less easy to be sure about the precise purposes of doing it, perhaps because there is always a temptation to make one scheme serve many purposes. The argument that it is unhelpful to have

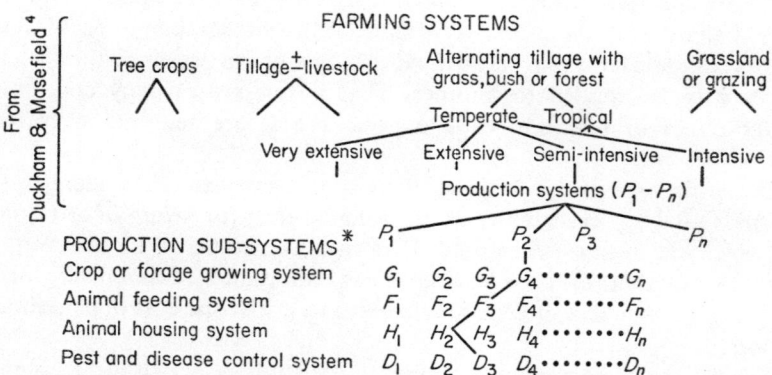

FIG. 1. Classification of agricultural systems (greatly abbreviated): from Spedding[21].
*These sub-systems are here considered as describable in terms of major components, different combinations of crop, cultivation method, fertiliser input, sowing date, etc., being represented by $G_1, \ldots, G_n$, for example.

COLLECTING — e.g. Wild oil palms in West Africa
Gum arabic in the Sudan

CULTIVATION
- *Shifting cultivation* (also classified by vegetation and by method of clearance)
  - Migration
  - Rotation
    - 45-year cycle
    - 30-year cycle
    - Semi-permanent farming
- *Semi-permanent cultivation*
  - on the fertile soils of the humid savannas
  - of perennial crops
  - with irrigation
  - unregulated leys in the drier savannas
  - unregulated leys at high altitude
- *Regulated ley farming*
  - Smallholders
  - Settlement schemes
  - Large farms
- *Permanent cultivation on rain-fed land*
  - Tropical highlands
  - Monsoon Asia
  - Hot humid tropics
  - African savannas
- *Arable irrigation farming*
  - Flooding
  - Basin irrigation
  - Furrow irrigation
  - Sprinkler irrigation
  - Underground infiltration
- *Perennial crops*
  - Cropping
    - Field crops
    - Shrubs
    - Trees
  - Exploitation
    - Estates
    - Smallholdings

GRAZING
- *Total nomadism*
- *Semi-nomadism*
- *Ranching*

FIG. 2. Classification of farming systems in the tropics (after Ruthenberg[5]).

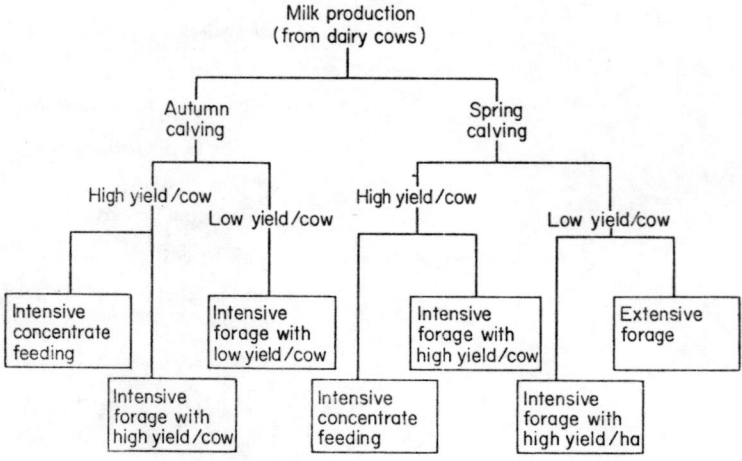

FIG. 3. A classification of milk production systems.

different classification schemes operating simultaneously is superficially reasonable, but it is rather like, for example, suggesting that there should be only one kind of cutting tool.

There is no simple answer to this problem: it requires very serious consideration and it would be useful to look at such schemes as have been proposed and to examine their purposes.

On a world basis, the work of Duckham & Masefield[4] and of Ruthenberg[5] laid foundations upon which others could build (see Figs. 1 and 2).

At a much more detailed level the UK Meat and Livestock Commission have classified production systems, for their purposes, and these can be schematically arranged. Figures 3 and 4 illustrate how such systems can be classified.

Related to the problem of classification is that of description.

1.4. THE DESCRIPTION OF AGRICULTURAL SYSTEMS

Descriptions may be more or less detailed. For identification purposes they may be very brief and often simple, especially if the object is a familiar one and has only to be distinguished from a small number of alternatives.

For classification, the minimum description consists of information about the features that characterise classes and about which combinations of such features are diagnostic.

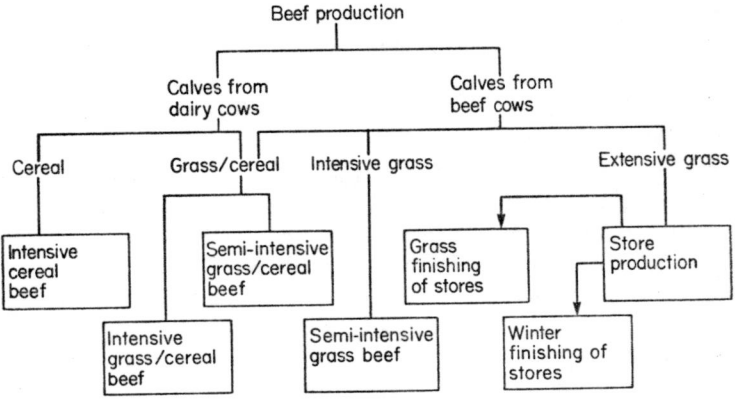

FIG. 4. A classification of beef production systems.

But for the purposes of copying, much more is required. It is only necessary to contemplate the difference between identifying and naming a person and providing enough information to paint a likeness.

Thus, if research results are thought to apply to all members of one class* of system, it is only necessary to be sure that the system to be modified belongs to the same class. If a new system is devised, however, or an existing system identified as highly successful, anyone wishing to set one up himself needs vastly more information, covering all the decisions to be made in doing so.

Descriptions of this kind involve foreseeing these decisions and assessing their importance.

In a system of lowland sheep production (see Spedding[6]), for example, some information is obviously required about the ram. But it is by no means obvious where to draw the line. The breed, size, age, weight and longevity are clearly important and it may even be hoped that 'breed' might summarise several of these. If breed is stated, however, it implies that it has to be this one, that others will not do; and yet it may often be the case that one breed is known to be suitable, that several others can be substituted without significant effect and that some other breeds are known to be unsuitable.

Similarly, the dates of mating and lambing must matter but they probably do not matter to a few days or even weeks.

* This assumption will only be justified if the experiment was actually concerned with diagnostic features.

Much more detailed considerations of ram management, such as the ratio of ewes to rams and how many share a paddock, may be of great importance but may be judged to be matters within the competence of any probable operator. Thus, the person for whom the description is prepared is also a factor in determining its content. Indeed, the data required for copying purposes depend greatly on what assumptions can be made about the skill, background knowledge and even common sense of the copier.

There are, nonetheless, a great many pieces of information about which decisions have to be made (the ram, after all, is only one component of a sheep system), as to their relevance, essentiality and most useful form of expression: often the last will be as a range within which something should lie.

There is one important consequence of the fact that descriptions need to be detailed. If such information is listed or written about, it does not necessarily add up to a picture of the system, in the reader's mind. Indeed, a long list of unrelated items simply means that the earlier items are forgotten before the later ones are considered.

Some framework is needed in order to relate items to each other and to the purpose of the system and there are many advantages if system descriptions can be diagrammatically represented (see Fig. 5).

Such pictures of systems are bound to be rather static, oversimplified versions of what occurs and/or what a system consists of: they are, in fact, crude models and it is worth considering to what extent more elaborate models can contribute to the understanding of agricultural systems.

1.5. THE USE OF MODELS

In their simplest forms, models are bound to be used: it would hardly be possible to communicate about agricultural systems or their components if different minds did not contain sufficiently similar pictures of them for the immediate purposes of discussion.

The questions to be answered, therefore, are what kinds of model are most useful for what kinds of purpose and to what extent can they assist in the purposes for which agricultural systems are studied.

Because a mathematical model can contain a vast amount of information (so does a long list), it does not follow that it provides a picture of a system: yet it is certainly not the case that pictures have to be simple and contain little detail. The value of mathematical models, in eliminating ambiguity and allowing an immense amount of manipulation and calculation, should not obscure the fact that they have to be preceded by a picture-model; they do not generate the basic pictures themselves.

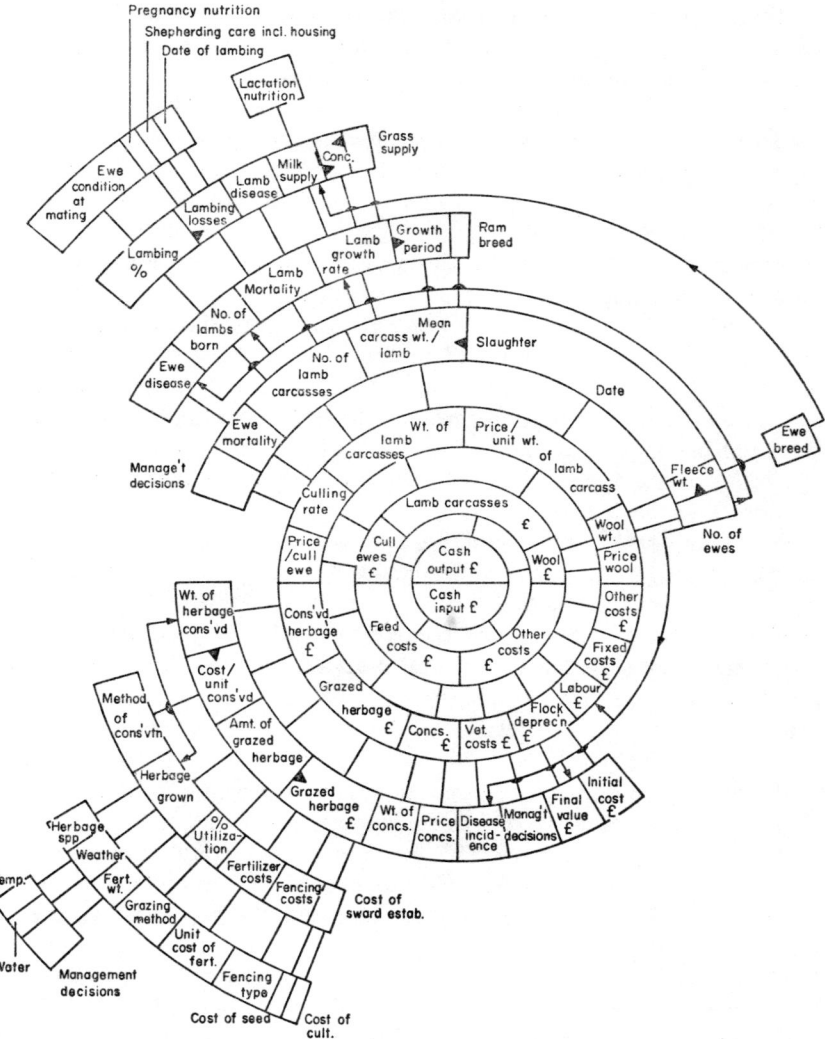

Fig. 5. Components of sheep production. A circular diagram used to describe the components of a sheep production system and the relationships between these components. The system objective (or purpose) is placed centrally and successive circles contain those factors that directly affect components internal to them, indicated by radial lines, all of which are directed inwards, and by circular lines with arrowheads. Only a few of the latter are illustrated in this diagram.

Furthermore, a picture-model has already involved decisions as to content: what must be included and what can be safely omitted are major issues in constructing models, as they are for any other form of description.

So, in effect, models are selective descriptions and cannot contain everything that is present in the system being modelled. When models are used to generate hypotheses, or as statements of hypotheses, the same argument applies. The need to test such hypotheses is to test the validity of the descriptions: not only of content or form, however, but also of function and relationships.

Since criteria for selective description vary with purpose, many different models may be constructed of the same system.

The well-known work of de Wit and his colleagues [7,8] on the physiology and growth of maize, of Holling [9] on hunger in the mantis, and of Watt [10] and Van Dyne [11] on resource management, illustrates the range of successful modelling that has been undertaken. Indeed a vast amount of work of this kind has now been completed and there can be no doubt about the feasibility of the approach in relation to a wide range of biological and resource management problems.

Two main problems remain, in addition to the technical problems concerned with errors in the input data and the approximations generally involved in simulation (Charlton & Street [12]). First, there are a few convincing examples of the use of complex models of agricultural systems being used as a basis for agricultural practice, though this situation may rapidly change (Morley [13]).

Secondly, there is insufficient guidance as to the worthwhileness of building any particular model. As with all other research, the amount that could usefully be done greatly exceeds the total capacity for doing it and priorities have to be arrived at in some systematic fashion.

It is not merely that innumerable workable models could be constructed, with present technical methods, of different subjects, but that a great many different models could be made of the *same* subject, expressed differently for different purposes.

The relative importance of models can only be determined by reference to a larger or more comprehensive model, the purpose of which is defined. Leaving aside the difficulty of ever having enough information to construct the larger models required and the obvious corollary that they would have to be at a different level of detail and thus not require the same information as must be contained in all the sub-models, a major problem remains.

This problem is whether it is *possible* to have a *complete* picture (this being the most comprehensive conceivable).

Consideration of any familiar system, such as a horse, shows that it is *not* possible. Innumerable views of a horse may be imagined but none that show all sides, inside and outside, in full detail, even in a static sense, quite apart from dynamic aspects, functions and physiological states, over even a short period of time.

Intellectually, the problem is met by rapid reviews of albums of pictures, giving a 'rounded' view in space and time.

It may be that comprehension of a whole agricultural system also requires an album of models and it is important to consider what these should be and in what form each model or picture should be presented. It is important to remember that 'comprehension' is probably essential at this stage, so it will be necessary to imagine holding such an album in the mind rather than holding a collection of data in a computer.

This is not to belittle the enormous capacity of the computer to encompass in detail what can only be comprehended in outline, but the danger of constructing an amalgam that cannot be visualised should be recognised.

If priorities between models are to be assessed, so that one part of an agricultural system may be said to be more important than another (for research, in practice, or for profit), the overall view must be bio-economic. Farm business analysis has used a variety of methods for improving procedures for selection of enterprises and combining resources, employing the computer to increase the speed and scope of the calculations.

Linear programming (see Lloyd[14] for a review) is an example of these techniques and its advantages and disadvantages are well documented. Simulation techniques, including Monte Carlo programming (Thompson[15]), have been developed to overcome some of these disadvantages and to extend the range of planning problems that can be solved.

Such techniques can undoubtedly aid decision-making and it has to be remembered that farmers and managers cannot defer their decisions until all the desired data are available.

The questions to be posed are what kind of decisions can be based on what kind of models. If little biological information is included in the model, decisions can only relate to the biological framework used. But some of the major changes that may be required in agricultural systems will relate to the underlying biology and, if this is not included, decisions cannot be taken about such changes. The debate about the nature and meaning of optimisation (see Anderson[16]) illustrates the difficulty. Only bio-economic models allow assessments of the economic results of manipulation of the underlying biological processes.

Clearly, if the whole can never be pictured, it can hardly be studied or subjected to experimentation. It is extremely unlikely that even the system represented by an 'album' can often be experimentally treated and it is essential to be able to extract from it parts that may usefully be studied separately. The appropriate terminology at this point is rather confused at the present time, but I have found it helpful to distinguish between parts, components and sub-systems.

'Parts' may be literally any portion of a system and, since their demarcation may be quite arbitrary, they have no integrity of their own. In this sense, it is possible to be working on or with a 'part' of a system in exactly the same sense that a painter engaged on a portrait may be currently working on one ear.

'Components' of, for example, a system of milk production by goats from scrubland, might be goats, or shrubs, or a relevant beetle. All of these can be legitimately studied as separate systems at a finer level of detail, but they are then independent of the whole from which they originated and the relevance to the whole of the results of such studies cannot be guaranteed. Similarly, 'component systems' can be identified, such as herbage growth or the life-cycle of a relevant insect, and studied separately.

Since all these examples are extracted from the original system without the lines of interaction that connected them all together, in no case can the results of separate studies be expected, necessarily, to lead to improved understanding of the functioning of the whole system. Yet this is what is often required.

'Sub-systems' are needed, therefore, that do have this property, that they possess an integrity of their own, such that when studied in relative isolation the resulting increased knowledge about them can be fitted back into the whole system from which they were derived and contribute to our understanding of the latter.

To do this, the following characteristics appear to be required. First, that a sub-system has the same end-point or main output as the whole system. Secondly, that its focus of interest is a component variable of the whole. Thirdly, it must contain all the important interactions between it and the main output, including those between intermediate components.

The problems of defining such sub-systems are substantial but the result appears extremely desirable, if not essential, to the study of agricultural systems.

First of all, it is necessary to be able to describe the whole system in terms that allow the extraction of sub-systems. One way of doing this (see Spedding[17]) is by circular diagrams showing all the components of a system and their relationships. This is illustrated in Figs. 6 and 7.

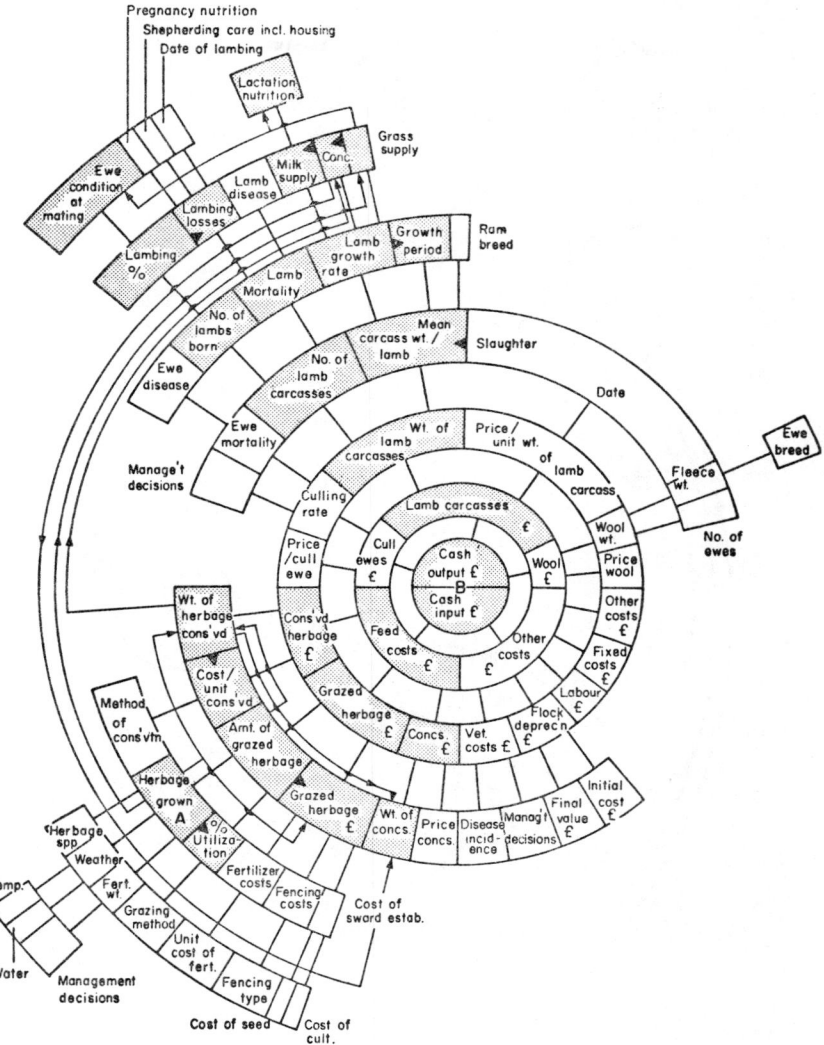

FIG. 6. This version of Fig. 5 shows those components (shaded) that are involved in the effect of the amount of herbage grown (A) on the cash output/cash input relationship (B) that is of central interest. This network of related components is considered to be the sub-system concerned with the effect of A on B.

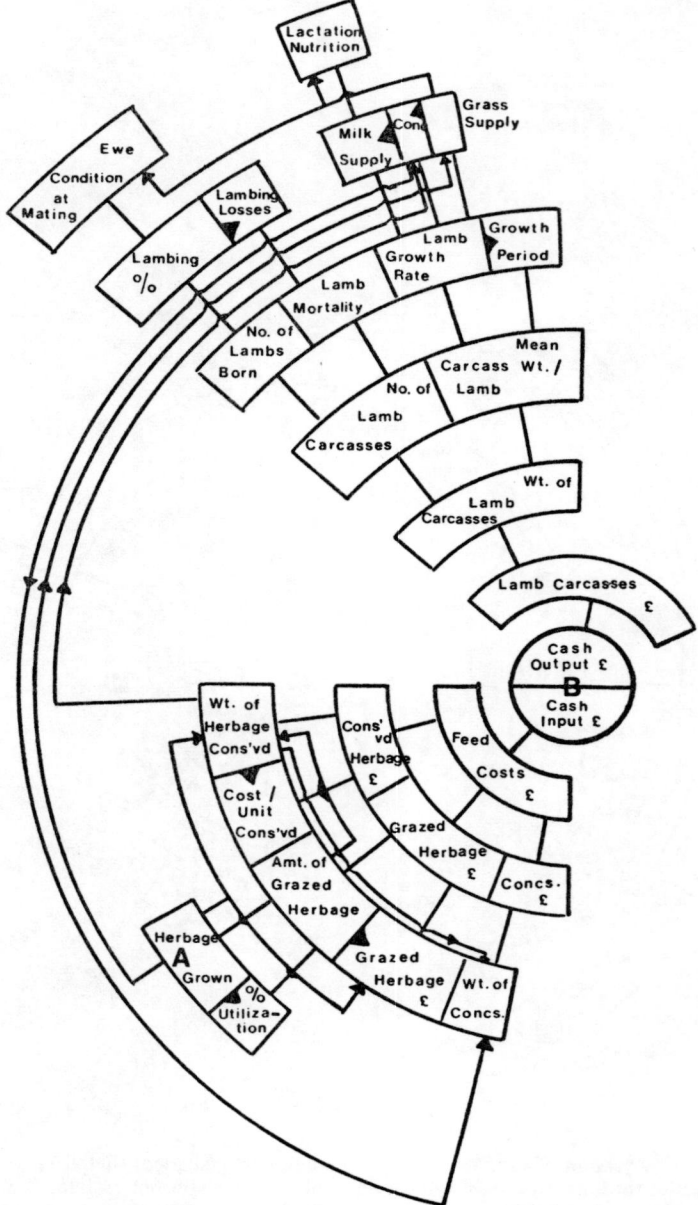

Fig. 7. The A/B sub-system extracted from Fig. 6.

If sub-systems can usefully be defined and extracted in this way, it seems likely that they would represent those parts of a system that may be studied separately, whether by modelling or by physical experimentation, whilst retaining relevance to the whole.

If such sub-systems are still too large or too complex for these purposes, they may be further simplified by adding restraints on the context of the system or by making some provisional assumptions. Apart from modifications of this kind, however, these sub-systems would be the minimum versions of the system worth studying *as integral parts of the system*; they can, of course, be studied in their own right and the results would be expected to apply in a more general fashion or at least to be relevant to a wider range of systems.

1.6. THE CONTENT OF AGRICULTURAL SYSTEMS

In discussing, at the outset, the special difficulties of studying agricultural systems, it was assumed that they were inevitably multidisciplinary. It is easily argued, furthermore, that economics is an indispensable component and it is, therefore, important to explore the possibilities and values of bio-economic models. Whilst the latter certainly appears to be valid, it is perhaps wise to reflect upon the many and varied purposes for which agriculture is undertaken (Bunting[18]; Spedding[19]).

It is quite conceivable that profit-making might not be the main purpose of an agricultural system, although it is difficult to see that economics could be totally ignored. This is so for most systems: if it appears otherwise, it is probably because only a part of the total system is being considered. In research terms, however, the biological feasibility of potential agricultural systems might ignore economics as not being a calculable component at that stage.

The point is that the methods of study must be relevant to the content of agricultural systems and this may vary greatly. This, however, simply adds to the reasons for a symposium of this kind, to discover what the main problems are, what solutions are needed and what methods have been found useful so far and in what circumstances.

1.7. ORGANISATIONAL IMPLICATIONS

The foregoing discussion has implications for the conduct of agricultural research and the organisation of research teams. It is customary to argue in favour of multidisciplinary teams (Watt[20]), and the number of disciplines commonly involved in agricultural systems certainly suggests that such teams will usually be required.

However, the essential point is that a research team should have in it the kind of staff needed by the subject of the research. The latter will vary greatly and it is misleading to generalise about the need for mathematicians or economists or entomologists: until the subject has been specified this need cannot be known, but once it is specified, it will be obvious whether, for example, an economist is needed or not.

The only useful generalisation is that a team intended to tackle a whole range of problems in agricultural systems will need within it (or be able to draw upon) all the disciplines ever likely to be required. This will certainly include a variety of biologists, mathematicians, systems analysts, economists, agriculturalists, ecologists, and sociologists.

It is no use arguing that such combinations cannot operate or that such multidisciplinary subjects should not exist. They exist in practice and there is no reason why a subject cannot be defined with reference to its focal point of interest and the relevant parts of all those disciplines that impinge upon it (Spedding[21]).

REFERENCES

1. Morley, F. H. W. & Spedding, C. R. W. (1968). Agricultural systems and grazing experiments, *Herbage Abstracts*, **38**, 279–87.
2. Spedding, C. R. W. & Brockington, N. R. (1975). Experimentation in agricultural systems, *Agricultural Systems*, **1** (in press).
3. Spedding, C. R. W. (1970). The relative complexity of grassland systems, *Proceedings XIth Grassland Congress*, pp. A126–31.
4. Duckham, A. N. & Masefield, G. B. (1970). *Farming Systems of the World*, Chatto & Windus, London.
5. Ruthenberg, H. (1971). *Farming Systems in the Tropics*, Oxford University Press, London.
6. Spedding, C. R. W. (1972). Systems synthesis in agriculture, *Fourrages*, No. 51, 3–18.
7. Wit, C. T. de & Brouwer, R. (1968). Über ein dynamisches Modell des vegetativen Wachstum von Pflanzenbestanden, *Angewandte Botanik*, **42**, 1–12.
8. Wit, C. T. de, Brouwer, R. & Penning de Vries, F. W. T. (1971). A dynamic model of plant and crop growth, in: *Potential Crop Production* (ed. P. F. Wareing and J. B. Cooper), Heinemann Educational Books, London, pp. 117–42.

9. Holling, C. S. (1966). The strategy of building models of complex ecological systems, in: *Systems Analysis in Ecology* (ed. K. E. F. Watt), Academic Press, New York, pp. 195–214.
10. Watt, K. E. F. (1968). *Ecology and Resource Management*, McGraw-Hill, New York.
11. Van Dyne, G. M., ed. (1969). *The Ecosystem Concept in Natural Resource Management*, Academic Press, New York.
12. Charlton, P. J. & Street, P. R. (1970). Some general problems involved in the modelling of economic systems on a digital computer, in: *The Use of Models in Agricultural and Biological Research* (ed. J. G. W. Jones), Grassland Research Institute, Hurley, pp. 50–62.
13. Morley, F. H. W. (1972). An evaluation of the symposium on systems analysis, *Proceedings of the Australian Society of Animal Production*, **9**, 137–8.
14. Lloyd, D. H. (1970). The development of farm business analysis and planning in Britain: a methodological review and appraisal, Department of Agriculture Study No. 6, Reading University.
15. Thompson, S. C. (1970). A user's manual for Monte Carlo programming, Department of Agriculture Study No. 9, Reading University.
16. Anderson, J. R. (1972). Economic models and agricultural production systems, *Proceedings of the Australian Society of Animal Production*, **9**, 77–83.
17. Spedding, C. R. W. (1973). Modern methods of sheep production in Great Britain, *Proceedings of the 'Greenweek' Symposium*, Berlin pp. 102–12; Moderne Methoden der Schafproduktion Grossbritannien, in: *Zur Agrarpolitischen Diskussion: Internationale Grüne Woche*, Berlin, 1973.
18. Bunting, A. H. (1971). Ecology of agriculture in the world of today and tomorrow, presented to a Symposium at the National Academy of Science, Washington, DC.
19. Spedding, C. R. W. (1973). The future development of agriculture, *Agricultural Progress*, **48**, 48–58.
20. Watt, K. E. F., ed. (1966). *Systems Analysis in Ecology*, Academic Press, New York.
21. Spedding, C. R. W. (1971). Agricultural ecosystems, *Outlook on Agriculture*, **6**, 242–7.

PART 1

Systems Methodology

2

Agricultural Systems Models and Modelling: An Overview

G. M. VAN DYNE and Z. ABRAMSKY

Natural Resource Ecology Laboratory, Colorado State University, Fort Collins, Colorado, USA

2.1. INTRODUCTION

Models are a form of synthesis and word, picture and mathematical models are commonly found in today's literature on agricultural systems. A synthesis is the combination of separate elements or parts into a whole. Models of various types, and particularly mathematical models, form a rather precise structure for the synthesis of information. But all mathematical models are incomplete. No one has yet developed a complete mathematical model for an agricultural system. The problem is one of time, tools and complexity. Yet mathematical models of various forms have been used effectively for many years in agricultural systems research and management.

Extensive development of agricultural systems models has awaited the advent of modern electronic computing machines. This is because of the complexity of the systems and of the nature of models used to simulate and optimise management of these systems. Agricultural systems include biological components. Biological components operate on biological and physical processes. The qualities of biological processes alone are such that one must consider great diversity in time and space, the existence of threshold processes, limiting values and discontinuities when representing the system mathematically. This level of complexity precludes the use of simple linearised models adapted from the classical tools of mathematical physics for all but the simplest situations. Social and economic considerations even further complicate the problem. Thus, although mathematical models are common for agricultural systems, there are few detailed textbooks on the building of these models or on detailed analysis of the utility of these models in the agricultural scene.

We have taken a rather broad view of agricultural systems. Thus some of our examples are drawn from economics as well as from conventional agriculture and related disciplines such as agronomy, animal husbandry, range management, wildlife biology and forestry. In effect, we have *included those renewable natural resource systems and artificial biological resource systems which are directly under the control of man.* Because the systems are under the control of man, we have included forays into the areas of economics, sociology, and regional planning.

Our definition of agricultural systems is somewhat broader than that of Arnold & Bennett[1] who speak of agricultural systems ranging from the intensive, e.g. pig and poultry system, to the extensive end, e.g. rangeland systems of the world. Alternately, Dent[2] describes four levels of agricultural

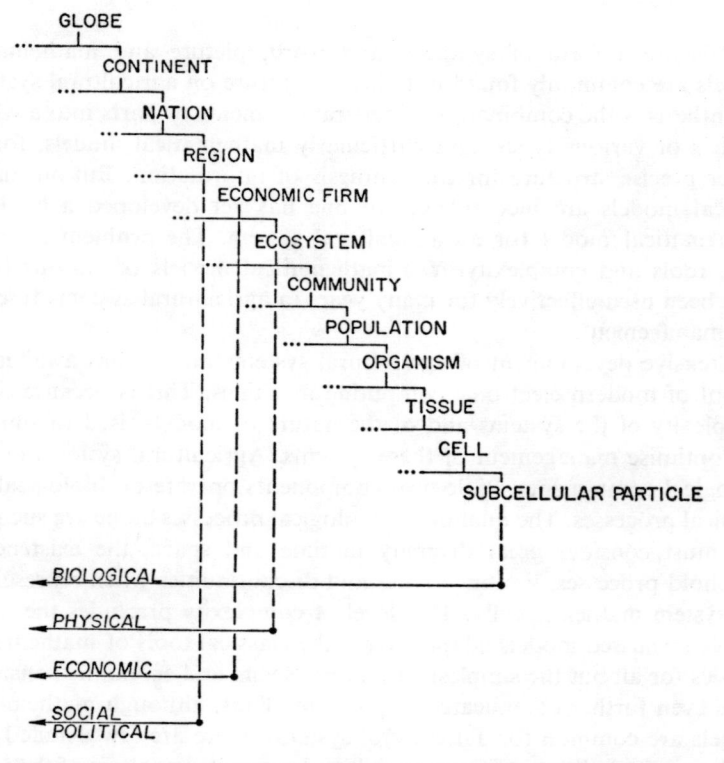

FIG. 1. Hierarchies of systems and levels at which biological, physical, economic, social and political components enter.

systems, with examples, including biochemical and physical, plant and animal, farm business and national and international.

The total range of systems of concern is illustrated in Fig. 1. We have shown a hierarchical arrangement in which we can nest one level of systems within another. A general principle in this diagram of architecture of complexity is that new principles or behavioural characteristics arise at each level of system nesting that cannot be predicted directly from the levels below. For example, when various populations or organisms are assembled to form a community, the dynamics of the community may not be predictable from the dynamics of the populations. It is the interactions of the individual components of sub-systems that help account for this difference.

In the hierarchical order used here, biological components enter at the lowest level and carry throughout all levels. Physical components enter at the ecosystem level, which has both biotic and abiotic components. Economic components enter when one couples ecosystems for human purposes, thus entering the market place. Even at the economic firm level, but certainly at the regional level, social and political considerations enter.

Most of the systems of concern in our paper range from that of the level of the community up to the region. Agricultural system models draw upon, of course, models at the population level and feed to models at the national level.

2.2. OBJECTIVES

Our objectives in developing this paper were:

(i) *To review typical agricultural systems models.* The review is not exhaustive; these models are only examples. We have restricted our review primarily to numerical models and primarily to areas in which we have not encountered review papers (see Section 2.4.1).
(ii) To examine the above papers and to *relate them to general systems concepts and approaches, including some of the art of modelling.*
(iii) *To criticise the design of and the method of reporting and publishing* on agricultural systems models.

We felt it more useful to make a broad review of agricultural systems models and modelling rather than to go into depth into systems analysis theoretical concepts. Thus, we do not treat in this review the important problems of model evaluation (including verification and validation,[3] sensitivity analysis[4,5] and others).

We have excluded from our review those papers related to water resource systems *in toto*. For example, we have not extended our excursion into the area of watershed management or river basin planning. We have, however, given examples of models of irrigation of agricultural lands.

Furthermore, our review suffers from that of most reviews; we over-represent those papers and languages which are readily available to us. We would appreciate learning of important agricultural system model papers we have overlooked.

2.3. GENERAL SYSTEMS CONCEPTS AND APPROACHES

2.3.1. Relation to General Systems Theory

A statement credited to Aristotle is '*the whole is more than the sum of its parts*'. In a sense, general systems theory consists of the scientific exploration of wholes and wholeness.[6] Systems thinking then follows a new paradigm contrasting to the previously predominant, elementalistic approach and conceptions. General systems theories have been elaborated and published by various workers including von Bertalanffy,[7] Wymore,[8] Klir[9] and Mesarovic *et al.*[10] More recent analysis of the trends in general systems theories shows the field is evolving, but the utility in day-to-day operations still needs to be demonstrated.[11] General systems theory is *more a philosophy of approach or paradigm* rather than an operational formula. Numerous definitions and interrelationships of techniques are provided by Ackoff.[12] Discussion of a 'systems approach' as distinct from 'systems theory' is provided by Jenkins[13] and Van Dyne.[14]

The number of practitioners of system science has increased, but it was only in 1954 that spreading interest led to the foundation of the Society for General Systems Research, initially named the Society for Advancement of General System Theory. In effect, this field of effort is the scientific exploration and theory of systems in the various sciences; and general systems theory is the doctrine of principles applying to all, or defined subclasses of, systems.

Dynamic system theory is concerned with the changes of systems in time as described either internally or externally. Internal descriptions express the system in terms of simultaneous differential equations. In external description, the system is considered a black box. Block and flow diagrams and input–output diagrams are used in description.

At least in simplified systems it is possible to interrelate mathematical descriptions according to the internal and the external procedures. For example, an input–output function can, under certain conditions, be developed as a linear nth-order differential equation. The terms of the differential equation can be considered the state variables. Thus, in some cases it is possible to provide a formal translation from one language into the other.[15]

The internal description of systems is essentially structural. The description of the systems behaviour is in terms of variables and their interdependences. The external description of a system is functional. The system's behaviour is described in terms of its interaction with the environment.[6]

A variety of systems theoretic approaches have evolved in the last 30 years including general system theory, cybernetics, theory of automata, control theory, information theory, set and graph and network theory, relational mathematics, game and decision theory, computerisation and simulation and so forth.[6]

These general system theory approaches, as different as they may seem, have in common that they relate to 'system problems', that is, problems of interrelation within a superordinate whole. Systems theorists are often concerned with giving mathematical descriptions to system properties such as wholeness, sum, growth, competition, allometry, mechanisation, centralisation, finality and equafinality. These properties often are derived from the system description by simultaneous differential equations.

2.3.2. Relationship to Operational–Conceptual Approaches

2.3.2.1. Simulation Models

Simulation expresses the dynamics of the system rather than the statics. Such an approach, as applied to agricultural system problems, utilises definitions of at least the following components:[16]

(i) *Driving variables* are those variables which affect the system, but are not affected by the system (e.g. precipitation). Included here is man's control over the system (e.g. fertilizer application, supplemental feeding).

(ii) A *system state variable* is one which changes over time, is a component of the system, and is usually expressed in measures of mass or energy in concentration units (e.g. aboveground live plant biomass).

(iii) A *rate process* accounts for the transfer of matter and energy within the system by physiological, physical, or sociological mechanisms (e.g. infiltration of water is a physical process, ingestion of food or photosynthesis are physiological processes).

(iv) *Parameters* are coefficients or terms in rate process functions.

The change, either continuous or incremented, of a system state variable is calculated in a differential or a difference equation as a function of one or more rate processes, each of which in turn may be a function of driving variables and system state variables. These dynamic models have time as at least one independent variable. In fact, in most agricultural system models, time is the only independent variable. Where there is more than one independent variable, it results in partial differential or difference equation systems.

There are a variety of simulation modelling approaches. Generally the models may be characterised as differential equations, difference equations, integral–differential equations, life table models and Markov models.

Frequently the system state variables are denoted by the vector $\mathbf{x}$ in systems engineering work, the driving variables by the vector $\mathbf{z}$ and the output variables by the vector $\mathbf{y}$.[15] A rather common approach of diagramming such systems is the use of flow diagrams, commonly following the notation of Forrester.[17] An example diagram adapted from Van Dyne & Anway[18] is shown in Fig. 2.

In a flow diagram of the system the state variables are denoted as boxes and are connected with arrows which represent the flows of matter or energy from part to part within the system. Solid lines represent the flow of matter and dotted lines represent the flow of information. The system may also consist of more than one subsystem. Within a subsystem it is possible to have both matter and information flows; between subsystems it is possible to only have information flows.

The most common form of simulation models is that of 'compartment models'; in flow diagrams the compartments represent the state variables in systems of difference or differential equations. Alternately, a 'transfer function' approach can be diagrammed in which the lines (arrows) represent the processes that transfer energy of biomass between its various forms. Each of these representations can be looked on as the dual of the other.[19] A comparison of flow diagrams with transfer function diagrams is given in Fig. 3. This is a compartmental representation of a model for energy flow through an ecosystem. The model at any time has energy distributed between nine 'compartments'. A crude, but running form was

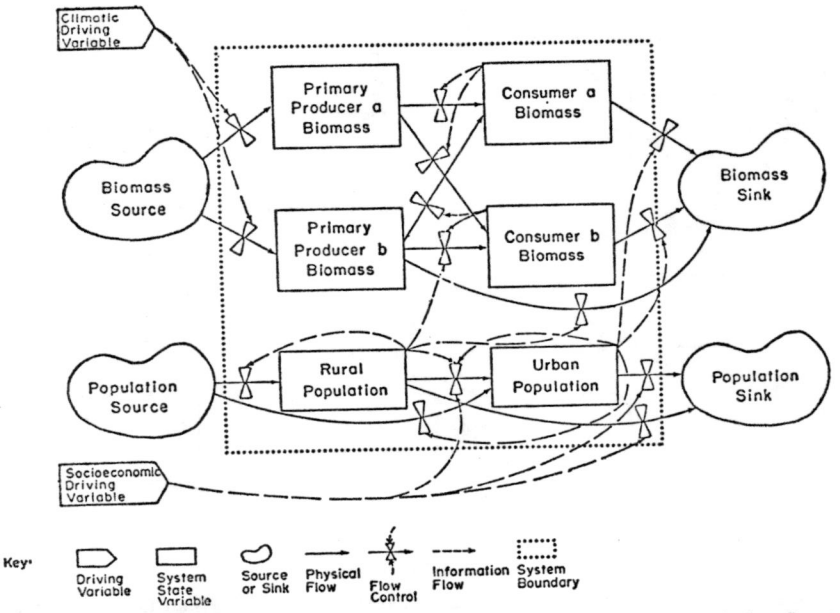

FIG. 2. A hypothetical, simplified flow model showing biomass and population flows (subsystems) through 6 system state variables, along 15 flow pathways or processes, from 2 sources and into 2 sinks, as affected by 2 driving variables. Note that a given flow may be controlled by a driving variable as well as by levels of state variables in one or both subsystems.

programmed and the code has been reported.[20] The processes of reproduction and growth, represented in Fig. 3, have been omitted in the transfer function approach. This was done because, in one form of conceptualisation, these processes do not represent absolute energy transfer between the compartments. Reproduction, for example, would require a subcompartmentalisation for those compartments in Fig. 3 representing live plants and animals. In Fig. 3(a) and 3(b) both temperature and precipitation are included as driving variables. These are not shown in the transfer function diagram because they do not represent energy inputs into the system. The effects of these inputs are, however, certainly important to photosynthesis and would be included as elements in the transfer function for photosynthesis when the diagram is mathematised. Although the transfer function approach (Fig. 3(c)) represents a coherent and instructive display tool for diagramming energy flow through a system, it has not been used frequently in agricultural systems modelling.

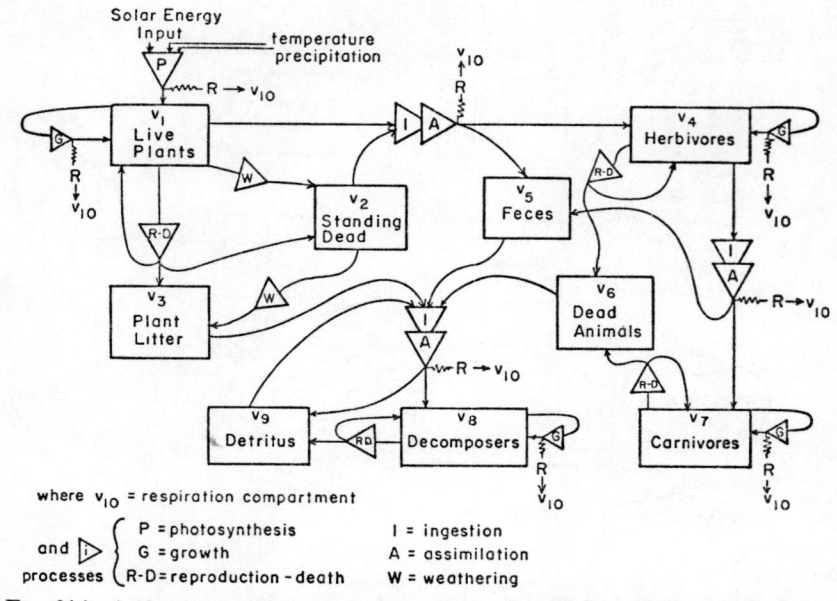

FIG. 3(a). A 'free-form' diagram for energy flow adapted from Van Dyne.[14] Energy levels are in compartments and flows are along arrows through labelled processes. Energy loss via respiration, R, dissipates into the environment, v_{10}.

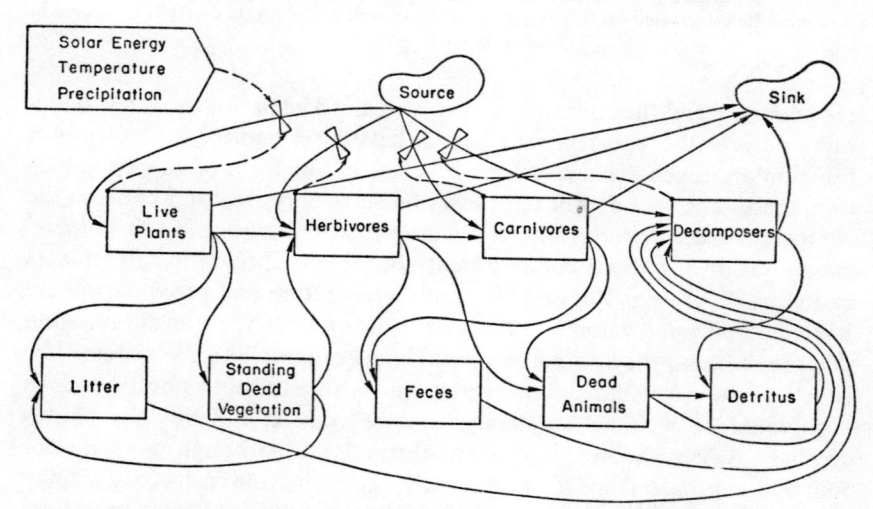

FIG. 3(b). A flow diagram approach for the same system. Flow controls have been shown in only a few instances.

Agricultural Systems Models and Modelling: An Overview

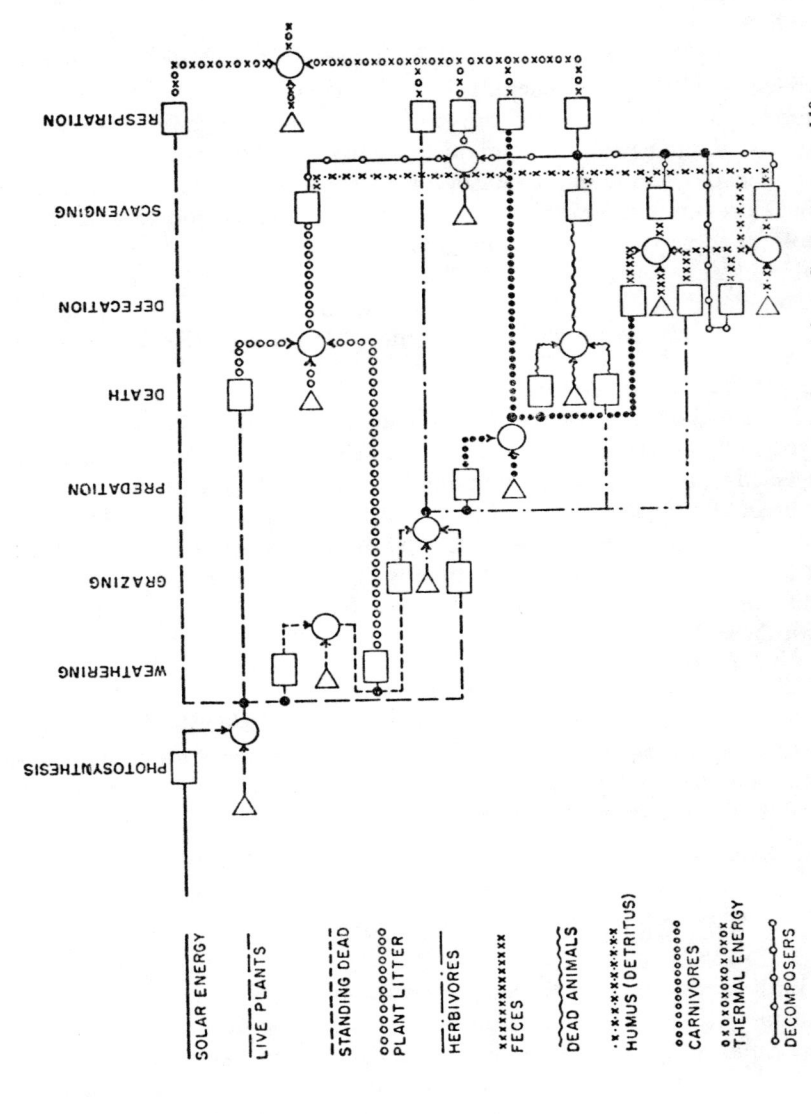

FIG. 3(c). A transfer function representation of the model adapted from Swartzman et al.[19]

2.3.2.2. Optimisation Models

Optimisation models discussed herein, generally, but not always, are static in nature. These techniques can be characterised as to whether they are linear or non-linear, static or dynamic and deterministic or stochastic. These classifications can be used as a gauge of difficulty of solution. Difficulty of solution generally increases continuously in moving from linear, static, deterministic models to non-linear, dynamic, stochastic models. Notable exceptions are queueing theory problems which, although non-linear, dynamic and stochastic, are sometimes easier to solve than some linear programming problems. In general, in optimisation techniques we look for those values of the controllable variables which give the maximal value of an objective function. The controllable variables are varied or manipulated to obtain the best combination of variables with regard to the selected objectives which are quantified in the objective function.

2.3.2.3. Statistical Models

A third class of models useful in many agricultural systems analyses is statistical models. In simulation modelling, statistical models, and especially non-linear regression models, are used in deriving functions wherein the dependent variable is a rate process and the independent variables are system state variables and driving variables. In an optimisation model, statistical equations, usually regression equations, are common and are often used to derive objective functions and constraints (e.g. Van Dyne[21]).

2.3.2.4. Model Notation Comparisons

A simplified comparison of simulation models (using differential equations as an example), optimisation models (using linear programming as an example) and statistical models (using multivariate multiple regression models as an example) is as follows:

Simulation:	Optimisation:	Statistical:
$\dot{x} = Ax + Bz$	max $c^T x$	$Y = BX + \epsilon$
$y(t) = Cx + Dz$	subject to $Ax < b$	
	$x > 0$	

In the above example for simulation models the vector x represents the system state variables, the vector z represents the driving variables and the vector y represents the output variables. The differential equations given by $\dot{x}$ are calculated as a function of the state variables and the driving variables at that given instant in time t. The output variables are calculated algebraically from the state variables and driving variables at a particular point in time. **A**, **B**, **C**, and **D** represent matrices of coefficients or parameters.

This is a simplified formulation and more complex and realistic generalisation approaches are provided elsewhere (Van Dyne[22]). For the simplified, linear, constant-coefficient deterministic form given above an analytical solution exists.[23,24]

In the optimisation example, the objective is given as the expression to be maximised. Here one seeks the vector of unknowns **x** to maximise the function subject to the inequalities given.

In the statistical model, the dependent variables are given by the matrix **Y** and the independent variables by the matrix **X**. A matrix of parameters is given by **B** (see also Section 2.3.2.6 for example of econometric model notation).

A comparison of these three forms of models follows:

(i) Everything is known. The solution of the model is initiated with values of the initial conditions for the **x** variables and with records or descriptions thereof of the **z** variables. From this information one may calculate the time course of **x**.

(ii) In the optimisation model the unknowns are **x**, but **c**, **A**, **b** are all known.

(iii) In the statistical model **Y** and **X** are known, and the values of **B** are obtained by least-squares or maximum likelihood methods. ϵ represents the error term.

In our tabulations in Section 2.4, we distinguish between differential equation simulation models, algebraic/difference equation models and algebraic equation models. A general comparison of these three types of equations is:[25]

$$\dot{\mathbf{x}}(t) = \mathbf{A}(\mathbf{x}, \mathbf{z}, t)\mathbf{x}(t) + \mathbf{B}(\mathbf{x}, \mathbf{z}, t)\mathbf{z}(t) \quad (1)$$

$$\mathbf{x}(nd) = \mathbf{A}'(nd)\mathbf{x}[(n-1)d] + \mathbf{B}'(nd)\mathbf{z}(nd) \quad (2)$$

$$\mathbf{x}(t_k) = \mathbf{A}''(t_k)\mathbf{x}(t_{k-1}) \quad (3)$$

In eqn. (1), $\mathbf{x}(t)$ is the vector of system variables and $\mathbf{z}(t)$ is a vector of driving variables. **A** and **B** are matrices whose elements may be functions of variables. Thus any a_{ij} in **A** or b_{ij} in **B** may be a function of any of the variables of **x** or **z** and time. This is a more general notation than the first-order linear form given previously.

When an approximate solution technique for differential equations is used, the resulting equations are similar in form to difference equations.

The general framework for these is shown in eqn. (2) above. Here the system variables $x(nd)$ are evaluated at discrete, prechosen time intervals d. Thus, $x(nd)$ are the values of the system variables at time nd, while $x[(n-1)d]$ gives these variables at time $(n-1)d$.

Markov chain models are equivalent to first-order ordinary differences equation models with coefficients independent of the variables. Markov models have the added generalities that the time step sizes need not be constant. The coefficients of the matrix may be time-varying, as in differential and difference equations. Usually no driving variables are considered in this approach. An example (eqn (3)) is given above. Here the variables are observed at times $t_0, t_1, \ldots, t_k, t_{k+1}$. The matrix A' is called the transition matrix and may be a time function but is never a function of x.

Note in eqns (1), (2) and (3) above, the *matrices* A, A' *and* A'' *are not the same, nor are* B *and* B'.

2.3.2.5. Events in Models

One further consideration in dynamic system models is that of including events. Events are particularly appropriate to include in agricultural systems models. In these cases the states of the system display little dynamic behaviour except at the occurrence of various events. In such modelling it requires a generation of a future-events chain which is then 'spun-out' in a simulation. It is an event-to-event model. The future-events chain need not consist of deterministically spaced events, but there may be probabilistic elements in them.[25]

An ecosystem modeller might use any one of the above approaches, or he might use several in combination. For example, one might combine an event structure with difference equations. Also the step sizes of the difference equation may be varied depending upon which event combinations are presently controlling the system.

2.3.2.6. Some Econometric Notations

It is instructive to compare the notations used in agricultural systems models (see Section 3.2.4.2) with that used by economists in macroeconomic modelling. One of the major approaches to macroeconomical modelling is the policy simulation approach. A review of the state of the art is given by Naylor.[26]

In such models the economy of the country in question is described by a set of simultaneous equations, simplified here to the linear form:

$$Ax_{(t)} + By_{(t)} + \sum_{j=1} B_j y_{(t-j)} + Cz_{(t)} + D = u_{(t)}$$

where $\mathbf{x}_{(t)} = m \times 1$ vector of exogenous variables; $\mathbf{y}_{(t)} = n \times 1$ vector of endogenous variables; $\mathbf{y}_{(t-j)} = n \times 1$ vector of lagged endogenous variables, $j = 1, \ldots, p$; $\mathbf{z}_{(t)} = q \times 1$ vector of policy instruments; $\mathbf{u}_{(t)} = n \times 1$ vector of stochastic distribution terms; and $\mathbf{A}, \mathbf{B}, \mathbf{C}, \mathbf{D}$ = coefficient matrices. Note that the vectors, $\mathbf{x}, \mathbf{y}$ and $\mathbf{z}$ *do not have the same usage* as in the state variable approach to engineering simulation models nor the usual mathematical programming notation. The notation here is more akin to statistical models.

The approach to using the above form of model as a simulation is to solve simultaneous sets of equations for $\mathbf{y}_{(t)}$ and the time path of $\mathbf{y}_{(t)}$ for as long a period as necessary. To do this, the exogenous variables $\mathbf{x}_{(t)}$ are read into the computer as driving data, the values of $\mathbf{y}_{(t-j)}$ generated in a previous period are fed into the model for period t, the policy variables are specified by the analyst, and the stochastic disturbances may be either suppressed or generated and added in. If the model is linear, the solution is of the following form:

$$\mathbf{y}_{(t)} = \mathbf{B}^{-1}\mathbf{A}\mathbf{x}_{(t)} + \mathbf{B}^{-1} \sum_{j=1}^{P} \mathbf{b}_j \mathbf{y}_{(t-j)} + \mathbf{B}^{-1}\mathbf{C}\mathbf{z}_{(t)} + \mathbf{B}^{-1}\mathbf{D} + \mathbf{B}^{-1}\mathbf{u}_{(t)}$$

where $\mathbf{b}_j$ is a vector in $\mathbf{B}$ and $\mathbf{B}^{-1}$ is the inverse of $\mathbf{B}$. If the system is not linear, as is frequently the case, the Gauss–Seidel method for solving systems of simultaneous non-linear equations frequently is used.[27,28]

2.4. OVERVIEW OF AGRICULTURAL SYSTEMS MODELS

2.4.1. Sources, Dates and Amounts of References

Our present review is uneven, for we have drawn heavily on previously published comparative reviews for various renewable resource system models:

(i) A recent discussion of the insect pest models and their uses is provided by Conway.[29]
(ii) Not so much for all agricultural systems, but for the insect population dynamics in forests, a workshop report is available.[30] Some seven papers concerned with the use of computer models in analysis of the sterile male technique in insect control and population dynamics are presented in an IAEA publication.[31]
(iii) Population ecology models recently were reviewed by Shoemaker[32] and Jaquette.[33]

(iv) A review of mathematical models in forest management with emphasis on linear programming and discussion of simulation and critical path analysis was published as the result of a meeting at the University of Edinburgh.[34]

(v) A thorough review of a topic highly related to agricultural systems, in the broad sense, is that of Patten[35] for plankton production models.

(vi) Paulik[36] summarises many papers on fisheries simulations.

(vii) An interesting series of papers utilising a matrix approach to calculate the dynamics of forest systems is provided by Usher[37-39] and Usher & Williamson.[40] This was an application of a model approach from population dynamics in animal studies into population dynamics of forest trees.

(viii) Couch[41] reviews the use of linear programming in ration formulation. Early work started in 1951, which is surprising in that the first publications on linear programming were by Dantzig in 1947.[42]

(ix) Van Dyne, Frayer & Bledsoe[43] review numerous optimisation problems in the natural resource sciences.

(x) Swartzman et al.[19,44] have provided general reviews of dynamic simulations models and of optimisation as related to land use planning.

(xi) Van Dyne, Innis & Swartzman[25] and Innis[45] have reviewed numerous grazing land system models.

(xii) Agricultural production functions have been summarised by Heady & Dillon.[46] Historically these have been mainly statistical regression functions. These functions are of primary value in agricultural systems modelling discussed herein in the structuring of optimisation models. There have been rapid developments in the last quarter century in agricultural economics in the usage of operations research tools. The greatest emphasis had been in mathematical programming, but other activity or decision analysis methods have received wide usage including game theory, Markov chain processes, inventory control, queueing theory, network analysis and simulation. Agrawal & Heady[47] provide a relatively readable textbook on these operations research methods as used in agriculture. They make the projective statement that perhaps, although there are developments and refinements in these methodologies still in progress, the number of new developments has slowed and the knowledge in the area appears to be stabilising.

Thus, their review may give the harried reader a chance to tarry awhile and try to catch up with a rapidly changing field!

(xiii) Arnold and Bennett[1] provide a useful review of optimisation models and Dent[2] reviews a number of simulation models, particularly for intensive agricultural systems. (Several individual chapters in Dent & Anderson[259] have been referred to but the reader may want to examine the entire book. Another symposium volume containing several papers of interest has been published by the National Institute of Agricultural Engineering.[260])

2.4.1.1. Bibliographies

Some abstract bibliographies of more strictly ecological models than several of those listed above are provided by O'Neill et al.[48] and particularly an extensive compilation of Kadlec[49]; a bibliography of selected statistical ecology and related references is provided by Schultz.[50]

2.4.1.2. Useful Conceptual Reviews

Two other reviews, while not emphasising mathematical models, but of major relevance to agricultural systems are those of Morley & Spedding[51] and Hutchinson.[52] The former emphasises grazing and the latter fodder conservation in addition.

2.4.1.3. Example Articles

We selected a collection of papers for this review non-methodically. The papers we selected and examined personally were those received from colleagues, those sought from the library by examining bibliographies of articles on hand, those we were referred to by talking to a few individuals, and a brief scan of only a few recent journals. We feel the example articles summarised below are sufficient to show the general trend in agricultural systems modelling. After our limited search of the literature, we were even more impressed by the extent of the articles in so many disciplines.

2.4.2. A Characterisation of Agricultural Systems Models

We have prepared several tables to list and briefly compare some of the models we examined in different areas of agricultural studies. In the review of these articles, we noted some generalities, both in problems and possibilities. We have grouped the references arbitrarily. We realise there is overlap within and among the categories in the tables (Tables 1 to 5). In the tables we have listed examples of typical references in a comparative

TABLE 1
*A review of agricultural systems models: example models primarily oriented to intensive agri-
Blanks represent not applicable*

Reference	General model	Specific model	Time stage	Attributes	Computer language
		DAIRY[a]			
Jenkins & Halter[114]	Optimisation				FORTRAN
Hutton[115]	Simulation		1 year		FORTRAN
Jensen[116]	Optimisation	Linear programming			
Lindner[117]	Optimisation	Linear programming			
Dean et al.[59]	Optimisation	Linear programming			
Dean et al.[60]	Optimisation	Linear programming			
Bath et al.[111]	Optimisation	Linear programming			
Chandler & Walker[118]	Optimisation	Linear programming		62 parameters	
Nelson & Elsgruber[119]	Optimisation	Dynamic	1 month	2 state variables	

Agricultural Systems Models and Modelling: An Overview 39

cultural management. An asterisk means we have estimated this information from the article. or unspecified information.

Model diagram	Program code	What it was built for	What it was used for	Comments
			DAIRY[a]	
No	Yes	The dairy cow replacement problems.		The authors suggest that the model might have practical use. The replacement policy specifies the productive unit that should be replaced, the time, and the age. A multistage stochastic replacement decision model.
Yes	Yes	To simulate a dairy herd for management decisions.	For making herd replacements.	Step-by-step explanation of the program. The model is capable of producing information on many aspects of the dairy farm. Good explanation about the data needed to run the model.
No	No	To develop optimum plans for tropical pastures dairy farm and to examine briefly the roles of labour, capital, and farm size in the development process.	To maximise income average cow dairy farm.	The programming results indicate that intensification of dairy farming based on tropical pastures is the appropriate strategy for dairy farmers in the Cooroy area. His suggestions that beans and pigs would be profitable sidelines have not been accepted.[123]
No	No	To examine the most profitable way in which the resources of a representative dairy farm can be used to increase beef production and raise farm incomes.		The results of this study confirm that increased beef production in conjunction with dairying can raise profitability, particularly of large farms. His suggestions have been applied (the number of beef cattle increased from 15 500 in 1965 to 26 000 in 1971.[123])
No	No	For maximising the profit.		The model selects the concentrate and roughage components of the ration, the roughage–concentrate ratio, level of feeding per cow, and quantity of milk production, that maximise income above feed costs. The program is adaptable to cows of different production abilities, different weights, and various economic situations.
No	No	To provide feeding programs which maximise economic returns over feed cost.	Income maximise above feed cost for dairy cattle.	The program was tested under field experimental and commercial conditions. It is a useful and flexible model but additional research is needed. The model is the same as in Dean *et al.*[59]
No	No	To formulate rations for dairy cows that will maximise income above feed costs.		Cows fed the computer-formulated rations returned $15–21 more income above feed cost per cow per year than cows fed by conventional practices. Application of Dean *et al.*[59] programme for actual milk production.
No	No	To facilitate formulation of least-cost rations through dynamic programming.	Dairy cows.	The model takes into account all the complex interrelationships among nutrients. Because nutrient specifications are generated from common terms associated with dairy cattle, people without expensive nutritional training can use it. The model is based on current National Research Council's suggestions.
No	No	Decision-making for beef feedlots.	To simulate beef feedlot and management decisions.	Adaptive multiperiod statistical decision model. The system requires further testing before more definite conclusions can be drawn and has three main subsystems: cattle price forecasting, beef feed formulation, and feedlot operations scheduling.

continued

TABLE 1

Reference	General model	Specific model	Time stage	Attributes	Computer language
Donaldson[120]	Simulation	Monte Carlo			
Dahlman & Sollins[121]	Simulation	Differential		9 state variables	FORTRAN
Cloud et al.[122]	Simulation				FORTRAN
		PEST CONTROL[b]			
Conway & Murdie[139]	Simulation	Difference	1 day		FORTRAN*
Watt[140]	Simulation	Difference		1 driving variable 30 parameters	FORTRAN
Shoemaker[32,141,142]	Simulation/ optimisation	Difference			
Pimentel & Shoemaker[143]	Minimisation	Linear			
Miles et al.[144]	Simulation	Differential	2 days	4 driving variables 6 state variables 9 flows	GASP-IV

—contd.

Model diagram	Program code	What it was built for	What it was used for	Comments
No	No	To take account of the effect of weather on cereal harvest.	Farm management decision.	In order to take account of the effect of weather, rates of combined work, harvest weather, and diurnal grain-moisture content is regarded as probabilistic with known distribution.
No	No			Synthetic model based on literature models for grassland in general. Generalised nitrogen cycle.
No	No	To aid farmers in planning their farm operation.	To analyse the effect of different harvesting systems on net farm income.	This study assessed the effect of the forage harvesting system; the date that harvest begins; the weather patterns; the method of grain feeding; and the size of herd on farm income. The simulation model consists of a series of five computer programs.
			PEST CONTROL[b]	
Yes	No	Teaching model.		A simple model of a hypothetical insect population on which we make control experiments; a simple reproduction model that was used as a basis for simulation studies aimed at understanding the different ways in which various pest control strategies affect population size through the reproduction component.
No	No	Evaluating various strategies of insect pest control.		The program simulates the effect of various dosages of insecticide, parasite release, and spraying of virus of any combination of these, weather and pest density on reproduction, dispersal and mortality in pest populations. The program predicts the pest and predator population fluctuations over a 35-year period. It calculates the total cost of the control; timber loss plus control costs for 15 control policies from which it chooses the best. Deterministic equations. Explanation of the model in appendix.
No	No	Pest management to develop a methodology for finding optimal pest control policies.	Model development.	The papers give theoretical considerations and the mathematical equations for building a model for pest control policies which are effective, economical and ecologically harmonious. The model is based on Hassel & Huffaker[146], but also suggestions are given for the future when more information on the interaction will be available; deterministic model.
No	No	To estimate the impact in crop prices and land use with and without insecticides, while producing enough of all crops to meet demand.	Corn and cotton production in year 2000 for management considerations.	Limitations: linearity, aggregation, and projected estimates of technological developments. A nationwide linear programming model for population of 280 million people which calculates crop costs and land use patterns that could be expected for several different restrictions regarding land and insecticide use. The model is based on Heady *et al.*[147]
Yes	No	To simulate alfalfa growth and use it for alfalfa pest management.		Simulated alfalfa yields agree with field data collected in 1967.

continued

TABLE 1

Reference	General model	Specific model	Time stage	Attributes	Computer language
Loewer et al.[145]	Simulation	Difference	1 day	3 driving variables 12 state variables	FORTRAN based
		IRRIGATION[c]			
Anderson & Maass[150]	Simulation	Optimisation			FORTRAN

[a] Other references:
Goudriaan & Waggoner[124] Duncan et al.[127]   Crabtree[130]   Jones et al.[133]
Walsh & Dugdale[125]   Deinum & Dirven[128]   Stapleton et al.[131]   Soribe & Curry[134]
Walsh et al.[126]   Visser[129]   Hathorn et al.[132]   Miles[135]

TABLE 2
A review of agricultural systems models: example models primarily oriented to extensive

Reference	General model	Specific model	Time stage	Attributes	Computer language
		GRAZING[a]			
Donnelly & Armstrong[89]	Simulation	Difference		10 'endogenous events' 1 driving variable 5 state variables	SIMSCRIPT
Freer et al.[90]	Simulation	Algebraic/ difference	1 day	8 state variables 1 driving variable 10 parameters	SIMSCRIPT
Armstrong[91]	Simulation/ optimisation	Algebraic/ difference	1 day	10 state variables 2 driving variables 13 constants 12 flows	FORTRAN/ SIMSCRIPT
Christian et al.[92]	Simulation/ optimisation	Difference	1 day	1 driving variable	FORTRAN

—contd.

Model diagram	Program code	What it was built for	What it was used for	Comments
Yes	No	To improve our understanding of crop-European corn borer ecology and to assist corn-pest managers in reducing crop damage by applying control measures when they will be most effective.		The model was validated and sensitivity analyses were conducted. It was found that the model is very sensitive to: (1) rainfall vs. the viable eggs laid per female; (2) temperature and rainfall in combination vs. the effect of each on the death rates of overwintering generation pupa; (3) temperature vs. the death rate of viable eggs and first-instar and second-instar larva.
			IRRIGATION[c]	
No	Yes	Decision-making in an irrigation system.	To determine how best to allocate irrigation water among crops and farms.	The model simulates decision-making at several levels of activity in an irrigation system, from the operations of water distribution system to the farms, and it takes into account the responses of different crops to irrigation water.

Miles et al.[136]
Baker & Horrocks[137]
Baker & Horrocks[138]

[b] Other references:
Southwood & Norton[148]
Ford-Livene[149]

[c] Other references:
Asseed & Kirkham[151]

agricultural management. Blanks represent not applicable or unspecified information.

Model diagram	Program code	What it was built for	What it was used for	Comments
			GRAZING[a]	
Yes	No	To simulate the grazing of summer pasture by sheep.	To predict the response of sheep weight to grazing subdivision, rainfall, growth rate of herbage, the amount of dry food available, and the efficiency with which it is grazed.	
Yes	No	Grazing of summer pasture by sheep.	To predict the response of sheep weight to changes in grazing subdivision, stocking rate, and growth rate of herbage.	Limited to the growing season; specific equations and diagrams and values of constants predictions reasonable. Plant growth and animal production.
Yes		Grazing management practices.	To compare the management strategies of set stocking and rotational grazing: model development.	Plant growth and animal production.
No	No	Selecting the best system of grazing management.	Model building for management decisions.	Limited to ewes during late summer. Examined a range of rotational management practices; based on Freer et al.[90], compared six rainfall patterns, eight stocking rates, two initial weights; calculated 'objective functions' based on value of live weight at start of mating, value of live weight change, cost of moving sheep, and value of food after 100 days and cost of supplementary feeding. A later version of the Freer et al.[90] model.

continued

TABLE 2

Reference	General model	Specific model	Time stage	Attributes	Computer language
Christian et al.[93]	Simulation/ optimisation	Difference	1 day	See Christian et al.[92]	FORTRAN
Morley & Graham[152]	Simulation	Algebraic/ difference	1 year	22 'input variables' 10 'input variables'	FORTRAN (KWIKTRA N)
Vickery & Hedges[57,58]	Simulation	Difference	1 week	11 state variables 3 driving variables 11 processes	FORTRAN
Morley[153]	Simulation/ optimisation	Difference	1 week		FORTRAN
Wright & Dent[154]	Simulation		1 week	18 state variables 2 driving variables	
McKinney[155]	Simulation			2 driving variables	FORTRAN CSMP
Chudleigh & Filan[156]	Simulation		1 month	2 driving variables	
Patten[157]	Simulation	Piecewise linear and stationary difference	1 week	40 state variables 3 driving variables	CSMP
May et al.[78]	Simulation	Differential	1 day	6 state variables 1 driving variable	Analog computer
Trebeck[158]	Simulation			1 driving variable	FORTRAN

—contd.

Model diagram	Program code	What it was built for	What it was used for	Comments
Yes	No	Farm producing lambs for meat.	Management decisions.	Limited to sheep and cattle; more complex than previous versions; animal management implemented through input code values; uses list-processing techniques in calculation; allows changing feed requirements for animals during the year.
No	No	Fodder conservation for drought.	Management decisions.	The model is restricted to comparisons between two of many possible policies.
No	No	Live weight changes and wool production for sheep on *Pholaris tuberosa* pasture.	Model development.	Plant growth and animal production limited in application because of the use of arbitrary assumptions; non-reproductive sheep simulated; used linear regression equations to interrelate observed and predicted data for four variables.
No	No	Evaluation of increase in growth rate of improved pasture.		The model is similar to this published by Vickery & Hedges.[57,58] Systems of young sheep production for meat and wool, or flock replacements, and young beef cattle grown for meat, or for herd replacement.
Yes	No	To represent a typical grazing system and to reach management decision.	Simulating 500 acre farm carrying 1500 mixed-age Merino ewes.	The grazing model was developed to explore the problems involved in developing and using models based on biological relationship for decision-making studies. The paper discusses some methodological aspects of simulation with specific reference to grazing system.
Yes	No	Estimating live weight change and pasture growth during 30 days of winter.	To predict the effects of availability and live weight in May and stocking rate from May to August on availability and live weight in August.	Soil water and nutrients were regarded as non-limiting. Pasture growth was estimated from a function of leaf area, temperature, and incoming radiation. The model was tested against data from different experiments. Sensitivity analysis techniques were used. Detailed model and runs are described by McKinney.[155]
Yes	No	A pastoral property in the Western Division of New South Wales.		Stochastic rainfall, lambing percentage, wool cut per head, wool quality, mean yield of wool and feed growth. The model was validated with the same data that was used for estimating a number of the parameters of the model.
No	No	To simulate the nominal and small perturbation behaviour of principal functional components of short grassland prairie as a total ecosystem unit.		Four submodel sections: abiotic, producer, consumer and decomposer.
Yes	Yes	To fit a model of a recycling sulphur network to these observations.	To predict the sensitivity of the system to hypothetical stimuli as a basis for further empirical tests of the validity of assumptions made.	A mathematical model to represent the sulphur cycle in a grazed pasture simulating the results of a published field experiment in which observations were made on the incorporation of radioactivity into wool following a single application of [^{35}S] gypsum to pasture.
No	No	To assist research into extensive beef production.		The model is general and capable of being used in a variety of situations. More research is needed in order to improve the model results.

continued

TABLE 2

Reference	General model	Specific model	Time stage	Attributes	Computer language
Arnold & Galbraith[159]	Simulation			4 driving variables 50 parameters	
Arnold & Campbell[160]	Simulation	Difference	1 day	10 state variables 3 driving variables 31 parameters 11 processes	FORTRAN
Arnold & Galbraith[159]	Simulation	Difference	1 day	7 state variables 4 driving variables 42 parameters 22 processes	FORTRAN
Arnold et al.[161]	Simulation	Difference	1 day	7 state variables 4 driving variables 20 processes	FORTRAN
Baldwin[162]	Simulation	Differential	Varies		Kinsum
Nelson[163]	Simulation	Algebraic/ difference	1 month	1 driving variable 75 parameters	FORTRAN
Edelsten et al.[164]	Simulation		1 day	30 state variables	
Singh[165]	Simulation	Linear/ differential	1 day	6 state variables 1 driving variable 6 processes	SIMCOMP
Morris et al.[166]	Simulation	Differential	1 day	9 state variables 4 driving variables	CSMP
Rice et al.[167]	Simulation	Difference	1 day	18 state variables 2 driving variables 7 parameters	CSMP
Smith & Williams[168]	Simulation/ optimisation		1 day		

—contd.

Model diagram	Program code	What it was built for	What it was used for	Comments
No	No		To simulate lupin growth and utilisation, and their use to predict likely animal modification responses.	Although the models are 'first generation,' they allow a wide combination of seasonal conditions, type of livestock, and stocking rates to be examined. The model of lupin growth was validated against data from different sources: live weight changes of Merino wethers and yearling cattle were simulated when grazed at three stocking rates.
Yes	No	Live weight changes and wool production.	Model development.	Limited to Merino wethers grazing on pasture; a submodel initially developed for a larger system model; equations presented with parameter values; agrees well with animal weight validation data, less well with wool data. Equations for plant growth on gravelly soils.
No	No	Crop and animal production.	To examine management strategy for crop use.	Limited to Lupin crops. In this modified version of Arnold et al.[161] model, the animal group is any stock of non-breeding sheep and cattle. Equations for plant growth on gravelly soils.
No	No	Submodel of a larger model.	Management decisions.	Restricted to one species of plant. Processes and variables affecting them listed but no equations.
		Energy balance at the tissue level.	Energy balance and animal production at biochemical level.	Deterministic non-stochastic linear equations; limited only for ruminants.
Yes	Yes	To test the feasibility of using a mix of native and domestic ruminants to rangeland.		Limitation: the model considers only green plants and ruminants (3 species). Far from reality; forage production is calculated on a per year basis and the rain is the only driving variable. The model is a simplified abstraction of a rangeland ecosystem. Excellent documentation of derivations of flow functions; relatively few runs made with final model.
Yes	No	To explore different systems of lowland fat-lamb production and to define the most profitable lowland fat-lamb production.	To explore the effect of management variables on the growth of lambs and to explore ways of controlling these factors.	Does not have an economic optimisation model; it simulates a block of sheep from when the lambs are born in mid-March to when they are sold in Sept.–Oct. The initial weights of the animals are given according to probability distribution (stochastic). The rest of the model is deterministic. Simulates individual animals.
Yes	No	To arrive at suitable transfer coefficient in order to approximate the pattern of changes in the compartment values and also to indicate the time intervals when these coefficients are likely to vary.	To depict intraseasonal changes in various primary producer compartment in three grasslands by adjusting some of the transfer coefficients.	A time-varying coefficient, linear compartment model of herbage dynamics for tropical grassland in India.
		Understanding the dynamics of nitrogen of ruminants.	Teaching and experimental design.	It is a compartment of subsystem model; limited for ruminants; requires validation by experiments. Deterministic.
Yes	No	To model the grazing ruminants.	To examine different grasslands—annual and perennial. Teaching.	Does not have a good representation of grazing on herbage growth; deterministic model with unstochastic rainfall. General in form but limited only to ruminants.
No	No	To postulate by computer simulation the probable importance of the length of deferment and initial plant density in relation to stocking rate.	To examine the probable response of a system of animal production to deferred grazing.	The combination of stocking rate and days of deferment that maximise the gross margin was approximated by use of a search routine. Because the model has not been validated against experimental data the results represent deduction from the model of biological processes.

continued

TABLE 2

Reference	General model	Specific model	Time stage	Attributes	Computer language
Hutchinson[169]	Simulation	Algebraic	1 week	4 state variables 4 driving variables 12 processes	FORTRAN
Swartzman[170]	Simulation	Differential		2 state variables	FORTRAN
Kelly et al.[171]	Simulation	Differential		6 state variables	FORTRAN
Wright & Van Dyne[172]	Simulation	Differential		8 state variables	FORTRAN
Goodall[72,85,86,88]	Simulation	Difference		7 state variables	FORTRAN
May et al.[78] May et al.[79] Till & May[81]	Simulation	Differential	Instant	41 state variables	Analog
Bledsoe et al.[173]	Simulation	Differential	1 hour	40 state variables	FORTRAN
Van Dyne[14,20]	Simulation	Difference		10 state variables	FORTRAN
PLANT GROWTH[b]					
Byrne & Tognetti[183]	Simulation	Algebraic/difference	5 days	2 driving variables	FORTRAN
Ross[184]	Simulation/game	Difference	1 day	10 state variables 2 driving variables 9 constants	
Ross et al.[185]	Simulation	Differential	1 day	2 driving variables 23 parameters	

[a] Other references:
Field & Hunt[174] Paltridge et al.[176] DeWit et al.[178] Smith[180] Jones[182]
Johns[175] Paltridge[177] Brockington[179] Fitzpatrick & Nix[181]

—contd.

Model diagram	Program code	What it was built for	What it was used for	Comments
Yes	No	To show how grazing experiments and modelling can be regarded as complementary activities.	Beef cattle fattening.	The model needs to be developed and improved with validation; example in management strategy to evaluate concept of fodder conservation, three stocking rates and six conservation levels examined; plant growth and animal production.
No	Yes		Prairie birds.	Parameters determined from experimental data.
No	Yes			Models for cool season and warm season grass communities, ungrazed; constant coefficient equations inadequate to represent dynamics. Time and system varying flows gave good results. Herbage dynamics.
No	Yes	General effort to look at ecological impacts of weather modification.	Simulate long-term dynamics of disused grassland.	Based on workshop exercise with realistic data inputs; incomplete; emphasis on semi-desert plant demography.
No	Yes			See text, Section 2.5.3.3; evolving model with spatial considerations. Sheep rangeland grazing.
No	Yes	Predictions of mineral cycle influence on sheep production.	Research guidance and synthesis.	Parts of an experimental–analytical programme; ^{35}S measured in various points in the system; sheep pasture sulphur cycle.
No	Yes	Research guidance and synthesis.	Experience in detailed total system model.	Part of a large-scale study to interrelate field and laboratory data and eventual predictions for system utilisation; one of the largest models of agricultural systems at the time.
No	Yes			Model a simplified grazing system for shortgrass prairie; output presented; unstable for decomposers.
			PLANT GROWTH[b]	
Yes	Yes	First attempt to establish the structure of a field system in a model.	Prediction of pasture response to an environmental change.	The work is to be regarded as a preliminary representation of pasture growth and a first attempt to establish the structure of a field system in models. Uses statistical trace or Monte Carlo technique.
No	No	To relate plant growth to moisture extraction.	Used to identify the major variables that control growth and moisture extraction of *Eragrostis eriopoda*.	Based on subtropical, arid climate; validation data on plant biomass taken by double sample technique, soil water by neutron probe; used in experimentally irrigated plots; model was planned to minimise number and site specifically of variables for predicting plant growth.
Yes	No	To describe the effects of nitrogen and light on the vegetative growth of grass–legume swards.		The mechanisms which were judged to be basic in determining the behaviour of a grass–legume sward are competition for light and supply of nitrogen. Proper validation will require more detailed information. The model consists of a set of first-order differential equations based on mechanisms involved.

Other references:
Lemon et al.[186] Stapleton et al.[188] Stapleton[190] Chen et al.[192]
Iwaka & Hirosaki[187] Stapleton et al.[189] Splinter[191]

TABLE 3

A review of agricultural systems models: example models primarily oriented article. Blanks represent not

Reference[a]	General model	Specific model	Time stage	Attributes	Computer language
Loucks[193]	Optimisation	Linear programming			
Leak[194]	Optimisation	Linear programming			
Nautiyal & Pearse[195]	Optimisation	Algebraic/linear			
Amidon & Akin[196]	Optimisation	Dynamic programming			
Myers[197]	Simulation	Difference	1 year*		FORTRAN
Myers[198]	Simulation	Difference	1 year*		FORTRAN
Botkin et al.[199,200]	Simulation	Difference	1 year		FORTRAN
Navon[201]	Optimisation	Linear			
Woodmansee & Innis[202]	Simulation	Difference		11 biomass and 12 potassium compartments. Several generated compartments to total of 40 210 year simulation ≥45 flows.	SIMCOMP

[a] *Other references:*
Liittschwager & Tcheng[203]   Teeguarden & Von Sperber[205]   Turnbull[207]
Navon & McConnen[204]   Kourtz & O'Regan[206]   Arvanitis & O'Regan[208]

Agricultural Systems Models and Modelling: An Overview 51

to forestry. An asterisk means we have estimated this information from the applicable or unspecified information.

Model diagram	Program code	What it was built for	What it was used for	Comments
No	No	To assist foresters in developing optimal continuous yield cutting schedules for a forest.	To illustrate by hypothetical example the usefulness as well as the limitations of these models.	
No	No	The estimation of maximum allowable timber yields.	Two hypothetical problems for illustration.	The model does not provide results that apply to any actual situation.
No	No	Management decisions.	To specify the economic optimum pattern of harvests from an irregular forest that will maximise the economic return.	The model is helpful in illustrating the economic interrelationship between the rotational length, the adapted period for conversion to sustained yield, and present worth of the forest. The model permits specification of the cutting programme that will maximise the present worth of expected yields and also can be used to define the optimum economic balance between present and future harvest.
No	No	Management decisions.	To determine optimal levels of growing stock in even aged stands during rotation.	A deterministic one-dimensional discrete dynamic programme.
No	Yes	Management of even aged timber stands for roundwood and saw logs.		Limited to even aged stands; main programme and 11 subprogrammes. Contains provisions for stand growth, thinning, harvest, cuts planting of non-stock areas, and other changes in forest conditions.
No	Yes	Management of even aged timber stands.		The original relationships and analytical producers (Myers[198]) have been modified and improved. Major modifications permit more general treatment of regeneration cuttings and more accurate estimates of intermediate cuttings on stands. Main programme and 21 subprogrammes.
No	Yes	To produce a dynamic model of forest growth.	IBM Programme.	Simulates the growth of the uneven aged mixed species stand of trees that occur in the northeastern US. Changes in the state of the forest are a function of its present state and random components. They avoid curve following a simulation that would not provide testable hypotheses about how trees grow, compete, and divide the resources. They used only standard library routines.
No	No	To generate cutting and reafforestation schedules for commercial forest lands under multiple management use.	Can be used to evaluate current management policies and goals and to evolve new ones.	It is a subsystem of a system. Uses computerised method for developing long-range management plans (up to 33 decades in the future). Limited to UNIVA 1108 computer with the ILONALP code. The programme and explanations are available by writing to the experiment station in Berkeley.
Yes	No	Lodgepole pine forest.	Evaluate impact of clear cutting on nutrient losses.	Simulated forest assumed to be in steady nutrient state over 70 year cycle before burning; biomass and submodels coupled; driven by literature-generated growth curves; flows presented and discussed individually.

TABLE 4
A review of agricultural systems models: example models primarily oriented to

References[a]	General model	Specific model	Time stage	Attributes	Computer language
Burt[209]	Optimisation	Dynamic programming	1 year	Maximise integral equation of economic return by varying length of renewal period.	FORTRAN
Nelson & Rittenhouse[210]	Simulation	Regression + budget calculation	1 year	Not a simulation *per se*, but deterministic budgetary calculations, data base eastern Oregon.	Not given
Van Dyne & Rebman[211]	Optimisation	Linear programming		4 variable objective function, 6 constraints	FORTRAN
Rae[212]	Optimisation	Linear programming		Essentially two variables and one constraint	Unspecified
Rae[213,214]	Optimisation	Discrete stochastic programming		50 activities, utility fit was non-linear relation of net income, 3 stages in planning period.	
Zusman & Amiad[215]	Simulation/ optimisation	Difference equation, steepest ascent search	3 periods in each year; about 2, 5, 3 and 6·5 months each	10-dimensional policy space; initial condition set to expected stationary organisation	Unspecified) (IBM 1620)
Johnson et al.[216]	Simulation/ optimisation	Transformation relation/ stochastic linear program	Steps of 1 year over 15 years	Model not clearly specified	Unspecified
Dalton[54]	Simulation	Algebraic	1 hour	Unspecified	ALGOL
Heifner[217]	Optimisation	Quadratic programming		~10 variable quadratic objective functions, 2 linear constraints	FORTRAN (Wolfes algorithm)

Agricultural Systems Models and Modelling: An Overview 53

agricultural economics. Blanks represent not applicable or unspecified information.

Model diagram	Program code	What it was built for	What it was used for	Comments
		Economic replacement problem.	Pinyon-juniper control analyses.	Variation of classical replacement problem in economics; deals with pinyon-juniper control in southwest US; determine optimum interval for control considering forage production rates and cost of control over time varying interest rates and forage prices.
		Economic replacement problem.	Sagebrush control in crested wheatgrass.	Net present value of resource calculated as function of initial brush cover, control effectiveness, potential forage value, cost of control, and cost of capital.
		Determining optimal animal combinations for grazing.	Hypothetical Edwards Plateau ranching management problems.	Select optimum combination of cattle, sheep, mohair goats and deer to maintain balance of grass, forbs and shrubs and give maximum economic return.
		Capital budgeting model.	Decisions on a New Zealand crop farm.	Criterion function based on tax-free cash on hand at end of planning period and value of assets at that time; subject to sufficient cash being withdrawn for financing the operators' personal requirements.
		Farm management decisions.	Fresh vegetable production.	Thoroughly described pair of papers: problem includes choosing decision dates (and thus stages in planning period), random events in nature at each stage (e.g. weather), probabilities of each state; possible activities and restraints need defined for each state of nature.
		Optimising organisational and managerial policies of a farm with uncertain rainfall.	Planning operation on a kibbutz in the northwest Negev with random weather impact.	Considered dryland crops and livestock management; policies related to herd size, average forage inventory, speculative inventories, hay acreage, sale of cows at varying ages under feed scarcity, winter grain–sorghum–Sudan grass planting related to weather, and ploughing under winter grain with crop failure; response surface developed by partial factorial uniform grid sampling (47 runs) and then fitting simplified response function.
		Farm growth.	Dryland cash crop farm, Texas Southern High Plains.	Specify initial asset state and their weights (constant dollars); develop a transformation relation to give dynamics (unspecified herein); show relation of accumulation of value to changes in assets and varying growth rates; develop criterion function to develop best sets of assets from all possible; provide for risk; make 20 runs for each stochastic situation and developed mean and variance.
		For farm investment plans for harvesting grain on UK farms.	To try out different investment strategies by changing single parameters one at a time.	The mathematical equations were derived mainly from experimental data. The price of computations limit widespread use by small firms; full equation not presented.
		To develop a method for improving the decisions made by firms.	To determine efficient grain inventories.	Limitations: (1) the technique results in only a partial solution to the problem of managing seasonal inventories; (2) the price of computations limits widespread use of quadratic programming by small firms.

continued

TABLE 4

Reference	General model	Specific model	Time stage	Attributes	Computer language
Stryg[218]	Optimisation	Linear programming			
Halter & Dean[219]	Simulation	Difference	1 day		DYNAMO
D'Aquino[220]	Optimisation	Linear programming		52 variables, number constraints unspecified	FORTRAN
Bartlett et al.[221]	Optimisation	Linear programming		Unspecified, but included limits of available land, limits of herbage use, limits on available capital	FORTRAN

[a] Other references:
Davis[222] Saunders et al.[225] LaFerney[227] Little & Doeksen[230] Burt & Johnson[233]
Tung et al.[223] Vandenborre[226] Sharples et al.[228] Charlton & Thompson[231] Crom & Maki[234]
Hardaker[224] Cloud et al.[122] Mo[229] Lin & Heady[232] Kuang[235]

TABLE 5
A review of agricultural systems models: example models primarily oriented from the article. Blanks represent

Reference	General model	Specific model	Time stage	Attributes	Computer language
		FISHERY[a]			
Larkin & Hourston[239]	Simulation	Algebraic/difference	1 year		FORTRAN II
Silliman[240]	Simulation	Differential	1 year*		Analog
Walters[241]	Simulation/Optimisation	Difference	1 year		FORTRAN

—contd.

Model diagram	Program code	What it was built for	What it was used for	Comments
		To maximise income. Model development.		Comparison of linear programming and Monte Carlo method by application to farm planning. Discussion of the advantages and disadvantages of the two methods.
Yes	Yes	Decision-making under uncertainty.	To improve the managerial decision on a large California ranch by using computer simulation under uncertain environments.	The two sources of uncertainty are weather and prices of factors and products.
		General resource management model for goods with supply and demand components.	Compare supplementation strategies on grassland cropland grazing system.	Based on data from Eastern Colorado Range Station; objective was profit maximisation; tableau not provided in detail; less 1% gain in income in nutrient supplementation strategy (from Ph.D diss.); no code provided.
		As above, but to handle multiple usage seasonally of grazinglands.	Determine hay production, class of stock, supplemental feed strategies, and season of use of four range sites.	Based on data from Central Plains Experimental Range; based on D'Aquino[220] approach but used discrete continuity equation to allow forage, animals and cash to flow seasonally; no code provided, no tableau provided.

Puterbaugh et al.[236]
Barker[237]
How & Hazell[238]

to fishery and wildlife. An asterisk means we have estimated this information not applicable or unspecified information.

Model diagram	Program code	What it was built for	What it was used for	Comments
			FISHERY[a]	
Yes	No	Simulation of the population dynamics of salmon spawning in a large river system.		Application of the models is illustrated by two examples, the reproduction is calculated from a series of theoretical reproduction curves operating in successive stages.
No	No	To simulate exploited fish populations.		The model is generally applicable to fisheries for which good measures of total catch, growth rate, natural and fishing mortality rates, and stock–recruitment relation are available. The model does not take account of immigration, emigration or environmental effect.
No	No	Fish population simulation and maximum yield determination.	Evaluate behaviour of fish population in response to fishing.	Deterministic model, using only biological variables such as mortality rates, fecundity rates etc. No objective test has been made as to, whether the model is able to predict actual data.

continued

TABLE 5

Reference	General model	Specific model	Time stage	Attributes	Computer language
		WILDLIFE[b]			
Beyer et al.[244]	Simulation	Difference	1 year*		FORTRAN
Davis[245]	Dynamic programming	Algebraic/ linear			
Parker[246]	Simulation	Differential	1 week	4 state variables 4 driving variables 9 constants	
Eberhardt & Manson[247]	Simulation	Difference	1 day	3 state variables 14 constants 3 flows	DYNAMO
Bledsoe & Van Dyne[248]	Simulation	Differential		6 state variables	Analog
Walters & Bunnell[249]	Simulation	Difference/ differential	1 year	181 parameters	FORTRAN
Walters & Gross[250]	Simulation	Differential	1 year		FORTRAN
Lobdell et al.[251]	Simulation		1 year		FORTRAN (GASP II)

[a] Other references:
Patten[242]
Rothschild & Balsinger[243]

[b] Other references:
Geier et al.[252]
Eberhardt et al.[254]

O'Brien & Wroblewski[255]
Martin & Turner[256]

Hacker et al.[357]
Mann[258]

—contd.

Model diagram	Program code	What it was built for	What it was used for	Comments
		WILDLIFE[b]		
No	No	To develop a Monte Carlo simulation of population interactions in a biome to provide a close fit to existing real data as well as a rational basis for quantitative predictions.	To simulate three species system over 15-year period.	A stochastic model employing an event (birth–death–predation), consequences technique: plant, moose, wolves.
No	No	For analysis of deer herd management over time, and deer and commercial timber production.	Deer management planning.	
No	No	To understand and regulate consequences of inorganic phosphate pollution and damming.	Management decision.	A deterministic model that was built to test if a relatively simple model could be used to simulate the observed behaviour of system components; aquatic ecosystem.
Yes	No	To represent the concentration of caesium-137 in a lichen–caribou–eskimo food chain.	Arctic food chain.	The model's principal function is considered to be analytical rather than predictive. A simple deterministic model of the food chain passage of caesium-137 in Alaska.
No	Yes			Based on literature studies for long-term changes of system on abandoned croplands.
No	Yes	To look at land use and big game populations in British Columbia.	As a teaching tool.	Plant production and succession wildlife habitat, food selection, and dynamics of wildlife herds. The model has proved useful in teaching students, and it may also be useful in research by helping to define critical interactions and data requirements. The model is a series of submodels. The model is general in a sense that no structure changes for application of the model to a different habitat are needed. The equations in the model are biotic interaction-oriented.
Yes	No	To gain insight to management decisions.	Suggestions of a management plan for white tail deer.	The mathematical derivations are described by Gross.[253] It is a single species population model that is based on interactions of density. Specific natality rates and specific mortality rates.
No	No	To analyse the long-term effect of addition of a spring gobbler hunt on the dynamics of a turkeys hypothetical population.	Simulation of a hypothetical population of wild turkeys.	Random variables: total annual mortality rate, the hunting mortality rate and reproduction rates were assumed to be normally distributed.

way; other papers we have reviewed in the same area are noted in the footnotes.

2.4.2.1. Description of the Tables

Generally, we have characterised the models as being either simulation or optimisation.

Under simulation we have included differential equation, difference equation, algebraic equation, and matrix equation systems. These simulation models have varying time steps with intervals of one hour, day, week, month, or, in some instances, year. We have attempted to list as attributes of simulation models the number of driving variables, state variables, flows or processes and parameters if this information could be readily ascertained from the article. We checked to see if the model was actually compared with field observations or if there was other validation.

For optimisation models we specify the type of technique if it was easily ascertained from the article. Generally, we reviewed linear, non-linear or dynamic programming optimisation models. In a few instances stochastic programming problems were reported.

In even fewer instances than in the above two types of models, there was a combination of some form of simulation with some form of optimisation. These models are identified as simulation/optimisation. Usually simulation/optimisation models couple a difference or differential equation simulation with a linear programming optimisation. Day,[53] however, couples time-series regression with linear programming.

We attempted to list the type of computer code where it was specified, such as FORTRAN and other compiler languages. We made a slight distinction as to what the model was built for and for what it was used. In the first instance, a broader or more general purpose often was the goal; but in the second instance a specific situation seemed to be the focus. Comments are provided on interrelationships of models, special characteristics, utility of findings, and some problems encountered.

2.4.2.2. Some Reflections on the Literature

Our general impressions of the models we surveyed are given as follows as a series of observations, critiques, and questions:

(i) By and large, we found the reporting of agricultural systems models inadequate for our purposes. We found it extremely difficult to assess the attributes of the model. We found it difficult to find exact and complete specifications of models. Thus, for example, we searched in the case of an optimisation problem (take linear programming as a specific) for the number of variables in the objective function, the number of constraints, and the number of non-zero parameters in the constant matrix. In simulation models we searched for a specific list of the driving variables, a list of the state variables, a list of the flows or processes as well as details on these items. We feel it is inexcusable now to report models in the vague way that they are reported in much of the literature. Perhaps it is understandable, however, at the time of publication considering the rapidly changing state of the art.

(ii) Many of the papers concerning simulation models would have greatly benefited by the inclusion of a suitable diagram of the model. Preferably, we would have liked to have seen flow diagrams of the Forrester[17] type, but any other well-described diagram would have 'been worth a thousand words'.

(iii) We were impressed that in most of the literature, models are simply built and reported but not analysed in detail. There was relatively, more analysis of optimisation models than of simulation models. Here such standard techniques as parametric analysis or examination of shadow prices frequently were made. For simulation models it would seem that only rarely did the authors test the sensitivity of the model to the initial conditions, the driving variables, or parameters. We feel that even qualitative sensitivity tests would be useful. More complete approaches such as that of Tomović & Vukobratović[4] would have been illustrative. Some workers are beginning to conduct experiments on the models to test sensitivity, such as the 2^n factorial design statistical analyses of the system as a method of sensitivity analysis.

(iv) We noted a distinct time trend in the papers reviewed in simulation models away from the constant-coefficient approach to specifying flows. Early papers frequently reported flows calculated as a constant proportion of the donor compartment. Such calculations seem rare nowadays.

(v) Some papers seemed to suffer from the possibility of 'circular reasoning'. The same sources of data that were used to build the model were used to evaluate (i.e. validate and verify) the model! In relatively few instances did we see a concentrated effort to get separate sets of data for developing models and for validating models.

(vi) Considerable evidence suggests that the potential use of the model is not clearly specified before the model is designed. That is, the objectives of the modelling exercise do not always match up closely with the model structure. It seems that the objectives are written last! Either this is the case, or else the authors become overly enthusiastic and make claims for their model such as 'with minor changes' the model could be applied to... situations.

(vii) Almost all of the models examined were deterministic. Most authors seem not to have recognised that the parameters have variances and covariances. An exception is provided by Dalton[54] who, although he only varied one parameter at a time, recognised the need of information about the covariances among parameters.

(viii) Closely related to (vi) is the problem of precision of model output. Particularly in the case of simulation models, very few statements are made as to the potential variance of the predictions from the model. Obviously, if the initial conditions, driving variables, or parameters in the flow functions are not known without error, then there could be considerable variability in the predictions from the model. The magnitude and direction of the variation could depend upon whether or not there are compensating effects among the parameters. This is an important problem when the time comes to apply model results. The interested reader could profit from examining Schutt[55] and Schweppe,[56] particularly the latter who treats the problem with rigour and humour!

(ix) Most authors say little about the computer implementation of their model. It would be valuable for potential users to know specifically what machine the model was run on or if it had been run on different machines. It should be noted that one cross-check on the code for a model may be obtained by attempting to run the model on various computers. We personally have encountered problems with code which had been 'triple checked' at one time on a given computer but was attempted to be run on another computer at the same time or on the same computer at a later time with a different compiler!

(x) Our analysis of the simulation articles suggests that there are increasing numbers of multiple-flow models. The early simulation models were generally concerned with only one type of material, primarily energy.

(xi) Our analysis of both simulation and optimisation models, and particularly the simulation/optimisation models, is that they have a 'once-only aspect'. It seems the models were of considerable value as a learning process; but after the 'new had worn off', relatively little was done with the model. We will review below, however, two examples where models evolved over time (see Section 2.5.3.3) and a record was made of the learning process for the modellers.

(xii) We see in the simulation models we received, a somewhat higher ratio of flows to state variables in recent as compared to early efforts. This indicates an increase in model complexity, i.e. a greater degree of coupling among the variables within the model.

(xiii) There is an almost complete lack of treatment of spatial problems in simulation models; but spatial problems are handled in a number of the linear programming models.

Following these criticisms, we feel it will be useful if we identify some type examples to which many of our objections above do not pertain. Our presentation of these type examples, however, should not be construed necessarily as what we visualise for a complete model presentation. We will discuss model presentation in Section 2.6, particularly in 2.6.4. We suggest the reader review the following modelling papers as useful guidelines: simulation, Vickery & Hedges;[57,58] optimisation, Dean et al.;[59,60] simulation/optimisation, Swartzman & Van Dyne.[61]

2.5 PERSPECTIVES ON AGRICULTURAL SYSTEMS ANALYSIS

2.5.1. Generality and Precision of Agricultural Systems Models

2.5.1.1. Realism, Precision, Generality, and Resolution

Four major properties of models important in their evaluation are *realism*, *precision*, *generality* and *resolution*. Any model must in effect compromise each of these properties, and the degree of compromise for each of them depends upon the modelling objectives. One of the major problems in modelling is the seeming incompatibility between obtaining precision and generality within the same model structure. Levins[62] argued the point emphatically some years ago that the two objectives are contradictory.

Assuming that the model hypotheses are realistic, precision then is the ability of the model to faithfully predict the time course of the model variables. Qualitative and quantitative aspects of a model formulation, respectively, might be termed the realism and precision.

Resolution of a model is more difficult to describe. In general, resolution is conditioned by the domain of the model. By domain we refer to the objects to be included in the model and the time span for which the model will predict. The objects to be included specify the extent and depth of the system, and the time span specifies the shortest interval and the overall duration of a prediction. In general, but not always, models with more state variables have more resolution; models with mechanistic, as compared to empirical, descriptions of flow functions have more resolution; and models attempting to abstract many qualities of a real system have high resolution while one which includes only a few attributes has low resolution.

Further consideration of model properties is given elsewhere.[63-65]

2.5.1.2. Need for Generalised Models

A skilful modeller can select or develop appropriate mathematical forms which will behave in a proper manner to follow the general properties of the system being modelled. Knowledge is required of the biology, physics, and economics of the situation being modelled and of the working of various mathematical forms to be able to choose a realistic form for the problem at hand.

There is considerable practical reason to hope that a general model can be built to preclude the necessity of building a specific model for each individual situation. A general model may not itself be adaptable to a wide variety of related situations for simulation and optimisation; it may not be widely adaptable as in the 'formulistic approach' for statistical models. There is also much more generalisation of models in optimisation than in simulation. For example, consider the generalised linear programming formulation (see Section 2.1.2). This is the most widely utilised mathematical formulation in business and industry. A general model may cover, nevertheless, the general rules and guidelines for developing specific models for a given situation.

Levins[62] indicated it is 'desirable to work with manageable models which maximise generality, realism, and precision toward the overlapping but identical goals of understanding, predicting, and modifying nature'. He concluded that this cannot be done! Some examples follow:

(i) One group of scientists have developed models related to the short-term behaviour of a particular organism, where fairly accurate measurements were made. The models are developed for the computer, and end with precise, testable predictions applied to those

particular situations. This was a *sacrifice of generality to gain realism and precision*.
(ii) Other workers used the Volterra predator–prey systems which omit time lags, physiological states, and the effect of a species population density on its own rate of increase. Such workers hope that many of the unrealistic assumptions will cancel each other and that small deviations from realism will result in only small deviations in conclusions. Thus, starting with precision they hope to increase realism, but in actuality they *sacrifice realism to gain generality and precision*.
(iii) Another general approach is taken when one is interested in the long run in qualitative rather than quantitative results. The justification here is that quantitative results are important only in testing hypotheses. If one is interested in the qualitative approach, then the models may or may not necessarily be implemented numerically. They may only be graphical; one is simply concerned whether the functions are increasing or decreasing, convex or concave, or greater or less than some value. Thus, one *sacrifices precision in order to attain realism and generality*. But the utility of such models for agricultural systems may be questioned, at least in the short run.

It probably takes all three types of models listed above to form a usable theory. Although the scale has increased greatly as compared to earlier models, models are still restricted by technical considerations to a relatively few components at a time, i.e. few compared to the total number of components in an agricultural system. Thus, a satisfactory theory of agricultural systems, as well as any other renewable natural resource system, is usually a cluster of models.

Holling[64] also questioned whether it is possible to develop realistic systems models with generality. He considers the *precise or detailed model* as '*tactical*' models and the general model as '*strategic*'.

An innovative approach which may overcome some of the problems of generality and precision is provided elsewhere.[66–68] These approaches are discussed in more detail with respect to hierarchies and modules of models in Section 2.6.3 along with problems related to time scale and interfacing. The basic concept is, however, that generalised model structures can be developed in which various submodels can be inserted to provide different levels or degrees of resolution. There are yet, to our knowledge, no detailed systems for general models of agricultural systems of this type.

2.5.2. Model Makers and Model Value

2.5.2.1. Model Use and Utility

It is illustrative to examine utility of models, as they are viewed by the model makers and by others.

Morales[69] outlines a series of general questions that one should ask about models: 'Are models useful? If so, for what? Do they produce more than what is put into them? What previous knowledge ... [is] ... necessary before the model can ever be constructed? And when the model is constructed and utilised, in what ways do its results differ from the results obtained without models?' He further suggests in modelling meetings that a clear distinction be made to the model question *per se* versus the model data.

Various models serve various purposes. A classical function of models in science is to state explicitly the simplified mental constructs of reality in the form of assumptions associated with a train of deductions leading to hypotheses.[70] If one follows this approach, they are led quickly to the mechanistic models (see Section 2.5.1, discussion of resolution). But there are various other ends to be served by models.[71]

(i) For some models, e.g. as for some in game management, one is not so concerned with an understanding of basic processes as they are with aiding practical decisions.

(ii) Other models, such as those sponsored by the Club of Rome, may not extend knowledge as much as they will increase public awareness of potential problems and thus hasten political action.

(iii) Even other models aid experimental analyses of data, analyses that cannot be approached either experimentally or analytically because of the complexity of the experimental situation.

All of these uses of models are valid. All are needed for advancement of science. Thus, field biologists and experimentalists who are arrayed against men who use computers, analytical mathematical models, or both should *at least clarify the purpose of the model before assessing its value!*

Agricultural system models have been developed for different purposes and are used in different ways. Most authors reporting agricultural system models still are at the stage of saying 'these results *could* be used....' Frequently the modeller emphasises potential uses of a simulation to evaluate the effect of some hypothetical and proposed treatment or set of treatments on the total system and not just those species which are primary subjects of control. Many authors have emphasised the economy of using models to examine the potential outputs of various treatments to minimise the actual number of treatments that might have to be examined experimentally. Examples follow.

Goodall[72] claimed that his model can potentially 'serve practical ends as a management tool, enabling the effects of disturbances ... to be predicted'. In discussing the construction of his model, however, Goodall noted that present experimental work conducted for other purposes usually has not provided data of the type required to estimate parameters needed in the model. Therefore, he concludes one of the values of an approach like his is that it can 'serve as a guide to future experimental work in indicating what types of observations are likely to be of greatest value for predictive purposes'.

2.5.2.2. Utility and Model Structure

Goodall[73] focused on the practical value of ecosystem modelling, which has direct analogy to agriculture system modelling. He emphasised that intuition needs aid and guidance in making resource management decisions in order to make reliable predictions. And this aid and guidance can be provided by modelling.

Extensive practical use of a model implies that a biologically sound conceptual framework has been developed for the model. This is a prerequisite. But a biologically sound conceptual framework, although essential, is difficult to achieve. Data must be available to show the form of functions relating the driving variable and state variables to rate processes and for securing the parameters in these functions. One must do some aggregating in modelling agricultural systems. Or else you would end with modelling each individual organism separately. Usually it is possible to do some aggregating without too much loss in information. But when modelling the system, one must balance precision and practicality.

It has been noted that *over time scientific groups tend to make models progressively more complex* mechanistically, i.e. with higher resolution.[5] Thus, over a period of time models tend toward less and less complex structural levels of organisation in the flow functions in order to explain changes in the state variables at a higher level of organisation. Complexity is noted here in the context of levels of organisation, such as shown in Fig. 1 in Section 2.1.1. For example, consider the levels of organisation in decreasing complexity as follows: ecosystem, community, populations, organism, tissue, and cells. If the state variables are mainly at the population level, the flow functions primarily incorporate information at the organism or the organ level. With increasing knowledge over time there is the tendency to break the state variables down to the next lower level of organisation. It follows, then, that the flow functions describe the changes in the state variables as they drop to an even lower level of organisation. This, of course, results in an increased number of flow functions and increased number of parameters to be estimated to be able to operate the model. DeWit[74] emphasises there is some basis for trying to contain the levels of organisation within the model within two or three.

No discussion on value of models of biological systems would be complete without consideration of the Lotka–Volterra type models. These models were introduced early and are characteristic of many fields of ecology. But they are more related to the basic science/theoretical approach than the applied science/practical approach. A detailed summary of these models has been published.[75] Initially, these were single differential equations showing change in population size for a species. Subsequently, coupled equations were developed, commonly for two species in a competing situation. And eventually an allowance was made for inclusion of external environmental effects. Usually, these models contained only system state variables and no external driving variables. Although only relatively simple models have been built, relatively complex analyses have been made of the Lotka–Volterra system models.

Because of the recent thrust on total system modelling the Lotka–Volterra equations have not been central to much of the present-day modelling. This is because they are primarily concerned with species interactions rather than such important considerations as nutrient cycling, energy flow, and management of the total system—biotic and abiotic.

2.5.2.3. Who are the Best Modellers?

'So far the best models are those prepared by biologists turned to mathematicians, and the more proficient the biologist is in mathematics, the better the models.'[76] Most of the models prepared to date have been prepared by one individual or very small groups of individuals. Most models tend to be of special design and adapted to particular problems. These models, developed primarily by biologists, may be satisfactory for the purpose at hand but frequently are formulated so that it is difficult to extend the model into new and unrelated situations. On the other hand, those models developed by mathematicians tend to have good general formulation. Because of their good formulation they are often very flexible and can be extended into new areas. But the mathematicians tend to oversimplify the biological world to such a degree that the models very often are unreal and perhaps unusable.

The logical alternative is that models should be prepared by teams of mathematicians and biologists. Within these teams the biologist must describe the biology of the situation. The mathematician must have described to him how the system operates. The mathematician's translation into a formulation must be reviewed by the biologists for appropriateness. Similarly, the biologist must examine the outputs to determine if they are logical and reasonable. *The biologist is the primary modeller.*

2.5.2.4. Data Base, Model Utility and Time Requirements

A general examination has been made of models published for agricultural systems to see the relationship between data and models. In most instances where there has been a strong relationship between data and models (see Section 2.4.2), the modelling work was either undertaken as part of the overall study or was based on the data collected by one of the authors. In a few instances the modelling group is part of a large experimental programme.[16] But to date, *most modelling reports do not include extensive references to the data* from the experiments in the programmes! An exception to this is a developing report on a total-ecosystem model (Innis *et al.*[77]) that is relatively strongly data based and is of high complexity.

Another example of a thorough, useful model based on considerable data is the model of May *et al.*[78] This series of papers, the most recent of which was available to us in manuscript form, is based on experiments which were initiated approximately seven years prior to the last publication.[78–81]

This phenomenon has implications for the organisation of modelling efforts in agricultural systems studies. Firstly, *modelling and model reporting should not be considered a once-only process.* Secondly, *the 'final' model may be published some five to ten years after the initiation of collection of experimental data.*

2.5.2.5. A Note on Speculations

It is interesting to speculate on the future of agricultural systems models. Particularly, at a symposium like this, surrounded by friends of the art and learning day by day of the increased availability of terminals, time-sharing systems and larger computers, one gets almost euphoric! But in dealing with computers, models, or any other tool one is brought back to ground by the interesting statement by Sterling & Pollack[82] recalling some predictions made shortly after microscopes were invented:

'One ambitious scheme for which the microscope was to be used was to count the number of generations that could be found in an ovum, in the naïve belief that in this way the time remaining until Judgement Day could be assessed. It is a credit to man's imagination that not only were some investigators able to differentiate the yet unformed eggs into 'male' and the 'female', but others were actually able to count the number of future generations contained in the confines of the ovum.'

Our friend, the late G. J. Paulik, notes, 'It is safe to predict that many of the results of early simulation modelling will not be any more reliable than the early microscopists' estimates of the time remaining until Judgement day!'[83]

2.5.3. Evolution of Models

Modelling is an expensive pastime. Many of the models summarised herein have had several, and some more than ten, man-years of modelling time spent in their development (see Section 2.5.4). Some of the models have developed over a long period of time involving a series of steps. Some overall steps and approaches in modelling are described below and examples of the evolution of different models summarised.

2.5.3.1. Top-down versus Bottom-up Approaches

There are probably two alternative strategies to be used in total system model development—*top-down and bottom-up modelling approaches*.[16]

In the top-down modelling approach one *starts with a series of objectives which dictate an overall total system model*. This model is continually improved by changing its structure in an organised stepwise manner. These changes are of two major types:

(i) Improvement can be made by increasing the mechanism in the flow function descriptions. It is possible to improve such mechanisms uniformly. More and more state variables and driving variables may be used to predict the change in a rate process. More and more complex relationships may be used for a given set of variables and a particular process, e.g. switching from a linear relationship to a non-linear relationship between the rate process and some particular state variables.

(ii) A second phase of development requires subdivision of some of the state variables, thereby introducing new variables.

No matter whether or not a method of logical approach is taken in the structuring of the modelling system at the outset, eventually one will reach 'ultimate units' of the system which must be treated as black boxes. The top-down philosophy is in a sense initially mechanistic in that it tries to subdivide the large system into a group of smaller components that can be handled adequately (see Section 2.6.3 regarding the hierarchical concept). Eventually, however, *basic units are treated in an empirical manner.*

The contrasting bottom-up approach is to *start with isolated and independent high-resolution and complex process models.* The procedure then is to couple together these complex process models eventually into subsystems, and the subsystems eventually into the system.

In most instances when one reviews evolution of models, the tendency appears for scientists to increase continually the complexity of the individual model, rather than to structure a framework to couple submodels and to elaborate remaining models after the initial few have been started (see Section 2.5.2.2).

There are advantages to either the top-down or the bottom-up approach. In the case of developing an isolated and independent high-resolution, complex process model one can have a simpler system with which to deal. Rather than dealing with 200 to 300 flow functions of the total system model, one may be considering only some 20 or 30 flow functions at a time. Rather than dealing with 75 to 100 state variables, one may be considering only 5 to 10. But difficulties often ensue when one tries to couple independently developed process models into a submodel, or submodels into a total system model. The advantage of being able to see and having a feel for the total subsystem being modelled may be overcome by the disadvantages of connecting subsystems.

A great deal of effort is required in conceptualising the overall model. This allows one to decouple the model into component parts which can be analysed independently. For example, upon examination one can see that each submodel requires certain inputs and gives certain outputs. Listing all the submodels with their inputs and outputs one can find the degree of coupling between submodels in the total system. If there is a high degree of coupling, it may be advantageous to do the model development *in toto* rather than piecemeal. If there is a low degree of coupling, then the driving variables from another subsystem which affect a process in the subsystem of concern may be read in as generated driving variables.

2.5.3.2. Evolution of Modelling Management

There are important managerial problems in modelling large systems because many individuals are involved. It is a doubly difficult problem. Some experience can be drawn from the work of Hamilton *et al.*[84]

When the number of modellers involved in an overall project is small, frequently the coordinator is a subject matter specialist himself. Generally, such a coordinator will have less interest or background in managing the work of others rather than participating himself. Assembling a team of competent personnel does not automatically lead to efficient research output.

Many of the scientists have neither the ability nor the interest to coordinate the work of others, and many of the scientists cannot be integrated into effective research teams! Secondly, personnel time is extremely expensive and budgets are seldom sufficient to the tasks outlined in programme proposals and cuts must be made. The consequences are clear; it is often considered too expensive to support an administrator–coordinator for only a small scientific staff.

An astute manager of modelling work may understand the system sufficiently well and spend enough time on the managerial aspects to be able to disaggregate the overall system and to treat different segments separately. A problem occurs in some types of agricultural systems models where there are large numbers of interconnections among the variables. Individual variables or flows or groups of flows may be the task of an individual modeller. An individual who sees his subtask as an end in itself may often work ineffectively.

Managing system modelling is an art. We are striving to find the minimum degree of interaction that must take place among all modellers, the minimum degree of structure that is needed to ensure coupling of all parts of the model and the kinds and number of personnel needed for these efforts.

A major management challenge in the top-down approach to organisation of models is in getting commitments from individual researchers. Starting with an overall systems model in the top-down approach, one soon finds there are components of the study which are of relatively little interest to many of the available researchers. This is particularly true if the researchers are drawn from academia and their main focus of attention on a year-long basis is on other matters. They are inclined to have relatively narrow research interests so that they can cope with the problem of keeping abreast of subject-matter techniques and knowledge in the limited time they have available to them. To some degree, in contrast, the researchers in federal or state agencies who are working full time on research have more opportunity for broader ranging. In any event, commitment from a suitable group of researchers is necessary if the models are to be improved upon in different revisions in the top-down approach.

The bottom-up approach isolates modelling activity into individual areas of research in which the investigators already have a strong interest and commitment. Unfortunately, as has been noted through experience, it is difficult or impossible to interface models developed in this approach into a larger-scale model.

2.5.3.3 Examples of Model Evolution

An example of an evolving model is that of Goodall.[72,73,85-88] This is a model concerning grazing management in an arid or semi-arid rangeland. Originally it was structured with a single herbivore, but subsequently with populations of different species of herbivores. This is essentially a difference equation model with early versions having about seven separate kinds of state variables. There were little or no data for validation of the model. The equations were reported in early versions. The model incorporated spatial aspects by a structure which enabled simulating a paddock with water supplies placed at different points and with different topographic features, such as slopes and flats. The model was driven by records of weather conditions, in the form of mean monthly records along with records of frequencies of different meteorological events such as daily totals of precipitation. This is allowed by a random sampling of meteorological conditions for a particular run. For drawing conclusions many such runs would be made to get an average response. By varying the numbers and kinds of species, by varying the numbers and placements of water points, and by controlling competing undesirable herbivores (such as rabbits), information can be gained from the model which would be a valuable aid in making a cost-benefit judgment on the potential value of the manipulation practice.

Another evolving model which has been reported several times is an Australian study of sheep on summer pasture, written in difference equations, originally with about five state variables, varying from SIMSCRIPT to FORTRAN coding, originally with relatively weak data relevance, but having the code reported.[89-93]

2.5.4. Costs of Models

It is difficult to assess the costs and benefits of models *per se*. Models are now deeply incorporated into many research and training efforts. A few authors have presented estimates of the specific time spent in the mathematical implementation of a model and sometimes they record the computer costs. Computer costs for a given job are generally smaller with larger, faster machines. Remote terminals and conversational compilers can also decrease development costs.

Most models at the usable agricultural system level require several man-years of scientific effort in model development and testing. But the amount of time required in model development depends a great deal on whether or not previous work was done that assisted the modelling effort. For example, Swartzman & Van Dyne[61] reported a simulation/optimisation model which was a combination of a difference equation simulation system having 94 state variables and 243 flows or couplings among the variables and an optimisation model with a linear programming solution with 36 decision variables and 36 by 47 original constraint matrix with 144 non-zero parameters. The development of the simulation portion of the model was greatly assisted through the use of a special compiler called SIMCOMP.[94] The implementation of the optimisation model was assisted through the use of a standard linear programming routine. These investigators spent about one scientist man-year in the model development and about US $1500 CDC6400 computer time for development, debugging, and making runs. Even so, the model was developed only to the running situation; it was not extensively experimented with or utilised.

Arnold & Bennett[1] report the cost of an agricultural system simulation model with which they have worked over a five-year period. The model was built in a step-wise manner with various submodels completed sequentially. They have invested about ten scientist man-years in the modelling and a computing time cost of about US $8000.

A large-scale, multiple-flow, total-system grassland ecosystem simulation model has been constructed as part of the US International Biological Programme Grassland Biome study.[18] Some 15 to 20 man-years of effort have already been expended in its implementation alone and this does not include development of earlier models which were the starting points. Probably not all the scientific manpower this model requires has yet been expended, for the major publications on the model are yet to come and it has yet to be tested widely over the range of grassland systems for which it was developed. It is difficult to estimate the amount of money spent in computer time in the development of the model to date, but a typical two-year simulation run with a two-day time step requires approximately 7 min of machine time (compiling and running but not input–output time) on a CDC6400.

The high costs of models are becoming increasingly well known. This leads many new modelling efforts into attempting to develop generalised models which can be adapted, by parameter changes, to a variety of situations.[18] This has introduced the concept of developing 'canonical' submodels for plants, consumers, etc. The utility of this approach has yet to be fully tested for agricultural and renewable natural resource systems.

2.5.5. Time Trends in Agricultural System Models

Agricultural systems deal with populations of soil organisms, plants, and wild and domestic animals and insects. Agricultural systems have the characteristics of multi-species systems, a type of system that Levins[62] not long ago considered fairly complex. He was discussing the models of single populations or species and considering the difficulties that would be encountered if multi-species systems were to be modelled. He indicated 'the naïve brute force approach' would be to set up a mathematical model which is a faithful one-to-one reflection of this complexity. This would require using 'perhaps 100 simultaneous partial differential equations with time lags', measuring hundreds of parameters, solving the equations to get numerical prediction, and then measuring these predictions against nature. He further went on to discuss that there were too many parameters to measure, 'the equations are insolvable analytically and exceed the capacity of even good computers'.

Since this time models have been built of and run for ecological, agricultural, and natural resource systems having 100 simultaneous equations (albeit not partial differential equations), models having time lags, non-linearities, and thresholds, and models having hundreds of parameters.

New journals appear each day with papers reporting the merits and advantages of simulation, systems analysis, and operations research techniques as applied to agricultural systems. Although a large number of papers are available in agriculture, still the majority of the published work on applications come from such fields as manufacturing, marketing, corporate structure, finance, inventory control and production. For the agriculturalist it would appear there is an acceptance lag.

Some five and a half years ago a symposium was held at the Grassland Research Institute at Hurley on the topic of models in agricultural and biological research.[95] It seems useful to make a brief comparison with the current symposium, and our review, with respect to trends. Some of our speakers and several of our participants attended and spoke at the earlier symposium. We are dealing with much the same topics as some five years ago—how to model, discussing example models, and illustrating deficiencies and needs. The present symposium places more stress on optimisation, less stress on training, less stress on various aspects of numerical analysis methods such as integration and of computer compiler systems and more stress on economics of the firm. In comparison with the literature we have sampled, the symposium five years ago contained several papers with clear flow-oriented diagrams, clearer specification of the computer language and compilers utilised but on the average poorer description of the flow equations. Several papers of the present symposium and the recent literature, as compared to the earlier symposium and literature, illustrate more post-development analyses of the models. This we feel is a healthy trend, and perhaps an outcome aided by the earlier symposium. The symposia and the literature are all alike in one main respect, all show there is much to be done to improve models and modelling and to convince those few who just will not seem to come around!

2.6. COMMENTARY ON DESIGN AND REPORTING OF MODELS

2.6.1. Steps in Hypothetical Model Building or Modelling Project

2.6.1.1. The Diagrammatic Stage

A first step in modelling, after definition of the overall subject matter, is to *develop a detailed word model* of the system of concern. This might be followed by what might be called a *relational diagram*. Such a diagram is free of many constraints but shows variables and relationships among variables without a rigid diagrammatic structure. The relational diagram would eventually be converted to the flow-oriented compartmental diagram. At this stage a 'degree of discipline' would be introduced in standardising the notational designation of flows, of variables, etc. (see Section 2.3.2.1 for examples).

After listing the individual flows one must determine what variables affect those flows and show graphically the relationships. At this stage, it is often possible to use least-square curve-fitting regression techniques to analyse data and develop equations to interrelate the flow, as a dependent variable, to different state variables or driving variables as independent variables. However, this analytical method may be more complicated than may be needed at early stages.[67] Consider the example where the modeller has observed in the literature a graphical relation between two variables which appear, in general, shaped like a parabola. One can simply scale three points from the graph to determine algebraically the constant for a quadratic curve to approximate the observed relation. This may be sufficient for initial purposes and allow the work to proceed.

2.6.1.2. Specifications for Submodels

It is important to define at an early stage a series of submodels. This is a practical consideration related to the development and the testing of the overall model. It is useful to break up the overall model into smaller submodels that can be, for development purposes, handled independently. Note, however, this is not equivalent to the 'bottom-up' approach discussed in Section 2.5.3.1.

Each submodel can have two types of inputs or factors affecting it: inputs from other submodels, i.e. levels of state variables in other submodels, or inputs from the driving variables, i.e. those variables which are external to all submodels. Each submodel can provide two types of outputs. These outputs are not to be confused with output variables in the context of driving-state-output variables in an overall model. The submodel in question may provide output information to other submodels, or the output information simply may be part of the total system result.

It is *not practical to combine all equations from all submodels into a single top-echelon model for the very first simulation run.* Yet it would not be a system model unless there was a fairly high degree of interaction between the different variables. For example, the question can be asked, 'How can a plant growth submodel be realistically operated without herbivore submodel interaction, and vice versa?'

For initial development runs, all the variables inputting to the submodel must be considered at least 'pseudo-extrinsic'. They are entered as sets of tabulated values or as functions having no dynamic interaction with the submodel. These extrinsic inputs should have tabulated values chosen with sufficient amplitude and combinations to give an adequate test to the response of the submodel. Care should be given to select tabulated values so that most possible situations which might arise in the higher-echelon model could be experienced. Systems engineers term this the 'open loop' response of the submodel; examination of models in this way can reveal important facts about the stability of the system 'closed loop' or whole system conditions.

The selection or specifications put on the submodels is extremely important. We have derived some general considerations about this problem from Clymer & Bledsoe[67] and Clymer.[96,97]

(i) The number of submodels in the overall system should be developed to ensure that the individual submodels have say between 5 and 15 differential or difference equations at the outset.
(ii) Each of the variables of concern in formulating the word model and subsequent diagrammatic models must be provided for in some submodel, either mechanistically or non-mechanistically.
(iii) The submodels must concentrate on the phenomena of chief interest. A minimum number of other phenomena should be

involved. The grouping of submodels, when run individually or collectively, should answer all high priority questions regarding the behaviour, composition and structure of the system.

(iv) One should minimise the number of time scales to be used in the submodels in the interest of simplicity and compatibility. If possible, nearly all submodels should be designed for the same time scale.

(v) The names and scopes of the submodels should be compatible with names and scopes of academic disciplines or established scientific specialties as much as possible.

(vi) The submodel should be designed with thought to the persons who will staff the modelling teams. The models must not be foreign to their experience and interests, i.e. not too large and complicated for different modelling skills, yet not so small and trivial as to be unmotivating.

(vii) When submodels are coupled together into a total system model, the inputs to a given submodel, which came as outputs from other submodels, become the coupling variables.

The uncoupled parts of the overall model are of value in the structuring process. But as a final product, uncoupled submodels are useful only on a short time scale because the lack of feedback loop closure is important. It is probable that appropriate driving variables cannot be generated short of modelling the associated systems, and even less likely that they will be realistic if the feedback loops are not included.

It is critical, on initial modelling, to design the model or submodels simply enough so that they are capable of representation for the human mind to comprehend, to encompass mentally, to hold in the mind at one time.[98] Although whole agricultural systems are extremely complex, they can be abstracted to simpler systems. We know that extremely complex physical systems are abstracted into simple models. *We should not be dismayed by attempted work on agricultural systems.* This is emphasised by Spedding[98] who indicates the 'apparent arrogance of dealing with whole grassland systems largely disappears if it is realised that a cell, a cow, a grassland system or a national industry are all systems with as many components and as much complexity as the investigator cares to recognise'. The modeller must be unafraid of either mathematics or economics. He must at times be willing to treat the boundaries between animals and plants as not particularly useful for the purpose at hand. The model builders should be intelligent and informed!

2.6.1.3. Calculation of Flows

There are three general ways in which flow rate might be represented:[16]

(i) The first and simplest way is to depict the flow using a *constant-coefficient approach*. For example, a given flow might be calculated as a constant fraction times the amount of the donor compartment. Here one assumes that the proportional amount flowing always remains a constant over all times and conditions.

(ii) A somewhat more mechanistic representation would be to allow the proportionality fraction to vary at different times or conditions, i.e. a *time-varying approach*. For example, the fraction might increase in one time of the year and decrease in another. Here we might make the fraction a sinusoidal function of time by making the coefficient vary with time.

(iii) The third and more general formulation is that any flow rate may be a function of all the state variables and all the driving variables in a system, i.e. in theory. The function representation can be non-linear and multivariate in nature. It is a *system-dependent approach*.

In the actual construction of the flow equations where some rate process is calculated to be a function of several state and driving variables, the question arises as how to combine these effects. A search of the literature will often show that important rate processes are functions of many different state variables or driving variables in the model. However, the same search frequently shows that no experiments had been conducted in which all possible combinations of the variables have been included. There are several alternative strategies in constructing the flow functions to include the impact of several variables. Two major approaches might be called the *maximum-reduction* approach and the *limiting factor* approach.

In the first instance, one sets the maximum rate available for a given process. This maximum rate is based on 'ideal' conditions. Then the maximum rate is reduced at a given time according to how far the driving variable and state variable conditions vary from the ideal conditions. The rate process is set at the maximum and then is multiplied by a series of terms all of which have a range of 0 to 1.

Several variables may interact, and the above simplified procedure does not take into consideration the interaction effect. This can be approached, as in a statistical model such as multiple regression analysis, where 'new' variables are regenerated as products of other variables, i.e. interaction terms.

The limiting factor approach reduces the optimum rate according to whichever factor is most limiting.

As one goes through increasingly complex approaches to generating flows for a simulation model, we are struck by the problems of complexity. O'Neill[99] made an interesting analysis of the *effect on systematic bias, measurement error and prediction error* as one changes from the constant-coefficient linear differential equation system model into a much more complex formulation of flow functions.

There are inaccuracies or errors in the information predicted by any model, probably due to two major components, namely, a systematic bias and a measurement error. Frequently, systematic bias may be decreased by increasing the resolution of the model, perhaps by increasing the number of state variables. This generally decreases prediction error, but at the same time it adds more parameters to be estimated in the model. This causes an increase in measurement error. Increase in measurement error can be expected to cause an increase in prediction error. One must balance the gain from increasing complexity and the loss from increasing measurement error. There are no hard guidelines, but it would seem that as a model increases in complexity up to a point there may well be great improvements in decreasing prediction error. Then, however, prediction error may begin to increase.

2.6.1.4. Comparison with Engineering and Economic Approaches

Engineers generally build models starting with detailed knowledge about the separate components. By building submodels they are able to combine these eventually into a system model. This procedure has had notable success for physical systems.

In economics an entirely different type of system is being examined, i.e. one which is noisy and complicated. In these situations generally long time-series data on different variables are obtained and the models are in effect constructed working backwards from the observed total system response results.

But neither of these approaches, the engineering nor economics approach, may necessarily work in the ecologically based agricultural systems. Instead, we must make more use of our considerable knowledge even though it may be descriptive in many instances. We need to segment the systems into their component parts and study the performance of the parts and recombine these into testable subsystems at higher and higher levels in the hierarchy (see Fig. 1).

2.6.1.5. Trends in Modelling Costs in Large Research Programmes

The above discourse shows clearly a possible role of modelling in an overall research effort. But what is the actuality, and what are the trends?

Watt[100] clarifies the magnitude and importance of the synthesis and modelling activities in what he considered 'typical ecological research programmes'. Comparing large-scale ecological-resource research programmes during two intervals (1920–35 and 1945–65), he notes *trends in research resource allocation:* a decrease in the proportion of effort in the systems measurement and systems description phase, an increase in the relative effort in systems analysis (i.e. data processing and statistical analyses) and little relative change in writing and systems simulation and optimisation. But simulation and optimisation accounted for less than 1% of the research effort!

Beginning in 1967 on a worldwide scale, a number of integrated, interdisciplinary ecological research programmes were initiated under the auspices of the International Biological Programme. In the United States a series of biome studies were developed as part of this effort. The largest of these to date has been the Grassland Biome study, which is described in detail elsewhere[5,16,18,22,101,102] and which can be considered typical of the other biome studies insofar as percentage allocation of funds is concerned. Over the period 1968 to 1974, some US $10·25 million have been spent in the grassland research and about 12% of this has been in development of simulation and optimisation models.

UNESCO is developing a new research programme called 'Man and the Biosphere'. Modelling will play a role in these researches. A panel of experts commenting on systems analysis and modelling approaches for this programme placed strong emphasis on model development in the overall research.[103] The proposed research format for this programme included successive phases of setting of objectives and preliminary synthesis, experimentation, management, evaluation and a final synthesis:

(i) In the initial phase, emphasis would be placed on establishing objectives and performing an *early synthesis* of existing information. Following setting of objectives, relevant literature should be assembled and synthesised in a preliminary manner and then the objective re-examined and clarified. This activity would require not only the collection of data and information but some degree of condensation and storage. The organisation of this information would have considerable value in the next phase.
(ii) As a second part of synthesis, there would be *model development, which still would precede experimentation.* Recommendations were made for successively forming word or logical models, followed by diagrammatic or graphic models and then by computer-implemented

numerical models. Models at this stage would be structured largely from the literature, from knowledge of experienced scientists and managers, and not necessarily on new experimental data.

(iii) The next research phase would involve *experimentation in the field and laboratory and on the mathematical models*. Specific suggestions were given for the collection of coordinated sets of driving variable data and of time series sets of state variable data for validation of simulation models. After some iterative improvements of the models, experiments would be conducted on the models to lead to new ideas of management of the systems being examined.

(iv) Eventually, suggestions were made as to how simulation and optimisation models of natural resource systems could be developed and combined and subsequently used in *management games*.

Thus, models would be a key part of the overall research effort, they would precede the detailed experimentation, they would interact with data collection, and finally they themselves would become objects of experiments.

2.6.2. Relationship of Model Objectives, Structure and Output

It has been indicated above (Section 2.5.4) that most agricultural system models for all but the simplest objectives require several man-years of modelling effort in addition to the effort for data collection and reduction. This is important when we consider that we want the models to be useful. The user has pressing problems to which he needs a solution today or tomorrow. Certainly he needs it before many, many years of effort. Those who build models have great enthusiasm for model utility. Yet model building of complex agricultural systems is still at an early stage in the art. Partly because of this, many practitioners couch their output in jargon and detailed excursions into esoteric theory! The potential user of models is busy making decisions and does not have time to 'cut through all the jargon'. As a result, he must proceed to make decisions the best way he knows based on experience and a cut-and-try process.

The implication of these phenomena is that the objectives of the model become increasingly critical because of the time and money required to complete the effort and the potential utility of the output. This has led to a great deal of modelling work to be directed to small problems (small in the sense of the total system) and specific questions. By narrowing the model to specific questions of interest to the resource manager, the model size is reduced to a degree that model development becomes more rapid. But, unfortunately, this could be akin to a 'quick technological fix' which has backfired so many times in development schemes.[104] A good answer may be developed quickly for the wrong question.

There is a strong need for developing models that encompass many parts of the system. This is discussed further in Section 2.6.3 with respect to the ability or inability to run submodels out of context of the rest of the system.

We need to condense our experience on how model objectives condition model structures; this is not well described in the literature. Much of the process of modelling is never described in the reporting of model results.[105]

It has been noted repeatedly that all models are simplifications of the real system and that no model is complete. The model objectives lead to simplifications. In studying the objectives there are two extreme situations, and each leads to obvious traps in modelling:[14]

(i) The first logical trap is being overly impressed by the complexity of the physical and biological processes operating within agricultural systems. For example, there are many stochastic inputs in such systems, both in the driving variables affecting the system (such as weather) and in the determination of parameters in the flow functions. Attempting to build a completely realistic and detailed model of all the physical and biological processes within a system results in an intractable model. Unfortunately, distribution functions for parameters may be virtually non-existent. Lucas[106] has given a valuable discussion concerning the source of randomness in data and the steps necessary to stochasticise a deterministic model. Only professional mathematicians have the tools and training to work with system models of such complexity. Lucas himself, competent mathematician as well as a good biologist, indicates the necessity for simplicity and thus a deterministic rather than a stochastic approach.[107] But realism may be more important than simplicity when modelling agricultural systems, so a balance of simplicity and complexity must be taken according to the judgment of the modellers and the objectives of the model.

(ii) The other logical extreme in modelling agricultural systems is to rely entirely upon conventional empirical models. In the case of simulation, no mechanisms are included in the flow functions and the resulting model lacks depth. Although this is attractive mathematically, it is equally unattractive biologically. This approach would, of course, lead only to deterministic models. Examples would be to use simple or multiple linear regression 'simulations' of system phenomena. In this approach it is difficult to attach much real-world meaning, in terms of biology or physics, to the coefficients of the equations. (Related points are discussed in Sections 2.5.2.1 and 2.5.2.4.)

2.6.3. Hierarchies and Modules in Models

Hierarchies of systems and levels at which biological, physical, economic, social and political components enter in the system model were presented in Fig. 1 and discussed in Section 2.1.1. In such a modelling structure, driving variables in one level or echelon may become state variables in another echelon model. State variables at one echelon may become output variables at another. For example, in the study or modelling of a given agricultural system, the surrounding agricultural systems (e.g. agricultural firms) may be considered exogenous or driving variables to the model for the system of concern. If one goes to the regional level, perhaps the surrounding systems then become state variables.

Similarly, a phenomenon which expresses a rate process at one hierarchical level may not be needed at another hierarchical level. Descriptors required in one hierarchical level may be changed at another or they may no longer be required. These phenomena relate to the concept of aggregation. As one develops hierarchies of models, it is possible to aggregate the system either organisationally or spatially over the heterogeneity of the system.[68]

An example of aggregating organisationally might be shown by reducing the number of plant taxa that would be carried as state variables in a given echelon as compared to the echelon below it. An example of aggregating spatially might be demonstrated by integrating over subsystems in a larger system model or a higher-level organisation. Instead of considering four or five different individual micro-watersheds, for example, the entire watershed might be modelled.

The hierarchical concept of modelling and the state variable approach and the flow-oriented conceptualisation and diagramming of models need yet to be combined in a thorough and unequivocable way. All of these concepts offer much to the agricultural systems modeller and should be given consideration in the next generation of development of large-scale models.

2.6.4. Needs in Reporting Models

2.6.4.1. Model Description

A clear and complete statement of at least the following seems critical for rapid advances for modelling efforts. Authors should clearly *specify the objectives* of their modelling effort, the *hypotheses or assumptions* used in the construction of their model, a statement of the *general mathematical form* of the model, for simulation models a list of the *specific driving variables* used, a list of the *state variables* used, a *diagram of the model* structure, a list of equations or *functional relationships of flows*, and, for full completeness, a listing of the *computer code*. Reporting a model, particularly a complex one, is no easy task, but is all too frequently omitted.

We prefer to see the equations for functional relationships of flows used in the model presented in algebraic format and notation, in as simple a form as possible, rather than presented as computer code-named acronyms. There is considerable value in including within an appendix or another article a full listing of the computer code.

2.6.4.2. Reporting the Model Code

Even if a number of journals begin to allow more complete articles, listing of model codes, and presentation of complex diagrams, many current agricultural system models are extremely complex and even with concise writing will take a great deal of space for the presentation. Perhaps the approach of Vickery & Hedges[57,58] is appropriate. This, in essence, is a joint publication. A condensed version of the model, still intelligible, is provided in a technical journal. As backup, however, the full computer code and details of the programme are provided in a report available on request and referenced in local libraries, at least, if not internationally.

Seldom are computer code listings sufficiently saturated with comment card statements. Conventional text will be required for a full description of the programme, usually because conceptualisation and assumptions must be spelled out in narrative.

The use of precompilers, which treat individual flows, help us focus attention on the actual components of the individual difference or differential equations of simulation models and thus assist in the organisation of thought and reporting on such information. It is useful to have appropriate graphs for each process to show the relationship of the process to the driving variables and state variables in a simulation model (see Section 2.6.4.4). Verbal description should be followed by mathematical representation of the flows or of the differential or difference equations in the model.

It would be useful to have a standard symbolism for the mathematical models. There seem to be two camps with respect to symbolism, for and against!

(i) Those for standard symbolism represent disciplines and situations in which there is a long history of modelling and in which there is considerable exchange of models between individuals. For example, there are rather standard formats and notations for representing linear programming optimisation models. In simulation modelling, the most likely standardisation is to refer to state variables as x, driving variables as z or v, and output variables as y. It is less clear how parameters should be noted.

(ii) The other camp represents those modellers who prefer to use highly variable acronym-like or mnemonic designations of variable names.

A complete listing of the model should include examples of the data used and the output of a test case. Further useful interpretative information includes the time required to run the programme on a particular computer and the storage requirements for the programme.

2.6.4.3. Journal Specifications

Rapid advances in agricultural systems modelling will be made only if authors give more concern to the form of their reports. At times, the *reporting is restricted by the mode of publication*. For example, as was noted earlier,[14] 'Our conventional journals and publications must allow inclusion of detailed lists of equations, FORTRAN or other computer code statements, and computer model output if we are to succeed in rapidly advancing the systems approaches to grassland analysis. This had seldom been the case in ecology and resource management sciences'.

The need for more complete listing of models has been recognised in recent journals. For example, in an editorial introduction for the *Journal of Environmental Management* it is noted:[108]

'Because of the current impact of modern mathematical and computing techniques on the environmental sciences, it is expected that many of the papers published by the Journal will have mathematical background... to meet the requirements of the kinds of studies needed to solve the complex problems of environmental management, a wide range of formats will be accepted, including computer-generated output and listings of relevant computer programs. Similarly, considerable flexibility in the length of papers published will be entertained so that relatively short notes will be accepted, as well as papers which are considerably longer than the standard usually set for scientific journals.'

2.6.4.4. Ideas on Standard Outputs and Model Diagrams

It is convenient for the reader if the authors use some separate notational format to characterise driving variables, state variables, output variables and parameters. It is also convenient for the reader if the parameters are included in a table showing the parameter code, the parameter value and the source of data used to estimate parameters, whether it be a scientific journal reference or a personal estimate or whatever.

Similarly, the functional forms of the relationships between a flow or rate process and specific state variables or driving variables should be cited.

Newton & Swartzman[109] presented a modified diagrammatic framework for showing the factors affecting the behaviour of the system elements in what they call a 'flow control diagram'. This technique of focusing on flows is useful and it presents a great deal of information in a clear and precise manner. Their approach is essentially to present a series of compartment diagrams showing the flow of each of the materials of importance in the model. With an overall model several submodels are identified, i.e. those submodels accounting for the separate flows such as water, carbon, money and land.

Each of the compartments in each of the subdiagrams is labelled and numbered with a number mnemonic combination relating to the submodel in which the compartment is found. Flows between compartments are identified by a mnemonic letter and two numbers separated by a comma. The first number denotes from where the flow emanates and the second is to where it is going. For each submodel a source and sink department is also identified. In addition to specifying driving variables, state variables and individual flows, decision rules are included, i.e. rules whereby control is effected over the system by an external manipulator, usually human.

This initial diagram is followed by separate graphs showing the relationship of the individual flows to the variables affecting them. A complete description of the model, if the model is reasonably complex, would generate many pages of individual flow descriptions.

By removing the factors affecting the flows from the overall compartmental diagram, the main diagram is significantly easier to read. The factors affecting the individual flows are then placed into perspective, a flow at a time.

At the present state of the art there are some possible disadvantages in attempts to standardise the approach to diagramming agricultural systems models. Some workers are concerned that rigidity of the choice of units, kinds of elements, and the ways the elements interrelate may limit the creative thinking process. Some time is required to learn any notation and symbolic procedure. Modelling is an evolving process (see Section 2.5.3), and during the process of modelling a diagram must be altered several times and this may be time-consuming.

Even considering these disadvantages, developing a thorough, standardised approach to model presentation and attempting to fit models into this brings to light missing information in model description. Any fairly standardised diagrammatic format facilitates communication, as would be the case also for a standardised notational format. The modularity of allowing individual flows to be presented separately assists when changes are made in the model during its evolution.

2.6.4.5. Reporting Model Limitations

In reporting models it is important for the authors to recognise and state the inadequacies of the models. Models are never complete, they are continually evolving. This provides difficulty in reporting models, as *many modellers are reluctant to publish upon their work until the 'final' model is completed*. The end result often is that the model is never published! If the model is published after a major stage of development, the author should indicate what the inadequacies of the model are. These inadequacies should be kept separate from features of the model which represent deliberate simplifications when the model was structured. The simplifications, of course, should be listed in the assumptions and the objectives of the modelling effort.

2.7. SOME OPPORTUNITIES IN AGRICULTURAL SYSTEMS MODELLING

2.7.1. Kind of Model Related to Field of Study

There is a much greater abundance of optimisation models in the agricultural economics fields than in ecology, for example. Part of this difference is related to the practicality issue discussed earlier.[62] The closer the field of effort seems to be to practical economic problems, the more likely it is that optimisation models get an early start. The closer the field seems to be to basic science, the more likely the early models will be those with differential equations or simulations of populations or system phenomena.

There appears to be a real need for developing more models that have both simulation and optimisation components, where optimisation is inserted in the modelling procedure in an analytical method rather than simply a trial-and-error method. It is common practice for builders of simulation models to play 'what if' or 'if then' games by varying initial conditions or parameter values or driving conditions to examine the response. But in any reasonably large and complex simulation model, there are extremely large numbers of strategies that may be imposed on the system by modifying driving variables or the state variables. There seems to be a time trend to more usage of simulation models in most of the fields of study reviewed, but only very recently have initial attempts been made at combining simulation and optimisation.

2.7.2. Lack of Use of Available Technology

Our review has been restricted primarily to simulation and optimisation models, and more specifically within optimisation to mathematical programming models and within simulation the use of differential or difference equations. Yet we were able to assess the usage of other optimisation techniques and other simulation techniques. Two common techniques which could be used more often appear in this review:

(i) There would be considerable opportunity for *more use of matrix approaches in empirical dynamic prediction* models. In some of these examples, a transition probability matrix is developed by analysis of large blocks of data. This transition matrix is used to change the state vector from one time step to the next time step. Although empirical in nature, the development of a transition matrix by data analysis results in the matrix having incorporated in a 'black box form' a relatively large degree of information. When used in similar situations and with appropriate initial conditions, the time course of the system, i.e. the trajectory of the system, can be calculated to an amazing degree of reliability.[110] One value of this approach is that the models are relatively simple to construct. They make use of large blocks of data which may be insufficient for the purpose of developing detailed flow functions for differential or difference equation simulation models.

(ii) A major lack of use of well-developed technology is in the general area of control theory. Most agricultural resource systems are feedback systems similar to those to which control theory techniques have been applied. This does not support the idea of forcing the problem into some predetermined mould, but of examining the general structures of the moulds to see if they are feasible to use for the problem at hand.

Dynamic programming approaches have had limited application in agricultural system modelling. To prevent confusion, it might be noted here that dynamic programming techniques, unlike other mathematical programming techniques, do not relate to specific solution algorithms, but rather they might be more adequately described as relating to a philosophy of solution techniques. Thus far, the technique of dynamic programming seems inapplicable to very large problems in agricultural systems modelling.

2.7.3. Transfer of Models to Users

A common problem in agricultural system modelling is that most of the development has been done by scientists in universities or in governmental research agencies. Frequently, it has not been the objective of the scientist to carry their models beyond the publication stage and, even so, that publication is an overly condensed version in a scientific journal.

In many instances it is the fault of the system rather than the scientist that this has taken place. Many scientists depend upon publications for advancement in rank and tenure. Were this not the case, they might spend more time in 'going to the field' with their models and demonstrating their utility or adapting the models to the level at which they can be used widely. There seems to be a slow trend in recognition of value of carrying the model from the drawing board through the computer and into the field, and perhaps we will see more of the imaginative effort put forward in modelling agricultural systems and carrying results to the potential user in the future. There are recent examples where agricultural extension or advisory uses are making significant inroads into the problem by *taking the models to the users and using them.*[111,112]

2.7.4. A Religion for Model Builders

The true believers who have read this far will recognise many elements of the modellers' doctrine have been interspersed throughout this paper. The faith has not been tacitly accepted; it has been questioned occasionally! It seems fitting to close with an overview of the religion for model builders, as adapted from writings of one of the priests.[113] Only the names of the specialists have been changed, to protect the ecologists!

(i) They are devout, their faith is unwavering; almost certainly, they are more devout than sane agriculturalists. They are imbued and driven by the faith that even though the approach may not have worked *yet*, it remains one of the few paths to truth and light!

(ii) They hold perhaps more faith than many agriculturalists in absolutes, in ubiquitous patterns, universal principles, or natural laws.

But agricultural modellers seek these grails in only an abstract and generalised sense. Further, they have unshaking faith in the model builders' ability to abstract such universals.

(iii) In seeming contradiction to his acceptance of generalised agricultural principles, the model builder generally does not believe that he cannot test these principles, or conclusively prove their existence. He recognises that his faith is pure and untainted by statistical exercise. He may state that 'the model is of most use when it is clearly wrong'.

(iv) Although supported by an imperturbable faith in general patterns, the model builder is convinced that only a small portion of the elements and patterns within a system are important (as opposed to the agriculturalists' teleological view). In the face of Pascal's statement that 'error comes from exclusion' the modeller believes that 'error comes from inclusion'. He adopts the teleological view of nature to the extent of recognising basic patterns, but remains convinced that there is a lot of garbage lying around with no real purpose—'noise' in the jargon of a communications engineer—containing little information.

(v) An outcome of these combined feelings is that the model builder assumes pattern through time and relative magnitude to be more important than absolute magnitude.

ACKNOWLEDGMENTS

We particularly thank Gene Nelson for providing bibliographies of important articles in the agriculture economics area. Kim Gilbert has been especially helpful in library assistance, and Susan O'Brien and Barbara Hendricks and crew in rapid processing of the manuscript. This paper reports on work supported in part by National Science Foundation Grant GB-41233X to the Grassland Biome, US International Biological Programme, for 'Analysis of Structure, Function, and Utilisation of Grassland Ecosystem'.

REFERENCES

1. Arnold, G. W. & Bennett, D. (1975). The problem of finding an optimum solution, in: *Study of Agricultural Systems* (ed. G. E. Dalton), Applied Science Publishers, London.

2. Dent, J. B. (1975). The application of systems theory in agriculture, in: *Study of Agricultural Systems* (ed. G. E. Dalton), Applied Science Publishers, London.
3. Mihram, G. A. (1972). Some practical aspects of the verification and validation of simulation models, *Operational Research Quarterly*, **23**, 17–29.
4. Tomović, R. & Vukobratović, M. (1972). *General Sensitivity Theory*, American Elsevier Publishing Co., New York, 258 pp.
5. Van Dyne, G. M. (1974). Problems in sensitivity analysis of ecological simulation and optimisation models, paper presented at NATO Conference on Mathematical Analysis of Decision Problems in Ecology, Istanbul, Turkey, July 1973 (in preparation).
6. Von Bertalanffy, L. (1969). *General System Theory: Foundations, Development, and Applications*, George Braziller, New York.
7. Von Bertalanffy, L. (1972). The history and status of general systems theory, *Academic Management Journal*, **15**, 407–26.
8. Wymore, A. W. (1967). *A Mathematical Theory of Systems Engineering: The Elements*, Wiley, New York, 353 pp.
9. Klir, G. J. (1969). *An Approach to General Systems Theory*, Van Nostrand-Reinhold, New York, 323 pp.
10. Mesarovic, M. D., Macko, D. & Takahara, T. (1970). *Theory of Hierarchical Multilevel Systems*, Academic Press, New York, 294 pp.
11. Klir, G. J., ed. (1972). *Trends in General Systems Theory*, Wiley-Interscience, New York.
12. Ackoff, R. L. (1971). Towards a system of systems concepts, *Management Science*, **17**, 11–18.
13. Jenkins, G. M. (1969). The systems approach, *Journal of Systems Engineering*, **1**, 1–17.
14. Van Dyne, G. M. (1969). Grasslands management, research, and training viewed in a systems context, Range Science Department Science Series No. 3, Colorado State University, Fort Collins, 50 pp.
15. DeRusso, P. M., Ray, R. J. & Close, C. M. (1965). *State Variables for Engineers*, Wiley, New York, 608 pp.
16. Van Dyne, G. M. (1974). Some procedures, problems, and potentials of system-oriented, ecosystem-level research programs, *Bulletin of the Swedish Natural Research Council, Barrskogslandskapets Ekologi* (in press).
17. Forrester, J. W. (1961). *Industrial Dynamics*, MIT Press, Cambridge, Massachusetts, 464 pp.
18. Van Dyne, G. M. & Anway, J. C. (1974). A program for and the process of building and testing grassland ecosystem models, *Journal of Range Management* (manuscript submitted).
19. Swartzman, G. L., coordinator (1970). Some concepts of modelling, US/IBP Grassland Biome Technical Report No 32, Colorado State University, Fort Collins, 142 pp.
20. Van Dyne, G. M. (1966). Ecosystems, systems ecology and systems ecologists, Oak Ridge National Laboratory Report ORNL 3957, Oak Ridge National Laboratory, Oak Ridge, Tennessee, 31 pp.

21. Van Dyne, G. M. (1966). Application and integration of multiple linear regression and linear programming in renewable resources analysis, *Journal of Range Management*, **19**, 356–62.
22. Van Dyne, G. M. (1972). Organisation and management of an integrated ecological research program—with special emphasis on systems analysis, universities and scientific cooperation, in: *Mathematical Models in Ecology* (ed. J. N. R. Jeffers), Blackwell, Oxford, pp. 111–72.
23. Kuo, F. F. (1966). Network analysis by digital computer, in: *Systems Analysis by Digital Computer* (ed. F. F. Kuo and J. F. Kaiser), Wiley, New York, pp. 1–33.
24. Pottle, C. (1966). State space techniques for general active network analysis, in: *Systems Analysis by Digital Computer* (ed. F. F. Kuo and J. R. Kaiser), Wiley, New York, pp. 59–98.
25. Van Dyne, G. M., Innis, G. S. & Swartzman, G. L. (1971). Some analytical and operation approaches to developing dynamic models of ecological systems, in: *Environmental Awareness, Second Annual Session, Colorado Chapter* (ed. M. Lillywhite and C. E. Martin), Institute of Environmental Science, Mount Prospect, Illinois, pp. 19–26.
26. Naylor, T. H. (1970). Policy simulation experiments with macro-econometric models: the state of the art, *American Journal of Agricultural Economics*, **52**, 263–71.
27. Evans, M. (1969). Non-linear econometric models, in: *The Design of Computer Simulation Experiments* (ed. T. H. Naylor), Duke University Press, Durham, North Carolina, pp. 369–92.
28. Klein, L.R. & Evans, M. T. (1969). *Econometric Gaming: A Kit for Computer Analysis of Macroeconomic Models*, Macmillan, New York, 44 pp.
29. Conway, G. R. (1973). Experience in insect pest modelling: a review of models, uses and future directions, in: *Insects: Studies in Population Management* (ed. P. W. Geier, L. R. Clark, D. J. Anderson and H. A. Nix), Ecological Society of Australia Memoirs No. 1, Canberra, Australia, pp. 103–30.
30. U.S. Forest Service (1969). Forest insect population dynamics: Proceedings of the Forest Insect Population Dynamics Workshop, US Forest Service Research Paper NE-125, US Forest Service, Washington, DC, 126 pp.
31. International Atomic Energy Commission (1973). *Computer Models and Application of the Sterile-Male Technique*, International Atomic Energy Agency, Vienna, 195 pp.
32. Shoemaker, C. (1973). Optimisation of agricultural pest management. I. Biological and mathematical background, *Mathematical Biosciences*, **16**, 143–75.
33. Jaquette, D. L. (1971). Mathematical models for the control of growing biological populations: a survey, Industrial and Systems Engineering, Technical Report 71-1, University of Southern California, Los Angeles, 16 pp.
34. Forestry Commission (1966). Mathematical models in forest management: Proceedings of the meeting held at the University of Edinburgh on 12–13 April 1965, *Forest Record*, No. 59, HMSO, London, 47 pp.

35. Patten, B. C. (1968). Mathematical models of plankton production, *Internationale Revue der Gesamten Hydrobiologie*, **53**, 357–408.
36. Paulik, G. J. (1969). Computer simulation models for fisheries research, management and teaching, *Transactions of the American Fishery Society*, **98**, 551–9.
37. Usher, M. B. (1966). A matrix approach to the management of renewable resources with special reference to selection forests, *Journal of Applied Ecology*, **3**, 355–67.
38. Usher, M. B. (1969). A matrix approach to the management of renewable resources with special reference to selection forests:—two extensions, *Journal of Applied Ecology*, **6**, 347–8.
39. Usher, M. B. (1969). A matrix model for forest management, *Biometrics*, **25**, 309–15.
40. Usher, M. B. & Williamson, M. H. (1970). A deterministic matrix model for handling the birth, death and migration processes of spatially distributed populations, *Biometrics*, **23**, 1–12.
41. Couch, J. R. (1966). Linear programming, *Proceedings of the Texas Nutrition Conference*, pp. 123–9.
42. Dantzig, G. C. (1963). *Linear Programming and Extensions*, Princeton University Press, Princeton, New Jersey, 625 pp.
43. Van Dyne, G. M., Frayer, W. E. & Bledsoe, L. J. (1970). Some optimisation techniques and problems in the natural resource sciences, in: *Studies in Optimisation. I: Symposium on Optimisation*, Society for Industrial and Applied Mathematics, Philadelphia, Pennsylvania, pp. 95–124.
44. Swartzman, G. L., ed. (1972). Optimisation techniques in ecosystem and land use planning, US/IBP Grassland Biome Technical Report No. 143, Colorado State University, Fort Collins, 164 pp.
45. Innis, G. S. (1975). Simulation models of grasslands and grazing lands, in: *Ecology of Grasslands and Bamboolands in the World* (ed. M. Numata), Springer-Verlag, New York.
46. Heady, E. O. & Dillon, J. L. (1961). *Agricultural Production Functions*, Iowa State University Press, Ames, 667 pp.
47. Agrawal, R. C. & Heady, E. O. (1972). *Operations Research Methods for Agricultural Decisions*, Iowa State University Press, Ames, 303 pp.
48. O'Neill, R. V., Hett, J. M. & Sollins, N. F. (1970). A preliminary bibliography of mathematical modelling in ecology, Oak Ridge National Laboratory Report ORNL-70-3, Oak Ridge National Laboratory, Oak Ridge, Tennessee, 97 pp.
49. Kadlec, J. A. (1971). *A Partial Annotated Bibliography of Mathematical Models in Ecology*, Analysis of Ecosystems, IBP, University of Michigan, Ann Arbor, approx. 200 pp.
50. Schultz, V. (1971). A bibliography of selected publications on population dynamics, mathematics and statistics in ecology, in: *Statistical Ecology: Many-Species Populations, Ecosystems and Systems Analysis*, Vol. 3 (ed. G. P. Patil, E. C. Pileou and W. E. Waters), Pennsylvania State University Press, University Park, pp. 417–25.

51. Morley, F. G. W. & Spedding, C. R. W. (1968). Agricultural systems and grazing experiments, *Herbage Abstracts*, **38**, 279–87.
52. Hutchingson, K. J. (1971). Productivity and energy flow in grazing/fodder conservation systems, *Herbage Abstracts*, **41**, 1–10.
53. Day, R. H. (1963). *Recursive Programming and Production Response*, North-Holland, Amsterdam, 226 pp.
54. Dalton, G. E. (1971). Simulation models for the specification of farm investment plans, *Journal of Agricultural Economics*, **22**, 131–42.
55. Schutt, T., coordinator (1974). Systems analysis within the Swedish coniferous forest project, Swedish Natural Science Research Council, 45 pp. (unpublished report).
56. Schweppe, F. C. (1973). *Uncertain Dynamic Systems*, Prentice-Hall, Englewood Cliffs, New Jersey.
57. Vickery, P. J. & Hedges, D. A. (1972). A productivity model of improved pasture grazed by Merino sheep, *Proceedings of the Australian Society of Animal Production*, **9**, 16–22.
58. Vickery, P. J. & Hedges, D. A. (1972). Mathematical relationships and computer routines for a productivity model of improved pasture grazed by Merino sheep, Animal Research Laboratory Technical Paper No. 4, CSIRO, Australia.
59. Dean, G. W., Bath, D. L. & Olayide, S. (1969). Computer program for maximising income above feed cost from dairy cattle, *Journal of Dairy Science*, **52**, 1008–16.
60. Dean, G. W., Carter, H. O., Wagstaff, H. R., Olayide, S. O., Ronning, M. & Bath, D. L. (1972). Production functions and linear programming models for dairy cattle feeding, Giannini Foundation Monograph No. 31, Giannini Foundation of Agricultural Economics, University of California, Davis, 53 pp.
61. Swartzman, G. L. & Van Dyne, G. M. (1972). An ecologically based simulation–optimisation approach to natural resource planning, *Annual Review of Ecology and Systematics*, **3**, 347–98.
62. Levins, R. (1966). The strategy of model building in population biology, *American Scientist*, **54**, 241–431.
63. Nooney, C. G. (1965). Mathematical models, reality and results, *Journal of Theoretical Biology*, **9**, 239–52.
64. Holling, C. S. (1966). The strategy of building models of complex ecological systems, in: *Systems Analysis in Ecology* (ed. K. E. F. Watt), Academic Press, New York, pp. 195–214.
65. Bledsoe, L. J. & Jameson, D. A. (1969). Model structure for a grassland ecosystem, in: *The Grassland Ecosystem: A Preliminary Synthesis* (ed. R. L. Dix and R. G. Beidleman), Range Science Department Science Series No. 2, Colorado State University, Fort Collins, pp. 410–37.
66. Goodall, D. W. (1973). Problems of scale and detail in ecological modelling, in: *Symposium of the International Institute for Applied Systems Analysis*, Vienna, 4 September, 12 pp.
67. Clymer, A. B. & Bledsoe, L. J. (1970). A guide to the mathematical modelling of an ecosystem, in: *Simulation and Analysis of Dynamics of a*

Semidesert Grassland: An Interdisciplinary Workshop Program Toward Evaluating the Potential Ecological Impact of Weather Modification (ed. R. G. Wright and G. M. Van Dyne), Range Science Department Science Series No. 6, Colorado State University, Fort Collins, pp. I-75 to I-99.
68. Van Dyne, G. M. & Innis, G. S., ed. (1972). Macrolevel ecosystem models in relations to man: a preliminary analysis of concepts and approaches, US/IBP Grassland Biome Technical Report No. 162, Colorado State University, Fort Collins, 61 pp.
69. Morales, R. (1972). Models: 1, *Limnology and Oceanography*, **17**, 499.
70. Popper, K. R. (1968). *The Logic of Scientific Discovery*, revised 3rd ed. (translation from German), Dowden, Hutchinson & Ross, Stroudsburg, Pennsylvania, 480 pp.
71. Stearns, S. C. (1972). Models: 2, *Limnology and Oceanography*, **17**, 500–1.
72. Goodall, D. W. (1969). Simulating the grazing situation, in: *Concepts and Models of Biomathematics: Simulation Techniques and Methods*, vol. I (ed. F. Heinments), Marcel Dekker, New York, pp. 211–36.
73. Goodall, D. W. (1972). Potential applications of biome modelling, *Terre et la vie*, **1**, 118–38.
74. DeWit, C. T. (1970). Dynamic concepts in biology, in: *International Biological Program: Prediction and Measurement of Photosynthetic Productivity*, Center for Agricultural Publishing and Documentation, Wageningen, The Netherlands, 632 pp.
75. Goel, N. S., Maitra, S. C. & Montroll, E. W. (1971). *On the Volterra and Other Non-linear Models of Interacting Populations*, Academic Press, New York, 145 pp.
76. Jameson, D. A. (1970). Basic concepts in mathematical modelling of grassland ecosystems, in: *Modelling and Systems Analysis* (ed. D. A. Jameson), Range Science Department Science Series No. 5, Colorado State University, Fort Collins, pp. 1–16.
77. Innis, G. S. (coordinator) et al. (1975). ELM: Version 2.0, US/IBP Grassland Biome Technical Report, Colorado State University, Fort Collins (in preparation).
78. May, P. F., Rill, A. R. & Cumming, M. J. (1972). Systems analysis of sulphur kinetics in pastures grazed by sheep, *Journal of Applied Ecology*, **9**, 25–49.
79. May, P. F., Till, A. R. & Downes, A. M. (1968). Nutrient cycling in grazed pastures. I. A preliminary investigation on the use of [^{35}S] gypsum, *Australian Journal of Agricultural Research*, **19**, 532–43.
80. Till, A. R. & May, P. F. (1970). Nutrient cycling in grazed pastures. II. Further observations with [^{35}S] gypsum, *Australian Journal of Agricultural Research*, **21**, 253–60.
81. Till, A. R. & May, P. F. (1970). Nutrient cycling and grazing pastures. III. Studies on labelling of the grazed pasture system by solid [^{35}S] gypsum and aqueous Mg^{35}SO$_4$, *Australian Journal of Agricultural Research*, **21**, 455–63.
82. Sterling, T. D. & Pollack, S. V. (1968). *Introduction to Statistical Data Processing*, Prentice-Hall, Englewood Cliffs, New Jersey.

83. Paulik, G. J. (1970). Digital simulation modelling in resource management and the training of applied ecologists, in: *Systems Analysis and Simulation in Ecology*, vol. 2 (ed. B. C. Patten), Academic Press, New York, pp. 373–418.
84. Hamilton, H. R., Goldstone, S. E., Milliman, J. W., Pugh, A. L. III, Roberts, E. B. & Zellner, A. (1969). *Systems Simulation for Regional Analysis: An Application to River-Basin Planning*, MIT Press, Cambridge, Massachusetts, 407 pp.
85. Goodall, D. W. (1967). Computer simulation of changes in vegetation subject to grazing, *Journal of the Indian Botanical Society*, 46, 356–62.
86. Goodall, D. W. (1970). Simulation of grazing systems, in: *Modelling and Systems Analysis in Range Science* (ed. D. A. Jameson), Range Science Department Science Series No. 5, Colorado State University, Fort Collins, pp. 51–74.
87. Goodall, D. W. (1970). Studying the effects of environmental factors on ecosystems, in: *Analysis of Temperate Forest Ecosystems* (ed. D. E. Reichle), Springer-Verlag, Berlin, pp. 19–26.
88. Goodall, D. W. (1971). Extensive grazing systems, in: *Systems Analysis in Agricultural Management* (ed. J. B. Dent and J. R. Anderson), Wiley, Sydney, Australia, pp. 173–87.
89. Donnelly, J. R. & Armstrong, J. S. (1968). Summer grazing, *Proceedings of the Second Conference for Applied Simulation*, New York, December 24, pp. 329–32.
90. Freer, M., Davidson, J. L., Armstrong, J. S. & Donnelly, J. R. (1970). Simulation of grazing systems, *Proceedings of the XI International Grassland Congress*, Surfer's Paradise, Queensland, Australia, pp. 913–17.
91. Armstrong, J. S. (1971). Modelling a grazing system, *Proceedings of the Ecological Society of Australia*, vol. 6, Canberra, Australia, pp. 194–202.
92. Christian, K. R., Armstrong, J. S., Donnelly, J. R., Davidson, J. L. & Freer, M. (1972). Optimisation of a grazing management system, *Proceedings of the Australian Society of Animal Production*, 9, 124–9.
93. Christian, K. R., Armstrong, J. S., Davidson, J. L., Donnelly, J. R. & Freer, M. (1974). A model for decision-making in grazing management, *Proceedings of the XII International Grassland Congress*, Part I, Moscow, USSR, pp. 126–30.
94. Gustafson, J. D. & Innis, G. S. (1972). SIMCOMP version 2.0 user's manual, US/IBP Grassland Biome Technical Report No. 138, Colorado State University, Fort Collins, 62 pp.
95. Jones, J. G. W., ed. (1970). *Proceedings of a Symposium on the Use of Models in Agricultural and Biological Research*, Grassland Research Institute, Hurley, England, 134 pp.
96. Clymer, A. B. (1969). The modelling and simulation of big systems, *Proceedings of the Simulation and Modelling Congress*, Pittsburgh, Pennsylvania, pp. 107–17.
97. Clymer, A. B. (1969). The modelling of hierarchical systems: keynote address, *Conference on Applications of Continuous System Simulation Languages*, San Francisco, California, 16 pp.
98. Spedding, C. R. W. (1970). The relative complexity of grassland systems,

Proceedings of the XI International Grassland Congress, Surfer's Paradise, Queensland, Australia, pp. A126–A131.
99. O'Neill, R. V. (1973). Error analysis of ecological models, in: *Proceedings of the Third National Symposium on Radio Ecology* (ed. O. G. Nelson), Atomic Energy Commission, Oak Ridge, Tennessee, pp. 898–90.
100. Watt, K. E. F. (1964). The use of mathematics and computers to determine optional strategy and tactics for a given insect pest control problem, *Canadian Entomologist*, 96, 202–20.
101. Van Dyne, G. M., principal investigator (1973). *Analysis of Structure, Function, and Utilisation of Grassland Ecosystems*, vol. 1: *A Continuation Proposal for 1974 through 1976*, US/IBP Grassland Biome, Colorado State University, Fort Collins, 317 pp.
102. Van Dyne, G. M., principal investigator (1973). *Analysis of Structure Function and Utilisation of Grassland Ecosystems*, vol. 2: *A Progress Report*, Colorado State University, Fort Collins, 305 pp.
103. UNESCO (1972). *Expert Panel on the Role of Systems Analysis and Modelling Approaches in the Program on Man and the Biosphere*, MAB: final report, UNESCO, Paris, 50 pp.
104. Farvar, M. T. & Milton, G. P. (1972). *The Careless Technology: Ecology and International Development*, Natural History Press, Garden City, New York, 1030 pp.
105. Mar, B. W. (1974). Problems encountered in multidisciplinary resources and environmental simulation models development, *Journal of Environmental Management*, 2, 83–100.
106. Lucas, H. L. (1964). Stochastic elements in biological models: their sources and significance, in: *Stochastic Models in Medicine and Biology* (ed. J. Gurland), University of Wisconsin Press, Madison, pp. 355–85.
107. Lucas, H. L. (1960). Theory and mathematics in grassland problems, *Proceedings of the VIII International Grassland Congress*, pp. 732–36.
108. Jeffers, J. N. R. (1973). Editorial introduction, *Journal of Environmental Management*, 1, 1–2.
109. Newton, J. E. & Swartzman, G. L. (1973). Frameworks for whole systems study with special reference to agricultural systems of temperate grassland, in: *Systems Workshop Report* (ed. J. E. Newton and G. L. Swartzman), Grassland Research Institute, Hurley, England, pp. 28–35.
110. Redetzke, K. A. (1973). A matrix model of a rangeland grazing system, Ph.D. dissertation, Colorado State University, Fort Collins, 145 pp.
111. Bath, D. L., Hutton, G., Jr. & Olson, E. H. (1972). Evaluation of computer program for maximising income above feed cost from dairy cattle, *Journal of Dairy Science*, 55, 1607–12.
112. Candler, W., Boehlje, M. & Saathoff, R. (1970). Computer software for farm management extension, *American Journal of Agricultural Economics*, 52, 71–80.
113. Bunnel, F. (1972). *Theological Ecology*, Faculty of Forestry, University of British Columbia, Vancouver, 4 pp. (mimeo).
114. Jenkins, K. B. & Halter, A. N. (1963). A multi-stage stochastic replacement decision model, Agricultural Experiment Station Technical Bulletin No. 67, Oregon State University, Corvallis, 31 pp.

115. Hutton, R. F. (1966). A simulation technique for making management decisions in dairy farming, Agricultural Economics Report No 87, US Department of Agriculture, Economic Research Service, Washington, DC, 143 pp.
116. Jensen, R. C. (1968). Farm development plans including tropical pastures for dairy farms in the Cooroy area of Queensland, *Review of Marketing and Agricultural Economics*, 36, 139–48.
117. Lindner, R. K. (1969). The economics of increased beef production on dairy farms in Western Australia, *Quarterly Review of Agricultural Economics*, 22, 147–64.
118. Chandler, P. T. & Walker, H. W. (1972). Generation of nutrient specifications for dairy cattle for computerised least cost ration formulation, *Journal of Dairy Science*, 55, 1741–9.
119. Nelson, A. G. & Eisgruber, L. M. (1974). A dynamic information and decision system for beef feedlots, paper presented at the Western Agricultural Economics Association Meetings, 12 pp.
120. Donaldson, G. F. (1968). Allowing for weather risk in assessing harvest machinery capacity, *American Journal of Agricultural Economics*, 50, 24–40.
121. Dahlman, R. C. & Sollins, P. (1973). Nitrogen cycling in grasslands, *Ecological Sciences Division, Annual Progress Report for the Period Ending July 31, 1970*, Oak Ridge National Laboratory, Oak Ridge, Tennessee, pp. 71–6.
122. Cloud, C., Frick, G. E. & Andrews, R. A. (1968). An economic analysis of hay harvesting and utilisation using a simulation model, US Department of Agriculture Station Bulletin 495, US Department of Agriculture, Washington, DC, 43 pp.
123. Bennett, D. (1974). Mathematical, economic and behavioural research in dairy farm decision-making, Australian Dairy Cattle Husbandry Conference, Ballina, New South Wales, Australia, 11–18 June, 10 pp. (unpublished manuscript).
124. Goudriaan, J. & Waggoner, P. E. (1972). Simulating both aerial microclimate and soil temperature from observations above the foliar canopy, *Netherlands Journal of Agricultural Science*, 20, 104–24.
125. Walsh, J. J. & Dugdale, R. C. (1971). A simulation model of the nitrogen flow in the Peruvian upwelling system, *Investigación Pesquera*, 35, 309–30.
126. Walsh, J. J., Kelley, J. C., Dugdale, R. C. & Frost, B. W. (1971). Gross features of the Peruvian upwelling system with special reference to possible diel variation, *Investigación Pesquera*, 35, 25–42.
127. Duncan, W. G., Loomis, R. S., Williams, W. A. & Hanau, R. (1967). A model for simulating photosynthesis in plant communities, *Hilgardia*, 38, 181–205.
128. Deinum, B. & Dirven, J. G. P. (1974). A model for the description of the effects of different environmental factors on the nutritive value of forages, *Proceedings of the XII International Grassland Congress*, Moscow, USSR, pp. 89–97.
129. Visser, W. D. (1969). Mathematical models in soil productivity studies, exemplified by the response to nitrogen, *Plant and Soil*, 30, 161–82.

130. Crabtree, J. R. (1969). Towards a dairy enterprise model, in: *Proceedings of a Symposium on the Use of Models in Agricultural and Biological Research* (ed. J. G. W. Jones), Grassland Research Institute, Hurley, England, pp. 35–42.
131. Stapleton, H. N., Buxton, D. R., Watson, F. L., Nolting, D. J. & Baker, D. N. (1973). Cotton: a computer simulation of cotton growth, Agricultural Experiment Station Technical Bulletin No. 206, University of Arizona, Tucson, 124 pp.
132. Hathorn, S., Jr., Stapleton, H. N. & Watson, F. L. (1973). Minimising costs in the cotton harvest–ginning system, Agricultural Experiment Station Technical Bulletin No. 204, University of Arizona, Tucson, 132 pp.
133. Jones, J. W., Wensink, R. B., Sowell, R. S. & Hesketh, J. D. (1974). Formulation of a decision model to select optimum crop production practices, paper presented at the 1974 Annual Meeting of the American Society of Agricultural Engineers, Oklahoma State University, Stillwater, 23–26 June, 24 pp.
134. Soribe, F. I. & Curry, R. B. (1973). Simulation of lettuce growth in an air-supported plastic greenhouse, *Journal of Agricultural Engineering Research*, 18, 133–40.
135. Miles, G. E. (1974). Developing pest management models, Agricultural Experiment Station Bulletin No. 50, Purdue University, West Lafayette, Indiana, 19 pp.
136. Miles, G. E., Peart, R. M., Bula, R. J., Wilson, M. C. & Hintz, T. R. (1974). Simulation of plant–pest interactions with GASP IV, paper presented at the 1974 Annual Meeting of the American Society of Agricultural Engineers, Oklahoma State University, Stillwater, 23–26 June, 11 pp.
137. Baker, C. H. & Horrocks, R. D. (1974). An overview of CORNMOD: a dynamic simulator of corn production, *Simulation* (in press).
138. Baker, C. H. & Horrocks, R. D. (1973). A computer simulation of corn grain production, *Transactions of the American Society of Agricultural Engineers*, 16, 1027–31.
139. Conway, G. R. & Murdie, G. (1972). Population models as a basis for pest control, in: *Mathematical Models in Ecology 1972* (ed. J. N. A. Jeffers), 12th Symposium of the British Ecological Society, pp. 195–213.
140. Watt, K. E. F. (1964). The use of mathematics and computers to determine optional strategy and tactics for a given insect pest control problem, *Canadian Entomologist*, 96, 202–20.
141. Shoemaker, C. (1973). Optimisation of agricultural pest management. II. Formulation of a control model, *Mathematical Biosciences*, 17, 357–65.
142. Shoemaker, C. (1973). Optimisation of agricultural pest management. III. Results and extensions of a model, *Mathematical Biosciences*, 18, 1–22.
143. Pimentel, D. & Shoemaker, C. (1974). An economic and land-use model for reducing insecticides on cotton and corn, *Environmental Entomology*, 3, 10–20.
144. Miles, G. E., Bula, R. J., Hot, D. A., Schreiber, M. M. & Peart, R. M. (1973). Simulation of alfalfa growth, American Society of Agricultural Engineers Paper No. 73–4547, 18 pp.

145. Loewer, O. J., Jr., Huber, R. T., Barrett, J. R., Jr. & Peart, R. M. (1974). Simulation of the effects of weather on an insect population, *Simulation*, **21**, 113–18.
146. Hassell, M. P. & Huffaker, C. B. (1969). Regulatory processes and population cyclicity in laboratory populations of *Anagasta kuhniella*. III. The development of population models, *Researches on Population Ecology*, **11**, 186–210.
147. Heady, E. O., Madsen, H. C., Nicol, K. J. & Hargrove, S. H. (1972). Future water and land use: effects of selected public agricultural and irrigation policies on water demand and land use, Report of the Center for Agricultural and Rural Development, Iowa State University of Science and Technology, prepared for the National Water Commission, PB–206–790 (NWC–F–72–031) NTIS, National Water Commission, Springfield, Virginia.
148. Southwood, T. R. E. & Norton, G. A. (1973). Economic aspects of pest management strategies and decisions, in: *Insects: Studies in Population Management* (ed. P. W. Geier, L. R. Clark, D. J. Anderson and H. A. Nix), Ecological Society of Australia Memoirs No. 1, Canberra, Australia.
149. Ford-Livene, C. (1972). Estimation, prediction and dynamic programming in ecology, Technical Report No. 72–17, University of Southern California, Los Angeles, 98 pp.
150. Anderson, R. L. & Maass, A. (1971). A simulation of irrigation systems, Technical Bulletin No. 1431, US Department of Agriculture, Economic Research Service, Washington, DC, 57 pp.
151. Asseed, M. & Kirkham, D. (1968). Advance of irrigation water on the soil surface in relation to soil infiltration rate: a mathematical and laboratory model study, Research Bulletin No. 565, Iowa Agriculture and Home Economics Experiment Station, Iowa State University, Ames, pp. 292–317.
152. Morley, F. H. W. & Graham, G. Y. (1971). Fodder conservation for drought, in: *Systems Analysis in Agricultural Management*, Wiley, Australia, pp. 212–36.
153. Morley, F. H. W. (1974). Evaluation by animal production of increases in pasture growth using computer simulation, *Proceedings of the XII International Grassland Congress*, Moscow, USSR, pp. 416–19.
154. Wright, A. & Dent, J. B. (1969). The application of simulation techniques to the study of grazing systems, *Australian Journal of Agricultural Economics*, **13**, 144–53.
155. McKinney, G. T. (1972). Simulation of winter grazing on temperate pasture, *Proceedings of the Australian Society of Animal Production*, **9**, 31.
156. Chudleigh, P. D. & Filan, S. J. (1972). A simulation model of an arid zone sheep property, *Australian Journal of Agricultural Economics*, **16**, 183–94.
157. Patten, B. C. (1972). A simulation of the shortgrass prairie ecosystem, *Simulation*, **19**, 177–86.
158. Trebeck, D. B. (1972). Simulation as an aid to research into extensive

beef production, *Proceedings of the Australian Society of Animal Production*, **9**, 94.
159. Arnold, G. W. & Galbraith, K. A. (1974). Predicting the value of lupins for sheep and cattle in cropping and pastoral farming systems, *Proceedings of the Australian Society of Animal Production*, **10**, 383.
160. Arnold, G. W. & Campbell, N. A. (1972). A model of a ley farming system, with particular reference to a sub-model for animal production, *Proceedings of the Australian Society of Animal Production*, **9**, 23–30.
161. Arnold, G. W., Carbon, B. A., Galbraith, K. A. & Biddiscombe, E. F. (1974). Use of a simulation model to assess the effects of a grazing management on pasture and animal production, *Proceedings of the XII International Grassland Congress*, Moscow, USSR, 7 pp.
162. Baldwin, R. L. (1972). Tissue metabolism and energy expenditures of maintenance and production, Tenth Brody Memorial Lecture, University of Missouri, Columbia, 24 pp.
163. Nelson, L. J. (1972). A model of competing range ruminants, M.S. thesis, University of California, Davis, 116 pp.
164. Edelsten, P. R., Newton, J. E. & Treacher, T. T. (1973). A model of ewes with lambs grazing at pasture, Internal Report No. 260, Grassland Research Institute, Hurley, England, 63 pp.
165. Singh, J. S. (1973). A compartment model of herbage dynamics for Indian tropical grasslands, *Oikos*, **24**, 367–72.
166. Morris, J. G., Baldwin, R. L., Maeng, W. J. & Maeda, B. T. (1974). Basic characteristics of a computer simulated model of nitrogen utilisation in the grazing ruminant, in: *Tracer Studies on Non-Protein-Nitrogen for Ruminants*, International Atomic Energy Agency, Vienna (in press).
167. Rice, R. W., Morris, J. G., Maeda, B. T. & Baldwin, R. L. (1974). Simulation of animal functions in models of production systems: ruminants on the range, *Federation Proceedings*, **33**, 188–95.
168. Smith, R. C. G. & Williams, W. A. (1974). Deferred grazing of Mediterranean annual pasture: a study by computer simulation, *Proceedings of the XII International Grassland Congress*, Moscow, USSR, pp. 646–55.
169. Hutchinson, K. J. (1974). Fodder conservation in grazing systems, CSIRO, Division of Animal Physiology, Armidale, New South Wales, Australia, 16 pp. (unpublished manuscript).
170. Swartzman, G. L. (1969). A preliminary bird population dynamics and biomass model, US/IBP Grassland Biome Technical Report No. 3, Colorado State University, Fort Collins, 16 pp.
171. Kelly, J. M., Opstrup, P. A., Olson, J. S., Auerbach, S. I. & Van Dyne, G. M. (1969). Models of seasonal primary productivity in eastern Tennessee *Festuca* and *Andropogon* ecosystems, Oak Ridge National Laboratory TM-4310, Oak Ridge National Laboratory, Oak Ridge, Tennessee, 296 pp.
172. Wright, R. J. & Van Dyne, G. M., ed. (1970). Simulation and analysis of dynamics of a semi-desert grassland: an interdisciplinary workshop program toward evaluating the potential ecological impact of weather modification, Range Science Department Science Series No. 6, Colorado State University, Fort Collins, 359 pp.

173. Bledsoe, L. J., Francis, R. C., Swartzman, G. L. & Gustafson, J. D. (1971). PWNEE: a grassland ecosystem model, US/IBP Grassland Biome Technical Report No. 64, Colorado State University, Fort Collins, 179 pp.
174. Field, T. R. O. & Hunt, L. A. (1974). The use of simulation techniques in the analysis of seasonal changes in the productivity of alfalfa (*Medicago sativa* L.) stands, *Proceedings of the XII International Grassland Congress*, Moscow, USSR, pp. 108–20.
175. Johns, G. G. (1974). A soil water: water use relationship for incorporation in models simulating dryland herbage production, *Proceedings of the XII International Grassland Congress*, Moscow, USSR, pp. 61–8.
176. Paltridge, G. W., Dilley, A. C., Garratt, J. R., Pearman, G. I., Shepherd, W. & Connor, D. J. (1972). The Rutherglen experiment on Sherpa wheat: environmental and biological data, Division of Atmospheric Physics Technical Paper No. 22, CSIRO, Australia, 155 pp.
177. Paltridge, G. W. (1970). A model of a growing pasture, *Agricultural Meteorology*, 7, 93–130.
178. DeWit, C. T., Brouwer, R. & Penning de Vries, F. W. T. (1971). A dynamic model of plant crop growth, in: *Potential Crop Production* (ed. P. F. Waring and J. P. Cooper), Heinemann, London, pp. 117–42.
179. Brockington, N. R. (1970). A simulation model of grass production in relation to water supply, 1969 Annual Report, Grassland Research Institute, Hurley, England.
180. Smith, M. W. (1974). Stochastic influences on the profitability of resowing rundown pastures with perennial species, *Proceedings of the XII International Grassland Congress*, Moscow, USSR, pp. 388–94.
181. Fitzpatrick, E. A. & Nix, H. A. (1969). A model for simulating soil water regime in alternating fallow-crop systems, *Agricultural Meteorology*, 6, 303–19.
182. Jones, J. G. W. (1970). Lamb production, in: *The Use of Models in Agricultural and Biological Research* (ed. J. G. W. Jones), Grassland Research Institute, Hurley, England, pp. 42–9.
183. Byrne, G. F. & Tognetti, K. (1969). Simulation of a pasture–environment interaction, *Agricultural Meteorology*, 6, 151–63.
184. Ross, M. A. (1973). Game: a model of growth and soil moisture extraction of *Eragrostis eripoda*, *Water in Rangelands Symposium*, Alice Springs, Australia, 15 pp.
185. Ross, P. J., Henzell, E. F. & Ross, D. R. (1972). Effects of nitrogen and light in grass–legume pastures: a systems analysis approach, *Journal of Applied Ecology*, 9, 535–56.
186. Lemon, E., Stuart, D. W. & Shawcroft, R. W. (1971). The sun's work in a cornfield, *Science*, 174, 371–8.
187. Iwaka, H. & Hirosaki, S. (1971). A compartment model of seasonal change in biomass of *Micanthus sacchariflorus* community, National Institute of Agricultural Science, University of Chiba, Yoyoi-cho, Chiba, Japan, 4 pp. (mimeo).
188. Stapleton, H. N., Cannon, M. D. & LePori, W. A. (1966). Cotton harvest schedule evaluation, paper presented at 20th Annual Cotton Defoliation-Physiology Conference, Memphis, Tennessee, January, 10 pp.

189. Stapleton, H. N., Cannon, M. D. & LePori, W. A. (1967). Cotton harvest-defoliation scheduling, *Transactions of the American Society of Agricultural Engineers*, **10**, 226–32.
190. Stapleton, H. N. (1975). A simulation program for cotton, authorised for publication as Arizona Agricultural Experiment Station Technical Paper No. 1309, University of Arizona, Tucson, 9 pp.
191. Splinter, W. E. (1975). Modelling plant–environment interactions: plant growth, *Agricultural Meteorology*, 14 pp. (in press).
192. Chen, L. H., Huang, B. K. & Splinter, W. E. (1968). Growth dynamics of small tobacco plants as affected by night temperature and initial plant size, *Transactions of the American Society of Agricultural Engineers*, **11**, 126–8.
193. Loucks, D. P. (1964). The development of an optimal program for sustained-yield management, *Journal of Forestry*, **62**, 485–9.
194. Leak, W. B. (1964). Estimating maximum allowable timber yield by linear programming, US Forest Service Research Paper NE-17, US Forest Service, Washington, DC, 9 pp.
195. Nautiyal, J. C. & Pearse, P. H. (1967). Optimising the conversion to sustained yield: a programming solution, *Forest Science*, **13**, 131–9.
196. Amidon, E. L. & Akin, G. S. (1968). Dynamic programming to determine optimum levels of growing stock, *Forest Science*, **14**, 287–91.
197. Myers, C. A. (1968). Simulating the management of even-aged timber stands, US Forest Service Research Paper RM-42, US Forest Service, Washington, DC, 32 pp.
198. Myers, C. A. (1973). Simulating changes in even-aged timber stands, US Forest Service Research Paper, RM-109, US Forest Service, Washington, DC, 47 pp.
199. Botkin, D. B., Janak, J. F., & Wallis, J. R. (1970). A simulator for northeastern forest growth: a contribution of the Hubbard Brook ecosystem study and IBM research, RC 3140 (No. 14356), IBM Research, Yorktown Heights, New York, 21 pp.
200. Botkin, D. B., Janak, J. F. & Wallis, J. R. (1970). The rationale limitations and assumptions of a Northeast forest growth simulator, RC 3188 (No. 14604), IBM Research, Yorktown Heights, New York, 39 pp.
201. Navon, D. I. (1971). Timber RAM: a long-range planning method for commercial timber lands under multiple-use management, US Forest Service Research Paper PSW-70/1971, US Forest Service, Washington, DC, 2 pp.
202. Woodmansee, R. G. & Innis, G. S. (1973). A simulation model of forest growth and nutrient cycling, *Proceedings of the 1973 Summer Computer Simulation Conference*, vol. II, Simulation Councils, La Jolla, California, pp. 697–721.
203. Liittschwager, J. M. & Tcheng, T. H. (1967). Solution of a large-scale forest scheduling problem by linear programming decomposition, *Journal of Forestry*, **65**, 644–6.
204. Navon, D. I. & McConnen, R. J. (1967), Evaluating forest management policies by parametric linear programming, US Forest Service Research Paper PSW-42, US Forest Service, Washington, DC, 13 pp.

205. Teeguarden, D. E. & Von Sperber, H. L. (1968). Scheduling Douglas fir reforestation investments: a comparison of methods, *Forest Science*, **14**, 354–68.
206. Kourtz, P. H. & O'Regan, W. G. (1968). A cost-effectiveness analysis of simulated forest fire detection systems, *Hilgardia*, **39**, 341–66.
207. Turnbull, K. J. (1963). *Population Dynamics in Mixed Forest Stands: A System of Mathematical Models of Mixed Stand Growth and Structure*, University Microfilms, Ann Arbor, Michigan, 186 pp.
208. Arvanitis, L. G. & O'Regan, W. G. (1967). Computer simulation and economic efficiency in forest sampling, *Hilgardia*, **38**, 133–64.
209. Burt, O. R. (1971). A dynamic economic model of pasture and range investments, *American Journal of Agricultural Economics*, **53**, 197–205.
210. Nelson, A. G. & Rittenhouse, L. (1974). Estimation of a performance function for evaluating range improvement investments, *Journal of American Economics* (manuscript submitted).
211. Van Dyne, G. M. & Rebman, K. R. (1967). Maintaining a profitable ecological balance, in: *Some Mathematical Models in Biology*, University of Michigan Report 4024-R-7 (5-Tol-GM01457-02) (compilers and ed. R. M. Thrall, J. A. Mortimer, K. R. Rebman and R. F. Baum), pp. PL6.1–69. See also *Economic and Management Planning of Range Ecosystems* (ed. O. A. Jameson, S. D'Aquino and E. T. Bartlett), A. A. Balkema, Rotterdam, pp. 158–60.
212. Rae, A. N. (1970). Capital budgeting, intertemporal programming models, with particular reference to agriculture, *Australian Journal of Agricultural Economics*, **14**, 39–52.
213. Rae, A. N. (1971). Stochastic programming, utility, and sequential decision problems in farm management, *American Journal of Agricultural Economics*, **53**, 448–60.
214. Rae, A. N. (1971). An empirical application and evaluation of discrete stochastic programming in farm management, *American Journal of Agricultural Economics*, **53**, 625–38.
215. Zusman, R. & Amiad, A. (1965). Simulation: a tool for farm planning under conditions of weather uncertainty, *Journal of Farm Economics*, **47**, 574–94.
216. Johnson, S. R., Tefertiller, K. R. & Moore, D. S. (1967). Stochastic linear programming and feasibility problems in farm growth analysis, *Journal of Farm Economics*, **49**, 908–19.
217. Heifner, R. G. (1966). Determining efficient seasonal grain inventories: an application of quadratic programming, *Journal of Farm Economics*, **48**, 648–60.
218. Stryg, P. E. (1967). Application of the Monte Carlo method and linear programming in farm planning, in: *Den Kongelige Veterinaer-Og Landbohojskole*, Arsskrift, Kobenhavn, pp. 195–217.
219. Halter, A. N. & Dean, G. W. (1965). Simulation of a California range-feedlot operation, Giannini Foundation Research Report No. 282, Giannini Foundation of Agricultural Economics, University of California, Davis, 125 pp.

220. D'Aquino, S. A. (1974). A case study for optimal allocation of range resources, *Journal of Range Management*, **27**, 228-33.
221. Bartlett, E. T., Evans, G. R. & Bement, R. E. (1974). A serial optimisation model for ranch management, *Journal of Range Management*, **27**, 233-9.
222. Davis, E. E. (1971). Future structure of the Texas cattle feeding industry as projected by transition probabilities utilising a convex program, *Southern Journal of Agricultural Economics*, **3**, 87-93.
223. Tung, T. H., Reu, L. & Millar, R. H. (1968). A location programming model of the Colorado dairy industry, Colorado State University Experiment Station Technical Bulletin No. 99, Colorado State University Experiment Station, Fort Collins, 22 pp.
224. Hardaker, J. B. (1967). The use of simulation techniques in farm management research, *Farm Economist*, **11**, 162-71.
225. Saunders, F. B., Braden, J. W., Fosgate, O. T., Worley, E. E., Cameron, N. W. & Hayes, D. D. (1970). An economic analysis of alternative dry-lot feeding systems for lactating dairy cows, College of Agricultural Experiment Station Research Bulletin No. 79, University of Georgia, Athens.
226. Vandenborre, R. J. (1967). An econometric analysis of the markets for soybean oil and soybean meal, Agricultural Experiment Station Bulletin No. 723, University of Illinois, Urbana, 54 pp.
227. LaFerney, P. E. (1969). A nonlinear model for evaluation of cotton processed by mills for specific end uses, Technical Bulletin No. 1401, US Department of Agriculture, Economic Research Service, Washington, DC, 25 pp.
228. Sharples, J. A., Miller, T. A. & Day, L. M. (1968). Evaluation of a firm model in estimating aggregate supply response, North Central Regional Research Publication No. 179, Iowa State University, Ames, 61 pp.
229. Mo, W. Y. (1968). An economic analysis of the dynamics of the United States wheat sector, Technical Bulletin No. 1395, US Department of Agriculture, Economic Research Service, Washington, DC, 55 pp.
230. Little, C. H. & Doeksen, G. A. (1968). An input-output analysis of Oklahoma's economy, Oklahoma Agriculture Experiment Station Technical Bulletin No. T-124, Oklahoma State University, Stillwater, 31 pp.
231. Charlton, P. J. & Thompson, S. C. (1970). Simulation of agricultural systems, *Journal of Agricultural Economics*, **21**, 373-89.
232. Lin, A. Y. & Heady, E. O. (1971). Simulated markets, farm structure, and agricultural policies, *Canadian Journal of Agricultural Economics*, **19**, 55-65.
233. Burt, O. R. & Johnson, R. D. (1967). Strategies for wheat production in the Great Plains, *Journal of Farm Economics*, **49**, 881-99.
234. Crom, R. J. & Maki, W. R. (1965). A dynamic model of a simulated livestock-meat economy, *Agricultural Economics Research*, **17**, 73-83.
235. Kuang, H. S. (1972). Allocation of random supply of tomatoes of varied quality produced in different areas among plants producing multiple product lines, *American Journal of Agricultural Economics*, **54**, 790-6.
236. Puterbaugh, H. L., Kehrberg, E. W. & Dunbar, J. O. (1957). Analysing

the solution tableau of a simplex linear programming problem in farming organisation, *Journal of Farm Economics*, **39**, 478–89.
237. Barker, R. (1964). Use of linear programming in making farm management decisions, US Department of Agriculture Bulletin No. 993, US Department of Agriculture, Washington, DC, 42 pp.
238. How, R. B. & Hazell, P. B. R. (1968). Use of quadratic programming in farm planning under uncertainty, *Agricultural Economics Research*, no. 250, Cornell University, Ithaca, New York, 25 pp.
239. Larkin, P. A. & Hourston, A. S. (1964). A model for simulation of the population biology of Pacific salmon, *Journal of the Fisheries Research Board of Canada*, **21**, 1245–65.
240. Silliman, R. P. (1966). Analog computer models of fish populations, *US Fish and Wildlife Service Fishery Bulletin*, **66**, 31–46.
241. Walters, C. J. (1969). A generalised computer simulation model for fish population studies, *Transactions of the American Fisheries Society*, **98**, 505–12.
242. Patten, B. C. (1969). Ecological systems analysis and fisheries science, *Transactions of the American Fisheries Society*, **98**, 570–81.
243. Rothschild, B. J. & Balsinger, J. W. (1971). A linear-programming solution to salmon management, *US Fish and Wildlife Services Fishery Bulletin*, **69**, 117–40.
244. Beyer, W. A., Harris, D. R. & Ryan, R. J. *A Stochastic model of the Isle Royale Biome*, Los Alamos Scientific Laboratory, University of California, Los Alamos, New Mexico, 28 pp.
245. Davis, L. S. (1967). Dynamic programming for deer management planning, *Journal of Wildlife Management*, **31**, 667–79.
246. Parker, R. A. (1968). Simulation of an aquatic ecosystem, *Biometrics*, **24**, 803–21.
247. Eberhardt, L. L. & Hanson, W. C. (1969). A simulation model for an arctic food chain, *Health Physics*, vol. 17, Pergamon Press, Baltimore, Maryland, pp. 793–806.
248. Bledsoe, L. J. & Van Dyne, G. M. (1971). A compartmental model simulation of secondary succession, in: *Systems Analysis and Simulation in Ecology*, vol. 1 (ed. B. C. Patten), Academic Press, New York, pp. 479–511.
249. Walters, C. J. & Bunnell, F. (1971). A computer management game of land use in British Columbia, *Journal of Wildlife Management*, **35**, 644–57.
250. Walters, C. J. & Gross, J. E. (1972). Development of big game management plans through simulation modeling, *Journal of Wildlife Management*, **36**, 119–28.
251. Lobdell, C. H., Case, K. E. & Mosby, H. S. (1972). Evaluation of harvest strategies for a simulated wild turkey population, *Journal of Wildlife Management*, **36**, 493–7.
252. Geier, P. W., Clark, L. R., Anderson, D. J. & Nix, H. A., ed. (1973). *Insects: Studies in Population Management*, Ecological Society of Australia Memoirs 1, Canberra, Australia, 190 pp.
253. Gross, J. E. (1970). Program ANPOP: a simulation modeling exercise on the Wichita Mountains National Wildlife Refuge, Colorado Cooperative

Wildlife Research Unit Program Report, Colorado State University, Fort Collins, 133 pp. (mimeo)
254. Eberhardt, L. L., Meeks, R. L. & Peterle, T. J. (1970). *DDT in a Freshwater Marsh: A Simulation Study*, Atomic Energy Commission Research Development Report, Battelle Memorial Institute, Pacific Northwest Laboratories, Richland, Washington, 63 pp.
255. O'Brien, J. J. & Wroblewski, J. S. (1972). An ecological model of the lower marine trophic levels on the continental shelf off west Florida, Department of Meteorology and Oceanography Technical Report, Florida State University, Tallahassee, 170 pp.
256. Martin, W. E. & Turner, F. B. (1966). Transfer of ^{89}Sr from plants to rabbits in a fallout field, *Health Physics*, vol. 12, Pergamon Press, New York, pp. 621–31.
257. Hacker, S., Billups, C., Wilkins, B., Jr. & Pike, R. W. (1970). Ecological modelling studies: I. Hydrodynamic model and shrimp population distribution model—systems analysis subproject of the LSU Sea Grant Program, Louisiana State University, Baton Rouge, 19 pp.
258. Mann, S. H. (1969). A mathematical theory for the harvest of natural animal populations in the case of male and female dependent birth rates, Research Memorandum No. 69-9, International Symposium on Statistical Ecology.
259. Dent, J. B. & Anderson, J. R., ed. (1971). *Systems Analysis in Agricultural Management*, Wiley, Sydney, Australia, 394 pp.
260. National Institute of Agricultural Engineering (1973). *Proceedings of the Symposium on Systems Applications in Agricultural Engineering*, National Institute of Agricultural Engineering, Silsoe, Bedford, England, 144 pp.

3

The Application of Systems Theory in Agriculture

J. B. DENT*

Department of Agriculture and Horticulture, University of Nottingham, England

3.1. STRUCTURE OF SYSTEMS RESEARCH IN AGRICULTURE

The proliferation of systems thinking among agriculturalists began about a decade ago, though by that time the concepts of general systems theory were already well established (Boulding[1]) and had been applied in advanced studies in other disciplinary areas (Bonini,[2] Cohen,[3] Shubick[4]). General systems theory was seen as the integrator of disciplines and simulation as the mechanism whereby systems could be modelled and studied. Since that time, systems theory has matured as outlined at this symposium by Van Dyne[5] and the power of simulation has increased through the availability of larger and faster computers.

Research in the application of systems theory to agriculture has been developed on a number of fronts: effectively at a number of 'levels' within the overall agricultural industrial system. This polarisation runs contrary to the very foundations of systems theory which emphasises the whole rather than the parts or elements of a system and underlines the consequent hierarchical framework (Wright[6]). Within this framework individual subsystems cannot be considered as having an independent function nor can they be fully understood without recourse to the complete system. The present lack of vertical integration in systems research in agriculture is a totally unsatisfactory condition and undoubtedly is a symptom of the intra-disciplinary bound nature of the personnel involved. The real world does not come to us in disciplinary packages and the fact that researchers are locked within disciplines (or at least 'levels' of disciplines) (Van Dyne[7])

* Present address: Department of Farm Management, Lincoln College, University of Canterbury, Canterbury, New Zealand.

could be a contributory factor in the general lack of impact of systems work in agriculture.

The concentration of systems research within levels is apparent from the following selected classification of published work in agricultural systems which is meant to be illustrative rather than exhaustive or taxonomic in nature:

Level 1. *Biochemical and physical systems*
  (i) Soil nutrient/plant growth relationships (Beek & Frizzell,[8] de Wit & van Keulen[9]).
  (ii) Studies in photosynthesis (Duncan et al.[10], Idso[11]).
  (iii) Animal metabolic studies (Baker,[12] Baldwin & Smith,[13] Smith[14]).

Level 2. *Plant and animal systems*
  (i) Plant and crop growth (Bryne & Tognetti[15]).
  (ii) Growth and development in livestock (Bywater[16]).
  (iii) Animal/pasture relationships (Donnelly & Armstrong,[17] Eadie,[18] Morley & Spedding,[19] Wright & Dent[20]).

Level 3. *Farm business systems*
  (i) Farm enterprise management (Anderson,[21] Blackie & Dent,[22] Halter & Dean[23]).
  (ii) Farm business management (Hutton & Hinman,[24] Eisgruber & Lee,[25] Maxwell et al.[26]).

Level 4. *National and international systems*
  (i) National agricultural demand/supply studies (McFarquhar & Evans[27]).
  (ii) International food supply models (Forrester,[28] Meadows et al.[29]).

This limited survey indicates the crystallisation of published work (at least) into specific levels and it is just this crystallisation which must limit the value of studies within any one level. System models in levels 1 and 2 in the above classification are essentially technical in nature, and the purpose behind their construction is partly to aid in the understanding of the various sub-systems involved (analysis) and partly to provide a feedback of information to guide further laboratory and field experimentation. The success of these models is debatable (and unquantifiable) but considerable benefits on both counts have been claimed (for example by Beek & Frizzell[8]). However, it seems inconceivable that much progress can be made in using models of this kind to aid the establishment of research priorities when the output from the model is expressed solely in biological and/or physical terms. Where limited finance for research is available it is important to make some socio-economic assessment of the benefits associated with the proposed research programme (Dent & Pearse[30]); and it is here, at a minimum, that liaison with systems of a 'higher' order is needed. Dillon[31] believes that, increasingly, physical and biological scientists will have their terms of reference set by social scientists.

Models of systems falling in levels 3 and 4 can be criticised in a general sense because they have been developed according to the interests of their constructors rather than the needs of their potential users. The result is that model builders and model users are mutually distrustful and the use made of systems models for management purposes in agriculture is correspondingly low. Furthermore, such models frequently have failed to represent adequately the technology (biology) of farming (or make satisfactory assumptions about it) due to poor communications with researchers at levels 2 and 1. Such a situation sets off a cycle of worsening communication between workers of different levels, a fixity of research within a given level, and a tendency to preoccupation with non-problems (Dobson[32]). It has led Anderson[33] to state that systems thinking and simulation methodology in agriculture has largely failed to fulfil its early promise.

The fledgling state of systems theory in agriculture has passed and it is now necessary to look for firm results and realistic application at all levels. To achieve this there is no doubt that some measure of vertical integration is required. The way ahead probably does not lie in the creation of large multi-disciplinary teams handling the range of problems from level 1 through to level 4 above, working under a single director, but rather in an improvement of communications between workers at different levels and a closer adherence to basic systems theory concepts. As Dillon[31] puts it, 'What's required is a structure which will facilitate a synthesising, integrative, team-oriented outlook rather than one that is analytical, compartmentalising and disciplinary.'

3.2. MODELS FOR MANAGEMENT OF FARM SYSTEMS

This paper is to be concerned largely with the application of systems concepts at the farm level at a point where research faces the felt problems of the real world. Models created at this level are particularly vulnerable to criticism since they rely to a large extent on biological data from lower-order systems in the hierarchy and are assessed according to socio-commercial criteria. Judged in this way, systems concepts and simulation models have had very little impact on the farming industry. This failure may be related to a number of factors:

(i) A lack of appreciation of the structure and function of the various biological sub-systems within the farm or enterprise model.
(ii) The lack of liaison between systems researchers and decision-makers.
(iii) The preoccupation of systems researchers with the model-building phase of their work without concomitant attention to the validation and application. Enthusiasm to stretch the new-found wings of computer model construction was understandable but the fledgling must sometime test its wings or perish.
(iv) The genuine uncertainty about how systems theory might find application in practical agriculture. Grappling with systems concepts in the heterogeneous and uncertain agricultural environment initially was taxing enough without considering the possibility of on-farm commercial application.

The last point is perhaps the most crucial and is still largely unresolved; yet systems concepts such as hierarchical structure, feedback and control mechanisms, the establishment of goals for systems, exogenous driving variables and dynamic behaviour are highly relevant to the management of the farm business. Management not only must generate plans for the business relative to specified objectives but also has the responsibility to implement the selected plan, evaluate its performance on a continuous basis and when necessary make adjustments to, or indeed completely change, the strategy involved (Churchman[34]). This control function of management (cybernetics) has been compared by Wiener[35] to that of a steersman who requires a constant flow of information about the effects of his actions in order to permit helm modifications to maintain his intended course (strategy). In farm management the type and flow of information required to 'steer' the business will depend on the type of farm or enterprise involved. In all cases an important function of management is to analyse the organisation of the farm and its possible control mechanisms and in the light of these to set up appropriate channels for the transmission of control information to the manager. Essentially, this amounts to establishing an information system for the individual farm manager, an aspect which has had scant research attention (Blackie,[36] Eisgruber,[37] Kennedy,[38] Nelson[39]) and yet is at the heart of the application of systems concepts to farming systems. A possible structure for such an information system is indicated in Fig. 1. Relative to Fig. 1 the management sequence would be:

(i) A comparison of alternative strategies via a planning package.
(ii) A detailed technical and financial forecast for the selected strategy.
(iii) Implementation of the selected strategy (using information in (ii)).
(iv) Recording of technical and financial performance.
(v) Analysis of records to be fed-back to the manager on a period-by-period basis.
(vi) Comparison of record analysis (actual performance) with forecast (planned performance) at corresponding time intervals. Deviation from the planned performance may necessitate an adaptive response by the manager (reorganising the plan, selection of an alternative strategy or modifying forecasts).

Conceptually, and in isolation, the various sub-systems of the complete information system (planning, recording, analysis of records, projection of targets, comparison between target and actual performance) are quite straightforward, but when they are brought together in an integrated

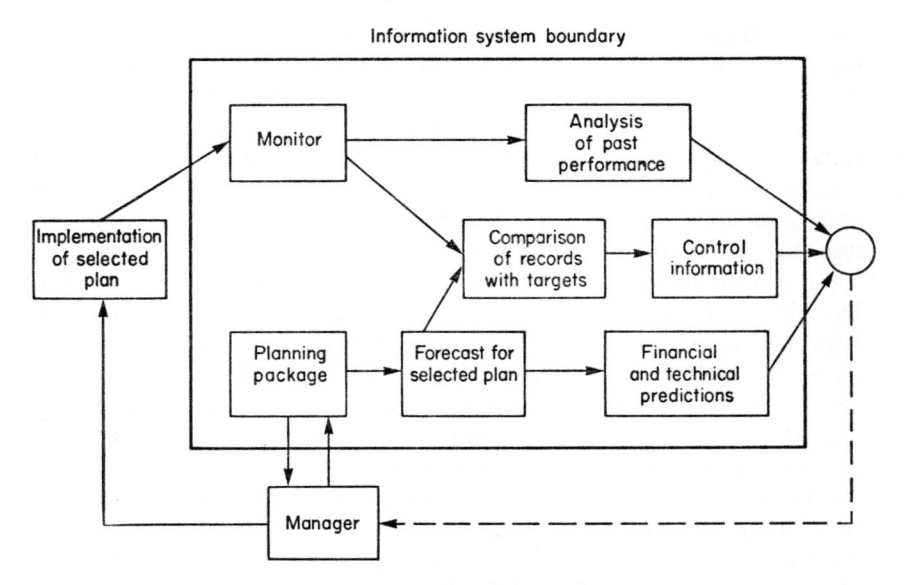

FIG. 1. The structure of an information system.

package their nature must be modified (Dent[40]). For example, the functioning of the planning sub-system within this context is quite different to the traditional role of planning in agriculture where emphasis has been placed on a statement of which enterprises should be incorporated into the farm and what their level should be in order to satisfy some financial criterion (profit maximisation, for example). In the systems context, planning has the function of providing data in a sufficiently comprehensive form about the expected outcome of alternative strategies to permit a realistic comparison between them. Selection of the desired strategy is left to management on the basis of the information generated in the planning package (Anderson and Dent[33]). Inevitably, the precise form of the sub-systems will depend on the nature of the farm or enterprise to which the information system is to be applied.

The sub-system dealing with planning is one which requires further consideration because this is the sub-system which must incorporate a model of the farm or enterprise system. In order to meet the needs of the information system the model must be capable of reflecting the timing of appropriate management decisions and to mimicking in detail the response expected from the real farming system in relation to these decisions. Technical adequacy in the various components of the farm system model is an obvious prerequisite, but technical inputs and outputs must be properly translated into cash terms so that financial as well as technical forecasts of business performance can be made in relation to any defined set of management decisions (strategy). The model must obviously be dynamic in form. Computer-based simulation models match up well to the requirement of providing time-dependent detail (Blackie & Dent[41]) and have the advantage that stochastic elements, often important in the correct specification of bio-economic systems (Mihram[42]), can be incorporated into the model structure with realistic ease.

On conceptual grounds the case for employing a simulation model in the information system is strong, but on operational grounds the gargantuan task of conceiving, building, programming and validating such a model for each farm or enterprise seems totally out of the question. The development costs for the vast majority of farms would far outweigh any managerial benefit that might be gained. The question is, can the benefits of using computer-based simulation models of production systems within the framework of an information system be made available without incurring the necessarily high costs associated with creating a model for every single farm? Three approaches which appear possible are the use of autonomous modules, the development of representative farm models and the construction of skeleton models. Each of these will be considered in turn.

3.2.1. The Use of Autonomous Modules

This approach, suggested by Anderson & Dent,[43] involves the development of a number of sub-models which are autonomous and which will remain stable over a wide range of farming conditions. Such sub-models might represent farming enterprises but more likely biological processes within farming enterprises. Obvious candidates among others for sub-modelling would be:

(i) The control of food intake by ruminants (Balch & Campling,[44] Baumgardp[45]).
(ii) The partition of energy in dairy cows between body fat deposition and milk production (Bywater[16]).
(iii) Grass growth (Wright[46]).
(iv) Grain yield in cereal crops (Donaldson[47]).
(v) Body development related to energy intake in non-ruminants (Fowler[48]).

There is no doubt that the construction of sub-models of this type would require a strong element of vertical integration between the levels of study set out at the beginning of this paper. Once the sub-models have been developed it could be anticipated that they will remain applicable over a reasonable period of time and to all farms within a region (on a certain soil type, etc.). The programmed versions of the modules would be held within the computer and called (effectively as sub-routines) in any combination relevant to the particular farm under study. This grouping would form the core of a full farm (or enterprise) model.

To this core the unique characteristics of the individual farm system (managerial preferences, capital structure, financial commitments, etc.) may be added. These additional sections may well be programmed relatively easily and linked with the selected modular core. The result would be a unique model for a particular farm. Intuitively, this approach is attractive and it ought to provide an integrative framework for all systems research in agriculture. Within this framework, research at all levels would be purposive, and justified relative to the end point of providing an effective management information system for farmers. Careful control over the level of detail specified in each sub-model would be essential if a balanced complete model is to be achieved. Difficulty can be expected in this regard from a multi-disciplinary team with individual interests limited to a single module.

3.2.2. The Development of Representative Farm Models

Individual farms and farm enterprises are characterised by the values taken for many factors (soil type, manager's age and experience, the availability of capital, etc.) and variation in only a few of these makes each farm unique. By the same token some farms will have features similar to other farms. The construction of a representative farm model involves two stages: first, the grouping together of similar farms, and secondly, the construction of a model typical of each group. Criteria upon which to base a suitable classification framework to permit the grouping of farms has been the source of much discussion (Bucknell & Hazel,[49] Millar[50]). Clearly, any grouping must be related to the structure, the operational efficiency and the development possibilities for farms (Day[51]). Provided factors can be isolated which adequately represent suitable criteria for classification, and provided these data are easily accessible for all farms in the relevant population, a number of algorithms have been established which permit clusters of farms to be produced related to these factors (Carmichael *et al.*,[52] Imbrie,[53] Parks,[54] Sokal[55]). However, farm and farm enterprise systems do not fall into obvious clusters, as for example some biological populations do, because the population of farms represents a continuum of change. The function of the clustering algorithm, then, is to locate relatively dense areas in the continuum, and to do this cluster analysis techniques rely to some extent on subjective judgement in the delineation of these areas. The larger the number of factors on which the clustering is based and the larger the sample of farms involved, the more time-consuming the clustering process becomes.

Requirements in terms of computer time and capacity vary from technique to technique but can be quite expensive (for example, an attempt to form clusters of farms described in terms of 10 factors from a sample of 274 farms took 12 min of 130K ICL 1906A computer time). Any reduction in the number of farms in the sample which have to be processed saves considerable time: a scan through the data to remove obvious 'lone' farms or to establish obvious first groups in the sample can be rewarding. Also any reduction in the number of factors used to describe the members of the sample can reduce the cost of computing and a review of the data may suggest that certain criteria may be eliminated without loss. Failing this, a group of mathematical procedures, including principal component analysis (Jeffers[56]), can be used to transform the original set of factors describing the sample of farms into orthogonal components. The linear transformation involved produces exactly the same number of components as original criteria, but it is generally found that a high percentage of the total variation is explained by a lesser number of components, thereby establishing an effective summary of the original data. The components need not necessarily bear an obvious similarity to the original factors. An example can be provided by a principal component analysis carried out on seven factors used to describe dairy farming systems in Great Britain (Billsdon[57]). These seven factors were forage acres per farm; herd size; yield per cow; concentrate feed per gallon of milk produced; percentage of annual milk produced in winter; units of nitrogen per acre of grass; and stocking rate. These data were available from 198 farms in the East Midlands Region of Great Britain, and Table 1 shows the components generated. Approximately 90% of the variation in the seven factors was explained by five components. Clustering can proceed on the basis of the components of which fewer may be used than the original number of factors with accompanied saving in computer time. Similar conclusions have also been derived (Crabtree[58]) from a sample of 251 herds which co-operated in the UK Milk Marketing Board's Low Cost Production Scheme.

TABLE 1

Principle components analysis on 198 dairy farms

	Percentage of variation explained				
Component number:	1	2	3	4	5
Individual factors	30·81	19·72	15·76	14·45	9·05
Cumulative	30·81	50·53	66·29	80·74	89·79

Having defined clusters of similar farms it may be possible to decide that one farm in the cluster is able to represent the rest, or, failing this, some synthesised mean farm for the cluster may be devised. A simulation model can be built for the representative farm in the knowledge that others in the group will be able to relate to it. There is no doubt that such a model could be used to explore the implications of alternative management possibilities and in general terms these implications will apply to all members in the group. In an educational sense, therefore, a representative model could be of considerable value. On the other hand it is unlikely that the forecasts of technical and financial performance of *individual* farms in the group would be specific enough for a model of this kind to find a place within an information system as previously described (Crabtree[58]). Operationally, therefore, representative models are likely to be of limited use for detailed management purposes.

Normative models of this ilk, though perhaps selected as being 'representative' in a less purposive manner, have been the most common type in farm business simulation (see, for example, Dent & Anderson,[59] Jones[60]), and form a suitable basis for multi-disciplinary team research. Unfortunately, in practice, this desirable movement has not materialised in any strength.

3.2.3. The Construction of Skeleton Models

A skeleton model by definition represents the logical structure of a real system and includes only the basic and unchanging parameters of real systems: it is literally a framework which will absorb the flesh of a whole range of real systems. The assumptions are that the defined logic is unchangeable from farm to farm and that the only data to be incorporated are those which can be considered constant between farms. Provided these assumptions hold, then when the skeleton model is combined with appropriate information from an individual farm the resultant coupled model can be considered as tailor-made to that farm and thus provide a reliable simulation of it. A coupled model, then, should be capable of reflecting actual management decisions and mimicking the real system in the nature of its response to these decisions. A skeleton model must be able to absorb the various details unique to individual farms or farm enterprises and these must affect the coupled model's performance in the same way as they affect the real system. In this manner the farmer can have access to an individually styled simulation model at a considerably lower cost than a model which is specifically designed for his own system, for although the development costs of a skeleton model may well be in excess of those of a conventional simulation model, it can find ready and immediate application on many farms.

An additional cost associated with the application of skeleton models in agriculture is the need to provide detailed coupling data. Although this detail may be comprehensive, many farms will undoubtedly be able to provide it readily from records or from experience, while on other farms it may be necessary for management consultants to be involved in assembling this data. This original set of data for every farm wishing to use the skeleton model is stored within a computer data file along with that from other farms to await its turn for coupling with the skeleton model. Obviously, arrangements must be made so that from time to time amendments can be made to these data held on file.

Skeleton models have recently been researched (Blackie & Dent,[22] Charlton,[61] Street[62]) but the emphasis has been on the simulation of individual farm enterprises rather than whole farm systems. Farm enterprises are simpler to model in this way since the variability within enterprises on different farms is less than that found between whole farm systems.

3.3. THE USE OF SKELETON MODELS

The coupled model is a symbolic representation of the real system and its main function is to permit the manager to compare the technical and financial implications of alternative strategies that may be feasible as developments from his present organisation. This is the planning sequence. Having decided on a specific new strategy (or indeed having decided to remain with the present organisation), the further function of the model will be to forecast the future outcome of the strategy in the form of time sequences of selected inputs and outputs. These predicted sequences represent in fact the working-out of a feasible strategy over time from the present organisation to one which fully incorporates the new strategy. In this way the model is providing guidelines for the implementation of a strategy. Selected parameters from the input and output forecasts can be used as targets (Anderson & Dent[33]). The inputs and outputs selected to act as targets should be those to which the performance of the business is particularly sensitive. A basis for management control is provided by the continuous comparison of the projected targets with the actual state of the system as it evolves under a particular strategy (Blackie & Dent[41]). If, for any reason, a significant divergence between actual performance and the targeted performance is observed, some adaptive behaviour may be necessary on the part of the manager. Managers will, of course, react uniquely to signs of divergence but will all be faced with the basic decision of whether to take some corrective action or to allow the system to run off target. Adaptive response may not necessarily involve an attempt to return the system towards the previous planned set of targets since some other alternative may seem more desirable at a particular point in time. The comparison between actual and target performance may not be straightforward since very often several parameters are involved.

It will be seen that the concept of a skeleton model links ideally with requirements of an information system for individual farm systems. The characteristic which permits a skeleton model to be integrated into a general information system while retaining the ability to mimic the individual farm production system distinguishes the skeleton model as a specific type (Blackie[36]). Within the context of an information system, skeleton models can form the basis for national, regional or industry-based schemes serving many farms. If this can be achieved, every production unit involved in the scheme has access to an information system related to its own particular management needs.

3.3.1. An Example of an Enterprise System Skeleton Model

A diagrammatic representation of a skeleton model for pig production systems is presented in Fig. 2. This, of course, is a static representation but shows the capability of accommodating all feasible methods of production. The computer version is constructed to allow the correct phasing and timing of inputs and outputs. The model is organised so that when it is coupled with actual data from a pig unit it provides a specific simulation of that unit and is able to reflect accurately the consequences of production decisions in respect of their sequencing and timing. The model is thus activated by data from individual farms relating to their performance, policy and status (Dent[40]).

The skeleton model is stored in a computer and each pig unit wishing to use it is allocated computer storage space on which its own input data is held and updated regularly. The computer is programmed so as to couple the data from each pig unit with the skeleton model in turn and in this way many farms can use the skeleton model at low unit cost. The output generated by the skeleton model provides detailed forecasts of cash flows and physical performance and status data for the coming twelve months: it also sets out targets of a physical and financial nature to be compared with monitored results of the actual units (records sent to the computer every four weeks by the farmer) (Dent[40]). These records are processed by the model to facilitate comparison and at the same time update the information held.

Experience with the present model suggests that target values remain reliable in respect of the phasing and timing of inputs and outputs up to about six four-weekly periods ahead and in respect to total inputs and outputs for up to a period of twelve months (Anderson & Dent[33]). In this case, therefore, it seems appropriate that detailed targets should be revised every four to six months.

In the pig herd information system outlined above, the farmer, every four weeks, receives updated forecasts for the coming twelve months on a period by period basis, an up-to-date analysis of his past performance and

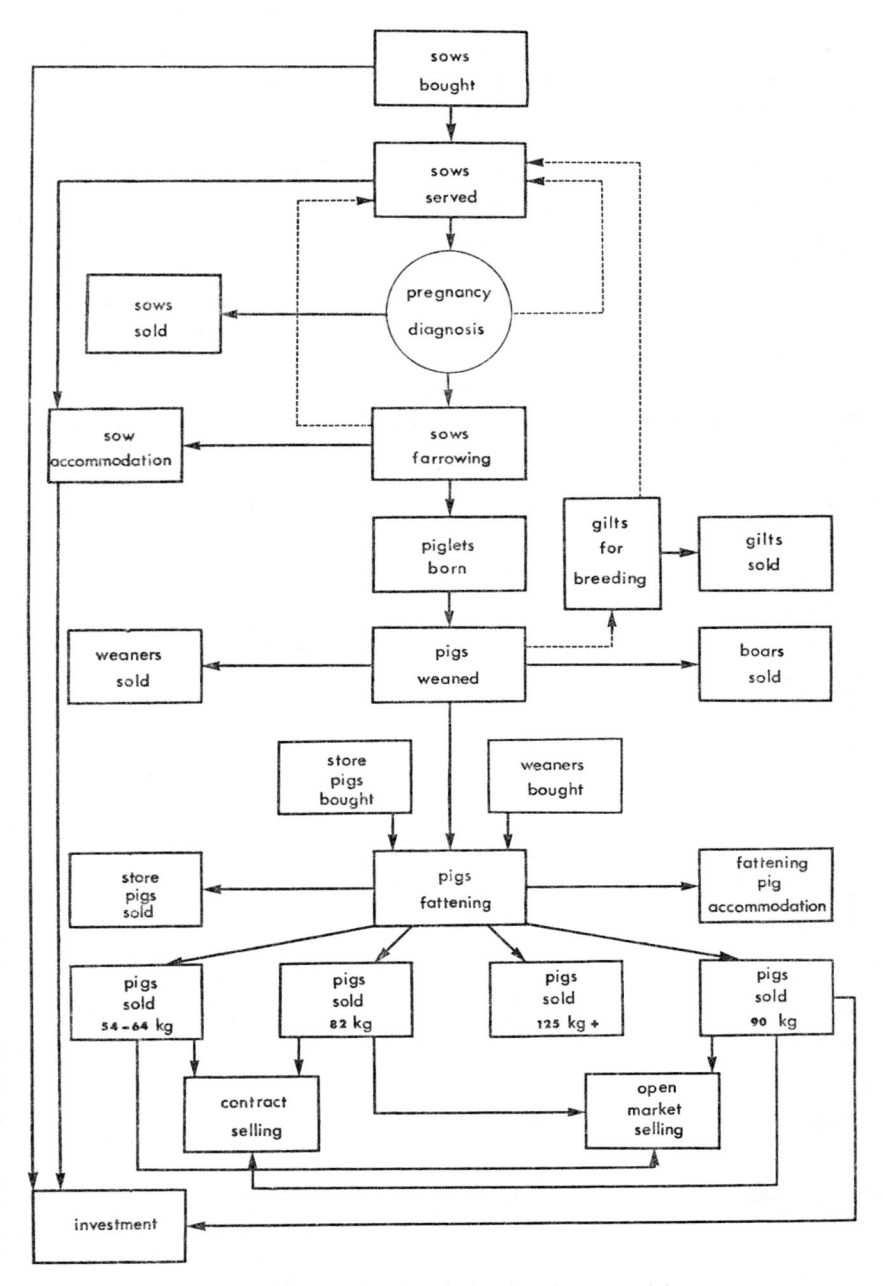

FIG. 2. The logic of a pig herd skeleton model.

a period-by-period comparison between actual and target performance up to the current stage. This procedure provides for the cybernetic loop required within an information system as well as establishing the means whereby individual farmers can access a specific model of their production system to 'experiment' with alternative sets of management decisions.

To be operative as an information system available generally to farmers, the skeleton model must be set in a computer organisation which checks recorded data inputs, stores them appropriately, organises the coupling with the skeleton, and provides output to the farm previously mentioned. Also, of course, an infrastructure must be created which directs input data from many farms to the central computer and output data from the computer to the farms. The basic design of such an operation has been explained by Blackie & Dent[41] and is summarised in Fig. 3.

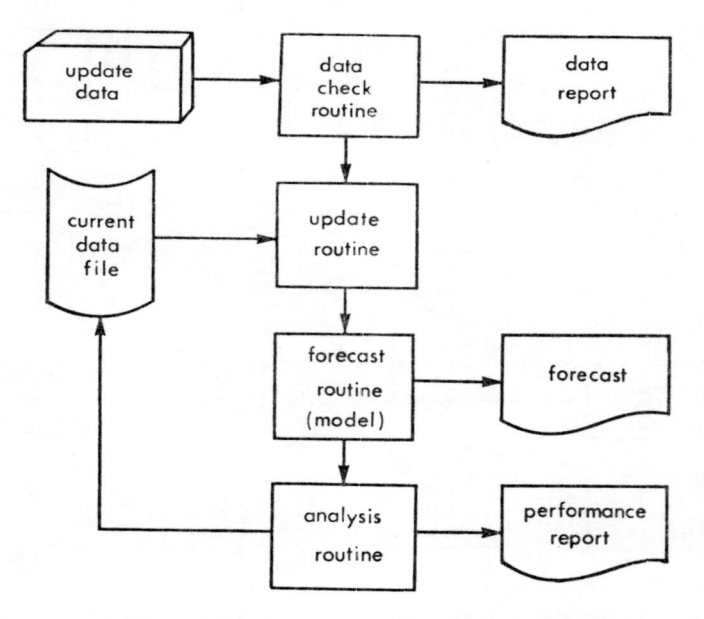

FIG. 3. An information system for a farm enterprise.

3.3.2. The Validation of Skeleton Models

There are basically two major sources of difficulty in attempting to validate a simulation model of a bio-economic system. First, there is the problem of the availability of suitable data against which model predictions can be assessed. In a grazing system model, dependent on climatic variables as main driving forces for example, validation must proceed by taking measurements of grass growth and livestock performance in the appropriate climatic sequence probably over several years. The technical difficulties involved in doing this are extensive and very often precise validation has been shirked (Jones & Brockington[63]). Secondly, the procedures used for comparison of actual and model data are by no means settled (Anderson[64]). Statistical procedures to aid this comparison do exist (Naylor,[65] Naylor & Finger,[66] McKenny,[67] Mihram,[42] Schrank & Holt[68]), but the complexity of the problem where a number of different and perhaps conflicting parameters have to be taken into account stretches standard statistical procedures. Often it has been found that subjective assessments coupled with experience in working with the model is the most satisfactory method of validation (Mihram,[42] Naylor & Finger[66]). However, subjective judgements can be vulnerable to criticism. A further problem can exist because some research may involve an attempt to model agricultural systems which do not currently exist. Validation of such models creates an environment in which controversy and debate will thrive (Anderson[64]).

Validation of skeleton models by contrast is relatively less complex (Blackie & Dent[22]). The skeleton model itself represents only the logical framework for a system and the interactive relationships between various parts of the system can be acceptably defined. There still remains the circumstance that real life processes are represented by mathematical mechanisms within the model and some testing is therefore essential, but this is more akin to mathematical verification than to validation (Blackie & Dent[22]).

The coupled model is intended to mimic existing real systems under precisely defined management circumstances and therefore real systems data are readily available to be compared with the model prediction. The problem therefore, with skeleton models is to determine the means whereby the accuracy of model predictions can be assessed rather than the provision of data to be used for comparison. Procedures for examining the accuracy of future event predictions from skeleton models against available historical time series data recorded on trial farms have been suggested by Blackie & Dent.[22] Two levels of assessment are obvious: one relating to precise event sequencing over time and one (less demanding) relating total event occurrence over time. In general, total event occurrences are likely to be more satisfactorily predicted by a model than event sequencing where an error of forecast is incurred whenever an actual event fails to coincide within the same time period as that forecast. For example, in relation to the performance of the pig herd skeleton model, predictions of the total number of piglets born in a herd during a year were generally more satisfactory than predictions of numbers of pigs to be born in each four-week period during the year. The difficulty with event occurrence predictions is underlined in this example, for if a sow actually farrows on the last day of a period when the forecast is that it should farrow on the first day of the subsequent period, this constitutes an error of prediction. In this sense, quite high errors of prediction by the method detailed by Blackie & Dent[22] can be tolerated. It is important to define for each particular model a satisfactory level of event scheduling error because in a model representing a dynamic system the ability to predict event sequencing accurately over the time ahead may be of major importance (Anderson & Dent[33]).

3.3.3. Skeleton Models in Research and Development

3.3.3.1. The Skeleton Model in the Assessment of New Technology

One particular advantage of the skeleton model is that it can be so organised to mimic any particular manifestation of an enterprise. It is then possible to use it as a 'test-bed' to explore the potential of any next technology within the defined system. Hence, provided a full information system is available, individual farmers may experiment with the model of their own unit to assess the value of, say, a new husbandry technique without putting the actual unit at risk. Dent[40] has illustrated this by the use of the pig herd information system. Extension officers can use the model in a similar way by creating a representative simulation model, via the skeleton model and the information system, of a type of unit with which they commonly have to deal. For the research worker it can also provide a means whereby the impact of any findings he may provide can be judged relative to certain types of production unit prior to the experimental programme being put into effect. Judicious use of the model in this way might well prove to be of some assistance to the research manager in determining the relative worth of alternative programmes (Dent & Anderson[59]). Certainly, the research worker/manager can be provided in this way with access to a proven model. The subsequent ease of operation of the skeleton model via an information system is a real advantage and an encouragement to prior testing of promising experimental programmes as they relate to different types of real-life enterprises. Exactly how objective use may be made of the results of this testing is not clear but is presently a subject of intense debate (Dillon,[31] Fishel,[69] Jeffers,[70,71] Pasternack & Passey,[72] Russell[73]).

3.3.3.2. The Skeleton Model in the Application of new Technology

For the farmer, even having had the advantage of being able to make a positive assessment of some new technology relative to his own unit, it is still no easy matter to introduce even what seems to be the simplest improvement into his unit. This is often a frustrating situation for research personnel convinced of the general worthwhileness of a new finding but not understanding the technical difficulty facing the farmer at its introduction into his system. In some cases, major changes of system are involved and an extended period of reorganisation from the existing to a new system must be undertaken. In other cases a constant review of the enterprise performance is essential. For both cases the use of the skeleton model within an information system can provide considerable assistance. Where major modification of the production system is involved, detailed forecasts of physical and financial targets throughout the transition period provides precise information on the timing of inputs and outputs required to implement the new system. The resultant cash flow predictions based on these forecasts gives guidance on the financial implications of the change. In the second type of situation, the continuous monitoring of the herd and the comparison of key performance characteristics between what is achieved and what was planned permits the manager to keep a close check on whether or not the new technology is meeting expectations.

3.4. CONCLUSIONS

This paper has attempted to indicate how, within the framework of an information system, system theory concepts can be applied in a direct manner to meet the needs of modern management on farms. The major sub-systems of planning, recording, record analysis and control have been expounded relative to the needs of the whole information system and it is clear that the integration of the sub-systems enforces change in their nature and function. The centre-pin to the information system is a simulation model for the production system which is designed in such a way as to be applicable, within an enterprise, to all farms. Such 'skeleton' models have recently been under development for a number of enterprises (Blackie,[36] Charlton,[61] Street[62]). To be operational, the information system, in which the skeleton model is embedded, requires considerable investment in carefully designed computer systems and in extension effort to introduce farmers to the information system and to oversee its operation.

The application of skeleton models for management purposes on individual farms can be confidently expected in the not too distant future.

However, the development of a skeleton model for an enterprise creates the opportunity for linking research (and various strata of research effort), development and extension in agriculture within that enterprise and so creates the possibility of more purposive selection of research programmes and permits a more efficient transfer of new findings to agricultural practice. This potential has not been fully explored but the pay-off from investigation could be considerable and would seem to justify the work involved.

REFERENCES

1. Boulding, K. E. (1956). General system theory: the skeleton science, *Management Science*, **2**, 197–208.
2. Bonini, C. P. (1963). *Simulation of Information and Decision Systems in the Firm*, Prentice-Hall, Englewood Cliffs, New Jersey.
3. Cohen, K. L. (1960). Simulation of the firm, *American Economic Review*, **50**, 534–40.
4. Shubick, M. (1960). Simulation of the industry and the firm, *American Economic Review*, **50**, 908–18.
5. Van Dyne, G. M. (1975). In: *Study of Agricultural Systems* (ed. G. E. Dalton), Applied Science Publishers, London.
6. Wright, A. (1971). Farming systems, models and simulation, in: *Systems Analysis in Agricultural Management* (ed. J. B. Dent and J. R. Anderson), Wiley, Sydney.
7. Van Dyne, G. M., ed. (1969). *The Ecosystem Concept in National Resource Management*, Academic Press, New York.
8. Beek, J. & Frizzell, M. J. (1973). *Simulation of Nitrogen Behaviour in Soils*, PUDOC, Wageningen, The Netherlands.
9. de Wit, C. T. & van Keulen, H., eds. (1972). *Simulation of Transport Processes in Soils*, PUDOC, Wageningen, The Netherlands.
10. Duncan, W. G., Loomis, R. A., Williams, W. A. & Hanan, R. (1967). A model for simulating photosynthesis in plant communities, *Hilgardia*, **38**, 181–205.
11. Idso, S. B. (1969). A theoretical framework for the photosynthetic modelling of plant communities, *Advance Frontiers of Plant Sciences*, **23**, 91–118.
12. Baker, N. (1969). The uses of computers to study rates of lipid metabolism, *Journal of Lipid Research*, **10**, 1–24.
13. Baldwin, R. L. & Smith, N. E. (1971). Application of simulation modelling techniques in analysis of dynamic aspects of animal energetics, *Journal of Federation Proceedings*, **30**, 1459–65.
14. Smith, N. E. (1971). Quantitative simulation analysis of ruminant metabolic function: basal, lactation, milk fat depression, unpublished Ph.D. thesis, University of California, Davis.
15. Byrne, G. E. & Tognetti, V. (1967). Simulation of pasture–environment interaction, *Agricultural Meteorology*, **6**, 151–63.

16. Bywater, A. C. (1973). Using technical information to aid farm decisions by means of the computer, *Farm Management*, **2**, 341–8.
17. Donnelly, J. R. & Armstrong, J. S. (1968). Summer grazing, *Proceedings of 2nd Simulation Conference*, New York, pp. 329–32.
18. Eadie, J. (1970). Hill sheep production systems development, Hill Farm Research Organisation, 5th Report, pp. 70–87.
19. Morley, F. H. W. & Spedding, C. R. W. (1968). Agricultural systems and grazing experiments, *Herbage Abstracts*, **38**, 279–87.
20. Wright, A. & Dent, J. B. (1969). The application of simulation techniques to the study of grazing systems, *Australian Journal of Agricultural Economics*, **13**, 144–53.
21. Anderson, J. R. (1971). Spatial diversification of high-risk sheep farms, in: *Systems Analysis in Agricultural Management* (ed. J. B. Dent and J. R. Anderson), Wiley, Sydney.
22. Blackie, M. J. & Dent, J. B. (1974). The concept and application of skeleton models in farm business analysis and planning, *Journal of Agricultural Economics*, **25**, 165–73.
23. Halter, A. H. & Dean, G. W. (1965). Simulation of a Californian range feed lot operation, Californian Agric. Exp. Stat., Giannini Res. Rep., No. 282.
24. Hutton, R. F. & Hinman, H. R. (1968). A general agricultural firm simulation, Dept. Agricultural Economics and Rural Sociology, Agric. Exp. Stat., Pennsylvania Univ., Rep. No. 72.
25. Eisgruber, L. M. & Lee, G. E. (1971). A systems approach to studying the growth of the farm firm, in: *Systems Analysis in Agricultural Management* (ed. J. B. Dent and J. R. Anderson), Wiley, Sydney.
26. Maxwell, T. J., Eadie, J. & Sibbald, A. R. (1973). Methods of economic appraisal of hill sheep production systems, *Potassium Institute Ltd, Colloquium Proceedings*, no. 3, pp. 103–13.
27. McFarquhar, A. A. M. & Evans, M. C. (1971). Projection models for United Kingdom food and agriculture, *Journal of Agricultural Economics*, **22**, 321–45.
28. Forrester, J. W. (1971). *World Dynamics*, Wright-Allen, Cambridge, Mass.
29. Meadows, D. H., Meadows, D. L., Randers, J. & Behrens, W. W. (1972). *The Limits to Growth*, Universe Books, New York.
30. Dent, J. B. & Pearse, R. A. (1973). Operations research, agricultural research and agricultural practice, in: *Operational Research '72* (ed. M. Ross), North-Holland, Amsterdam.
31. Dillon, J. L. (1973). The economics of systems research, *Proceedings of Systems Research Conference*, Massey University, New Zealand.
32. Dobson, W. D. (1970). Computer simulation models: the need for an applications orientation, *Canadian Journal of Agricultural Economics*, **18**, 175–8.
33. Anderson, F. M. & Dent, J. B. (1974). Planning and control of human food chains, in: *Human Food Chains and Nutrient Cycles* (ed. A. N. Duckham and J. G. W. Jones), North-Holland, Amsterdam (in press).
34. Churchman, C. W. (1968). Systems, in *Systems Analysis* (ed. S. L. Optner, 1973), Penguin, Harmondsworth.

35. Wiener, N. (1948). *Cybernetics*, Wiley, New York.
36. Blackie, M. J. (1974). A management system for the pig enterprise, unpublished Ph.D. thesis, Nottingham University.
37. Eisgruber, L. M. (1974). Managerial information and decision systems in the USA: historical developments, current status and major issues, *American Journal of Agricultural Economics*, 55, 930–7.
38. Kennedy, J. O. S. (1971). The design of integrated planning and control systems for agricultural enterprises with particular application to beef production, unpublished Ph.D. thesis, University of London.
39. Nelson, A. G. (1969). The feasibility of an information system: the beef feed lot case, unpublished Ph.D. thesis, Purdue University.
40. Dent, J. B. (1974). Application of system concepts and simulation in agriculture, Dept. Agric., University of Aberdeen, Misc. Pub.
41. Blackie, M. J. & Dent, J. B. (1973). A planning and control system for the small firm: a development of operations research in agriculture, *Zeitschrift für Operations Research*, 177, 173–82.
42. Mihram, G. A. (1972). *Simulation: Statistical Foundations and Methodology*, Academic Press, New York.
43. Anderson, J. R. & Dent, J. B. (1972). Systems simulation and agricultural research, *Journal of Australian Institute of Agricultural Science*, 38, 264–96.
44. Balch, C. C. & Campling, R. C. (1969). Voluntary intake of food, in: *Handbuch der Tierernährung*, vol. 1 (ed. W. Lenkett, K. Breirem and E. Craseman), Paul Parey, Berlin.
45. Baumgardp, B. R. (1970). Control of intake in the regulation of energy intake, in: *Physiology of Digestion and Metabolism in Ruminants* (ed. A. T. Phillipson), Oriel, Newcastle.
46. Wright, A. (1970). Systems research and grazing systems: management orientated simulation, *Farm Management Bulletin*, no. IV, University of New England, Australia.
47. Donaldson, G. F. (1968). Allowing for weather risk in assessing harvesting machinery, *American Journal of Agricultural Economics*, 50, 24–40.
48. Fowler, V. R. (1966). The application of current carcase knowledge to pig production, *Proc. 2nd Conf. Agricultural Research Workers and Agricultural Economists*, Brighton, PIDA, pp. 72–80.
49. Bucknell, A. E. & Hazel, P. B. R. (1972). Implications of aggregation bias for the construction of static and dynamic linear programming supply models, *Journal of Agricultural Economics*, 23, 119–34.
50. Millar, T. A. (1966). Sufficient conditions for aggregation in LP models, *Agricultural Economics Research*, 51–7.
51. Day, L. M. (1963). Use of representative firms in studies of inter-regional competition and production response, *Journal of Farm Economics*, 45, 1438–45.
52. Carmichael, J. W., George, J. A. & Julius, R. S. (1968). Finding natural clusters, *Journal of Systematic Zoology*, 17, 144–50.
53. Imbrie, J. (1963), Factor and sector analysis programs for analysing geologic data, Techn. Rep. No. 6, ONR Task No. 389-35.
54. Parks, J. M. (1966). Cluster analysis applied to multivariate geologic problems, *Journal of Geology*, 74, 703–15.

55. Sokal, R. A. (1961). Distance as a measure of taxonomic similarity, *Journal of Systematic Zoology*, **10**, 70–9.
56. Jeffers, J. N. R. (1967). Two case studies in the application of principal components, *Applied Statistics*, **16**, 225–36.
57. Billsdon, S. (1974). Unpublished material, University of Nottingham.
58. Crabtree, J. R. (1971). An assessment of the relative importance of factors affecting criteria of success in dairy farming using component analysis, *Farm Economist*, **12** (1), 17–30.
59. Dent, J. B. & Anderson, J. R. (1971). *Systems Analysis in Agricultural Management*, Wiley, Sydney.
60. Jones, J. G. W., ed. (1970). The use of models in agricultural and biological research, Grassland Research Institute, Hurley.
61. Charlton, P. J. (1972). Financing farm business growth, *Farm Management*, **2**, 60–70.
62. Street, P. R. (1973). Nottingham University dairy enterprise simulator, University of Nottingham, Dept. of Agric. & Hortic. Pub.
63. Jones, J. G. W. & Brockington, N. R. (1971). Intensive grazing systems, in: *Systems Analysis in Agricultural Management* (ed. J. B. Dent and J. R. Anderson), Wiley, Sydney.
64. Anderson, J. R. (1974). Simulation: methodology and application in agricultural economics, *Review of Marketing and Agricultural Economics*, (in press).
65. Naylor, T. H. (1973). Simulation and validation, in: *Operations Research '72* (ed. M. Ross), North-Holland, Amsterdam.
66. Naylor, T. H. & Finger, J. M. (1967). Verification of computer simulation models, *Management Science*, **14**, B92–B101.
67. McKenny, J. L. (1967). Critique of: Verification of computer simulation models, *Management Science*, **14**, B102–B103.
68. Schrank, W. R. & Holt, C. C. (1967). Critique of: Verification of computer simulation models, *Management Science*, **14**, B103–B104.
69. Fishel, W. L. (1971). *Resource Allocation in Agricultural Research*, University of Minnesota Press, Minneapolis.
70. Jeffers, J. N. R. (1972). Project planning and research administration, *Conference of the Advisory Group of Forest Statisticians*, IUFRO, 3rd, Jouy-en-Josas, Institut National de la Recherche Agronomique.
71. Jeffers, J. N. R. (1972). Summary and assessment: a research director's point of view, in: *Mathematical Models in Ecology* (ed. J. N. R. Jeffers), Blackwell, Oxford.
72. Pasternak, H. & Passey, V. (1973). Bicrieterian functions in annual activity planning, in: *Operations Research '72* (ed. M. Ross), North-Holland, Amsterdam.
73. Russell, D. G. (1973). *Resource Allocation System for Agricultural Research*, University of Stirling Press, Stirling.

4

The Problem of Finding an Optimum Solution

G. W. ARNOLD and D. BENNETT

CSIRO, Division of Land Resources Management, Perth, Western Australia

4.1. INTRODUCTION

For this paper we have defined the optimum solution as that path of investigation, recommendation and action which leads to the greatest gain in welfare. The welfare of the model user is of major concern, but one should also be aware of consequences to the community from decisions made on the basis of a systems model. These consequences, if adverse, are likely to lead to some constraints being applied to the farming system. A systems model is itself part of a system and the relevance of optimising the model can only be viewed with reference to the system it represents.

4.2. THE NATURE OF AGRICULTURAL SYSTEMS

Any consideration of optimisation of management for an agricultural system must be preceded by an analysis of the nature of the system being studied. There are, however, some general principles applicable to all systems which must first be appreciated before variations in the nature of specific systems are considered. Charlton & Thompson[1] briefly outlined these principles which are enlarged on below.

All agricultural systems are time-dependent and dynamic in the sense that they are in a constant state of change and evolution, whereby events which occur at the present time affect the way in which the system performs both financially and biologically in the future. The dynamics of a system will vary, depending on the specific system. The finances of a farm are always dynamic, with future profitability being affected by current investments and by unpredictable external events which influence the biological efficiency or level of production. For example, the current capital investment in plant and buildings will influence both the availability of credit and the choices of alternative enterprises in the future. Thus the purchase of a combine harvester in the current financial year will tend to favour cropping enterprises that will use that harvester in this and the following years.

In enterprises which are climate-dependent, weather conditions act as unpredictable external forces on farm finances. In all systems fluctuations in markets, changes in interest rate and removal of subsidies act as partially or wholly unpredictable external forces. The biological components of the agricultural system are also dynamic, being dependent on weather conditions and on the interactions between various enterprises on the farm. There is, however, a wide variation in the control that can be imposed on the production system, though controllable but unforeseen biological events may always occur (e.g. disease, adverse animal behavioural responses to density, etc.).

In an intensive pig or poultry system many of the biological components of the system can be manipulated: environment, nutrition, reproduction and through these the level of production. Although genotype will be a constraint and the timing of reproduction may be difficult, the control over the biological components will be far greater than for less intensive systems.

In a cropping or mixed farming system, management strategies such as crop variety, soil drainage, levels of fertiliser and pesticides, can be used to reduce climatic influences. There are often effects of previous events on the land, e.g. type of previous pasture or crop, crop rotation; and type and amount of previous fertiliser. These will influence the potential yield of the crop through effects on nutrient supply, weeds and plant diseases in ways which might be inadequately recognised. Much of the agriculture of Northern Europe and North America involves long periods in winter when ruminant stock are housed and this has the effect of temporarily preventing the natural feedbacks between the grazing stock and the pastures which they graze—allowing some reparation if necessary to adverse effects of the previous grazing season before a new one is commenced.

At the other end of the scale are the extensive rangelands systems of much of the world, in which often the only controls that can be imposed are in stock type and numbers, sometimes the timing of reproduction and, possibly, position of water supplies. In these systems there are continuous feedbacks between climate, pasture growth, nutrient cycling and the grazing animal. The effects of overgrazing at some critical times may alter the flora within the ecosystem for years to come and if erosion occurs this may alter productivity for centuries.

The message to be gained for optimisation is that the greater the number of natural feedbacks in a system and the fewer the opportunities for intervention (with fences, fertilisers, seed, etc.), the more constraints of a non-monetary nature will be required. (For example, if all the birds in a battery house die they can be replaced. If pasture density is decreased by continuous cropping of the ley, the rotation can be changed, but if the soil and vegetation are damaged as often occurs in arid grazing lands, then the cost of reparation is likely to be uneconomical or physically impossible.) In such systems there is the possibility that the financial time scale of the lessee decision-maker is shorter than some of the biological feedback mechanisms—leading, as is the case for many Australian rangelands, to decay. In such systems intervention by the State is becoming more frequent, and it is likely that the ground rules will be set by systems models.

A vital characteristic of many but not all agricultural systems is its sociology. Farming is practised for a variety of reasons, not all of which are directly profit-motivated. Farming as a recreation is increasingly practised in near-urban areas in much of the Western and developing world. Business investment in agriculture to avoid taxation cannot be overlooked; neither can farming as a way of life, a life in which the standard of living is matched to the individual or family's aspirations for life in a rural environment, or as an inescapable life through lack of suitable job training or job opportunities elsewhere.

Even where farming is largely profit-motivated, the personal idiosyncracies of the entrepreneur must be taken into account in developing an optimum strategy for the agricultural system being used. If a man dislikes cows, or lacks the skill to handle them, then he is unlikely to accept dairying as a major enterprise even if it were clearly the most profitable with the resources available.

These social constraints as well as the financial and biological nature of the system must be taken into account in deciding how to approach optimisation. However, decision-makers interested in optimising a system in which there are many controls on production, such as with broilers, are more likely to be satisfied with a single solution and, to them, market situations will be of major significance. Where there are complex biological feedbacks in the production system, such as in rangelands, the decision-maker is unlikely to be satisfied with a single solution, but a range of choices. With rangelands the time scales are such that market situations can seldom be taken into account and biological constraints dominate. The need is to ascertain that any proposed strategy is resilient, and does not lead to disaster in times of drought, flood, fire or monetary depression.

4.3. THE NATURE OF AGRICULTURAL SYSTEMS MODELS IN RELATION TO OPTIMISATION

Operational research techniques to study agricultural systems were developed by economists. Their models usually use static response surfaces to describe the biological outputs as inputs into an econometric model. Few biologists are attracted to the optimising models offered by operational research, primarily because models such as linear programs do not adequately describe the system so far as the biologist is concerned. Simulation modelling allows him to do so. A model that describes both the biology and economics of an agricultural system to the satisfaction of both biologist and economist may well be extremely difficult to optimise and thus be expensive to use widely.

The agricultural extension worker, on the other hand, needs an aid to decision-making that is easy and cheap to operate and easy to explain. For example, if the model is needed to predict optimum fertiliser strategy for a particular crop or pasture, it is likely that the only available input data will be a paddock history. Neither the farmer nor the extension worker can detail the water-holding capacity of the soil and its nutrient status can only be approximated by chemical tests. Inputs have to be simple and complex output is of little use because it is difficult to explain to the farmer.

There is a need to understand the uses of an agricultural systems model before any construction is attempted. Many 'cheap and nasty'* models can be very useful and contribute more to improved agricultural practice and farmers' profits than 'expensive but accurate' models. What is lacking is any experimental and extension comparison of the efficacy of simple and complex models.

What must be stressed is that optimisation must be tailored to the needs of the customer, be he farmer, extension worker, research worker or politician.

4.4. WHY OPTIMISE?

There seem to be three reasons for optimising:

(i) To surround a conjectural answer with so much mathematical garbage that sceptics are afraid to challenge it (Dillon, personal communication).

* 'Nasty' to the biologist because they oversimplify the known characteristics of the system and give a poorer fit to the real world than more complex models.

(ii) To satisfy the model builder that he has an important message.
(iii) To help answer questions posed by decision-makers, which are aimed at improving their management, and so improve their welfare.

4.4.1. Mathematical Garbage

This is a field in which there has been considerable work but which has been more productive in theses (especially in schools of agricultural economics) than useful decision-making. It has been spearheaded by work in demonstrating the usefulness of operations research techniques. In most cases the argument goes: someone (Dantzig, Bellman or one of their followers) has formulated a new algorithm which neatly optimises a mathematical problem. Can we bend a real-life problem to fit the straitjacket enforced by the technique? Examples are numerous and embarrassing to mention. (Many millions of pounds have been saved by the use of these operations research techniques in agriculture and elsewhere, but early illustrations in the literature have often been very poor practical examples.)

4.4.2. To Satisfy Model Builders

Most model builders in agriculture at present are specialists dissatisfied with previous methods of demonstrating the relevance of their research; most would consider their research was relevant but a few conscientious scientists have been concerned as to whether this was true. Research workers normally construct their models by expanding from the functions they have been investigating (e.g. crop response to fertiliser, or animal performance as a function of stocking rate) into a simulation model of a larger system (e.g. nutrient flow or grazing enterprise). Efforts to expand beyond this point become frustrated by the rapidly increasing size and complexity of the model and by increasing problems of financial* and hardware support. Whether they will ever be useful in solving agricultural problems is still, we think, an article of faith rather than an accepted fact. For these the optimisation becomes a distillation of the model output, a way to nominate certain combinations as being more important than others. If one considers the bias that is likely to enter into such a procedure, then its form will probably be an emphasis on system disturbances away

---

* Thus obeying the third law of simulation as stated by Dillon[2] that 'Once started, simulation of a system will continue until available funds are exhausted'. The first and second laws state: '(a) simulation like statistics cannot prove anything and (b) simulation like statistics can nearly prove anything'.

from present norms towards some doomsday or increased profitability position. Most agricultural simulation models so far constructed unfortunately fall into this category.

4.4.3. Answering Questions

What sort of questions will a decision-maker ask? This depends on the decision-maker and especially on his objectives. Models can serve to formulate government policies on agricultural subsidies, marketing policies for agricultural supply firms, etc., but mostly we think of farmers and they and their attitudes and objectives are more diverse than the conditions they control: from the broad-hatted Texans and Australian Pastoralists to Scottish crofters, from the businessman in Sydney with a play-farm at Bowral (or his equivalent in London living at Henley) to the family farmer of the World's grain belts with almost total dependence on his farm for economic survival, from the industrial poultry farmer whose attitudes are little different from the factory manager to the stud stockbreeder who has a personal relationship with each of his animals. Each farmer has a set of objectives, but the degree to which he could describe them would depend not only on his farming system but also on his educational background. Let us list a few: economic survival and preferably acquisition of some luxuries; a continuous (not necessarily dramatic) increase in security and status with age; a work load that satisfies (is interesting, humane and not too boring), that does not overtax either mind or body and declines with age; a feeling of community within his family and/or within his social environment which is generated by a behavioural pattern acceptable to that community. This list would probably satisfy most farmers, but it is still a long way from a cardinal objective function, with appropriate weightings and constraints, which is the way a modeller wishes to receive it.

So much for the input. Now what about the output? How do you convince a decision-maker that the pieces of paper on which you have printed the outcome of your computations constitute a plan which will improve his welfare? There is first a need to generate some level of belief in the decision-maker's mind that the model which you have produced has a behaviour similar to its real-world counterpart, so that meaningful conclusions can be made about management in the real world using the model. The method which satisfies the analyst in this regard is usually obtained from his simulation model by approximating the behaviour of a set of time series data obtained in the real world. A decision-maker may need different forms of testing before he is satisfied. We have found that setting the model up as an interactive management game and letting the decision-maker explore the model's behaviour can make him more confident because he can see the model conforms to his own experience of reality in situations with which he is familiar. (More likely it baffles him with science, since it is normally the first time he has seen a computer terminal, let alone operated one!)

4.4.4. Is Optimisation Necessary?

Having obtained the belief of the user in model performance, is it necessary to optimise? If the process of exploring the behaviour of the model is relatively simple then the decision-maker might be able to do it himself and provide all the answers he needs (Charlton & Street[3]). What the decision-maker might look for are the changes which have the greatest effect on the value of his objective function. Aids, such as a sorting routine which finds the parameter to which the objective function shows greatest sensitivity at time t, might be just as useful, if not more so to a decision-maker operating the model in iterative mode, than providing a total optimiser.

Having obtained the decision-maker's cardinal objective function, having obtained belief by the decision-maker that the model is a reasonable analogue of reality, and having been asked by the decision-maker, 'What set of decisions will maximise the value of my objective function?', then you can proceed to optimise.

4.5. ECONOMIC THEORY AND OPTIMISATION

In essence optimising is the same process as choosing: you choose a (set of) decision(s) which optimises your objective function. Farming, as it has so often been stated, is a game against nature (Dillon & Heady[4]). Nature is not an antagonist in that it is indifferent to the strategies adopted by the farmer and so the farmer's normatively optimal strategy is to select that strategy which maximises (over all possible strategies of nature) his objective function.

At this point we leave the world of pure economic theory and enter a grey world between economics and behavioural theory. The theory of the firm states that the objective of a firm in a perfectly competitive market is to maximise profit. A justifiable inference of the economic theory is that, to survive, a firm must act in this way. Luckily, world agriculture does not constitute a perfectly competitive market. Agricultural firms have sufficient organisational slack (Cyert & March[5]) to operate in a less competitive mode. Their behaviour can only be described and not dictated by the economic consultant. Economists have absorbed one of the more notable behavioural reactions, the avoidance of risk. If a choice exists between two enterprises which have the same merit value, then most farmers would choose the less risky of these two choices (note the use of the word 'most', not all farmers necessarily avoid risk). Attitudes to risk are normally summarised by the determination of the decision-maker's utility function (Officer et al.,[6] Officer & Dillon,[7] Dillon[8]). There is a degree of optimism amongst some agricultural economists about the accuracy and stability of utility functions (Officer & Halter[9]). This is not shared by the behavioural psychologists (Edwards & Tversky[10]). There is no guarantee that the individual decision-maker's utility function, gauged at any point before, will be a true reflection of his utility function at the time that the decision is made.

The other aspect emphasised by recent economic papers is the use of the decision-maker's own probability of outcomes of events rather than any more remote source of data (regional experiments, time trends, etc.). These subjective 'prior' probabilities can then be modified in the light of specifically derived information into a 'posterior' probability from actual observations observed prior to the decision by the use of Bayes' theorem (Officer & Dillon[7]).

Cyert & March[5] emphasise that the aspiration level on goal dimensions essentially set by three variables: the organisation's past goal, the organisation's past performance, and the past performance of other 'comparable' organisations. These variables are reflections of the factors quoted by Daw[11] as being important to farmers, namely compatibility with present farm pattern and farmers' own preferences about activities to be included or excluded.

One interesting economic problem arose when we accepted the standard normative economic model for fertiliser response, which fails to consider the residual value of phosphorus in soil. Future incomes from present applications are both real and one of the factors a Western Australian farmer considers when deciding on his rate of application. The local term used is to 'put it in the super bank.' Papers quoted by Anderson[12] use the sum of the discounted incomes of future crops with no further applications of fertiliser (Pesek et al.,[13] Heady & Dillon,[14] Doll et al.,[15]). Not only is the procedure unrealistic in that future crops will receive further applications in normal farming practice, but it is clumsy and not in accordance with good accounting practice, which is to value inventory at purchase rather than possible sale value. Thus the inventory of residual phosphate in the soil should be costed at its purchase or replacement cost rather than the possible yields of as yet ungrown crops.

The objective of most firms carrying an inventory can be expressed in the following equation:

$$OF = \text{Max} \left[(\text{Outgoing inventory})/(1 + R) - (\text{Incoming Inventory}) + ((\text{Profits})/(1 + R) - (\text{Costs}))\right] \quad (1)$$

where OF = objective function and R = rate for return from alternative investment.

For the residual fertiliser problem this can be translated to:

$$OF = \text{Max} \, (P_x(G_t(X + Z)/(1 + R) - Z)) + (P_y F(X)/(1 + R) - P_x X) \quad (2)$$

where P_x = price of applied fertiliser, $G_t(\,)$ = the time-dependent function which determines fertiliser carryover, X = quantity of fertiliser applied, Z = fertiliser carried over from previous decision period, P_y = price of product and $F(\,)$ = crop production function.

The optimum is obtained by differentiating with respect to X, setting the derivative to zero, and solving for X^* (the optimum). For a Mitscherlich representation of

$$F = A \, (1 - B \exp (-CX)) \quad (3)$$

and a constant discount representation of

$$G_t = VX \quad (4)$$

the solution is

$$X^* = \ln \left[ABC/((P_x/P_y)(1 + R - V))\right]/C \quad (5)$$

where V is the derivative with respect to X of the residual value function G_t. Having developed this approach we found it already documented by FAO.[16]

Thus the incorporation of a residual value changed the criterion for the optimum to better accord with both the chemical and farmers' conceptual aspects of the situation.

4.6. OPTIMISATION TECHNIQUES AND THEIR USE

4.6.1. The Techniques

Our assigned task is not to describe in detail the ever-increasing number of optimisation methods available but to give an account of our

TABLE 1

Optimisation techniques (after Nicholson[17] and Swartzman[18])

Method	Objective	Form of objective function	Form of constraints (linear or non-linear)	Static or dynamic	Deterministic or stochastic	Local or global optimum
1. Benefit–cost analysis (B/C)	To compare various selected management procedures with respect to their benefit–cost ratios.	Non-linear	Non-linear	Static	Deterministic	Local
2. Calculus of variations (Calc)	To find the optimal control which maximises the objective function when the variables are given by differential equations.	Differentiable	Simple and differentiable	Dynamic	Deterministic	Local
3. Linear programming (LP)	To maximise or minimise objective function subject to constraints.	Linear	Linear	Either	Deterministic	Global
4. Quasi-linearisation (QL)	To linearise a non-linear problem to a series of linear problems. It is often used in conjunction with dynamic programming or calculus of variations to linearise non-linear functions.	Converts non-linear to linear	Converts non-linear to linear	Dynamic	Deterministic	Local
5. Quadratic programming (Q.P.)	To maximise or minimise objective function subject to constraints	Quadratic	Linear	Static	Deterministic	Local
6. Geometric programming (GP)	To maximise or minimise objective function subject to constraints	Generalised polynomials	Polynomial equalities or inequalities	Static	Deterministic	Local
7. Non-linear programming (NLP)	To maximise or minimise objective function subject to constraints.	Linear or non-linear	Non-linear	Static	Deterministic	Local

The Problem of Finding an Optimum Solution

8. Stochastic programming (SP)	To maximise or minimise objective function subject to constraints.	Linear or non-linear	Either	Either	Stochastic	Local
9. Dynamic and recursive programming (DP)	To maximise or minimise objective function over time subject to constraints and recursion relations which relate to states earlier in time.	Linear or non-linear	Either	Dynamic	Either	Global
10. Optimal control (OC)	To maximise or minimise an objective function over time subject to constraints and recursion relations which relate to states earlier in time.	Well-behaved	Either	Dynamic	Either	Local
11. Simulation optimisation (Sim-opt)	To maximise or minimise objective functions at discrete stages over time with constraints supplied by a simulation.	Linear or non-linear	Either	Static optimisation with dynamic simulation	Either	Local (over time) Global (instantaneous)
12. Operational (simulation) gaming (GAME)	To test various management procedures using a simulation of a system to find the one that gives the highest value of an objective function over time.	Any—may even be poorly defined	Either	Dynamic	Either	Local
13. Experimental optimisation (Exp-opt)	To determine the optimal value of an objective function by performing a series of experiments on the pertinent variables. *Approaches* Random Factorial Single step Gradient	Any	Either	Either	Either	Local

experiences in optimising models and implementing the solutions in agricultural situations. However, we list the available methods in Table 1 in increasing order of complexity. For those who want more details of these techniques, for each of which there are several variations, we suggest they refer firstly to the source references from which we built this table (Nicholson,[17] and Swartzman[18]), and if these prove inadequate to the following four books.

For the dilettante who wishes to read an interesting, amusing and concise book on optimising procedures there is Kaufmann & Faure's *Introduction to Operations Research*.[19] A more advanced text is Wagner's *Principles of Operations Research*.[20] Tabak & Kuo[21] deal specifically with optimal control theory, while a mathematical treatment of the problems of optimisation can be found in Wilde & Beightler's *Foundations of Optimisation*.[22]

The first ten techniques listed in Table 1 normally require the problem to be formulated in a manner suitable for the optimising technique, while the last three can be used in conjunction with a simulation model developed without much prior consideration of the optimising technique to be used.

Theoretically, with simulation/optimisation any one of the first ten techniques can be used. In actual practice, as might be expected, linear programming (LP) has been almost the sole method of optimising simulation models used to date. Various methods of transfer of parameter values from the simulation to the optimising routine and back have been developed. One hopes that various optimising packages will shortly be provided as standard subroutines for simulation languages.

The general statement can be made that one can attempt to obtain an optimum solution for any constructed model. But increasing model complexity leads to increasing cost of finding the optimum solution. Our problem is a three way trade-off: to formulate a model which is an adequate representation of the agricultural system yet does not require too great an effort to optimise and in which the total cost is not too great for the (farm advisory) market to bear. Our advice to novices is to stick to simple models. More complicated models can always be substituted later if necessary.

4.6.2. The Use of Optimising Techniques

Table 2 lists a number of examples of optimising models in the literature. The list is not exhaustive and is biased towards Australian animal husbandry models. As can be seen from this table and Table 3, the rate of increase in methods of optimising still appears to be greater than the rate

TABLE 2
Examples of optimising models (after Swartzman[18,24])

Management Areas	Author(s)	Type of Programme [a]	Objective	Comments
Intra-farm	Bennett & Ozanne[25]	Calc	Optimise fertiliser rate within year.	Applied to a number of farms and in submissions to maintain a fertiliser subsidy.
	Richards & McCarthy[26]	LP	Maximise revenue by allocating cattle, sheep and crops among selected soil types on large farms.	Program was implemented.
	Rae[27]	LP	Maximise farm firm growth subject to annual income constraints.	Applied to the case study farm. Solution implemented.
	McFarquhar[28]	QP	Maximise utility on a cropping farm.	Hypothetical example of real problem.
	Carlsson et al.[29]	SP	Optimise a mixed farming operation.	Hypothetical solution to a real problem.
	Rae[30]	SP	Optimise the operation of an intensive horticultural enterprise.	Hypothetical solution to a real problem.
	Throsby[31]	DP	Maximise present value of an income from pasture improvement.	Results applied.
	Kennedy et al.[32]	DP	Find optimal crop and fertiliser policy through time.	Hypothetical solution to real problem.
	Kingma[33]	Sim-Opt	Maximise farm firm growth through time.	
Animal husbandry	Bird[34]	LP	Find new enterprises which will increase dairy farm income.	Found opportunity for diversification limited.

continued

TABLE 2—contd.

Management Areas	Author(s)	Type of Programme	Objective	Comments
Animal husbandry —contd.	Jensen[35]	LP	Find new enterprises which will increase dairy farm income.	Farmers found other alternatives not included in model.
	Sinden[36]	LP	Evaluate the role of poplar production on dairy farms.	Rust (*Melampsora* spp) ruined poplar production economics.
	Lindner[37] BAE[38]	LP	Find alternative enterprises for dairy farmers.	Farmers found same enterprises before study was completed.
	Bracken[39]	QP	Minimise radionucleotide intake of cattle while satisfying their diet needs.	A real problem implemented in the laboratory.
	Van de Panne & Popp[40]	SP	Minimise cost of feed while satisfying animal nutrient requirements.	A real problem implemented in a feeder lot.
	Smith[41]	DP	Optimise replacement of dairy cows within a herd.	Hypothetical solution to a real problem.
	Swartzman & Van Dyne[42]	Sim-Opt	Maximise long-term profit of a pastoral property.	Hypothetical solution to a real problem.
	Trebeck & Hardaker[43]	Sim-Opt	Find the cattle stocking rate which maximises utility.	Hypothetical solution to a real problem.

The Problem of Finding an Optimum Solution

	Christian et al.[44]	Exp-Opt	Find the schedule of stock movements which optimises a fat lamb enterprise.	Hypothetical solution to a real problem.
Pest control	Mann[45]	DP	Minimise loss of cropland to pests over given time period by eliminating pests to a certain level at each decision stage.	Hypothetical example assumes complete control over pest elimination.
Resource use	Heady & Whittlesley[46]	LP	Allocate crops to regions in the US to minimise production and transportation costs.	The problem was formulated with real data, its large size precluded application.
	Van Dyne[47]	LP	Optimise protein production in foothill range.	Hypothetical solution to real problem.
	Kennedy et al.[32]	LP	To predict aggregated changes in ouput of Australian agriculture due to changes in price, etc.	Model development still proceeding.
	Maruyama & Fuller[48]	QP	Decide on where to produce milk and how to transport it between competing markets to maximise net revenue.	Real data used (not applied).
	Bennett et al.[49]	QP	Optimise catchment land use.	Data collection and model development still proceeding.
	Townsley[50]	DP	Maximise revenues from sale of New Zealand butter in UK.	Hypothetical solution to real problem.

[a] See Table 1 for explanation of abbreviations.

TABLE 3
Application of programming techniques in resource management (after Swartzman[18])

Management region	Calculus of variations (Calc)	Linear programming (LP)	Quadratic programming (QP)	Stochastic programming (SP)	Dynamic programming (DP)	Simulation optimisation (Sim-Opt)	Experimental optimisation (Exp-Opt)
Intra-farm	R	R, H	R	H	H, R	H	
Animal husbandry		R	R	R	H	H	H
Pest control					H		
Regional resource use		H, R	H, R		H		

R = Real-life application of the optimal solution. H = Hypothetical application to real-life problem.

of increase in the use of these techniques in hypothetical agricultural situations, which is itself greater than the rate of increase in usage to solve real agricultural problems. (Unfortunately the bias introduced by both scientists' need to publish and by scientific editorial policy makes it easy to publish new algorithms and hypothetical examples, but extremely difficult to get any picture of the extent of usage of models in real-life decisions. From personal correspondence we know that ICI runs about 100 MASCOT LPs for farmers in the UK and that the Agricultural Business Research Institute at the University of New England, in Australia, has provided a large number of least-cost feed mix LP formulations as well as LP farm plans, regional matrices, etc.) It would appear that LP is used for farm planning on 0·05–0·5% of farms each year, and most other techniques are still in the 'once-off' stage. This is some but not much progress from the situation of a decade ago (Hutton[23]). More research effort should be assigned to studies of the use, and reasons for lack of use, of optimising models in agriculture. For example we would appreciate help in locating a generalised dynamic programming package.

A common use of LP documented in the literature is the generation of optimal strategies for 'average' district farms. There have been a number of such studies associated with finding opportunities for diversification by the depressed Australian butterfat industry. And in some cases there has also been enough time to evaluate the LP proposals. As can be seen from the comments in Table 2, the results do not indicate a very good record in this particular problem, partly from a failure to provide guidelines soon enough (Bureau of Agricultural Economics[38]), partly from a failure to recognise possible enterprises and price shifts (Jensen[35]) and partly from a failure to foresee biological problems (Sinden[36]).

To what degree can these procedures cope with maximising utility, rather than profit and the use of the decision-makers' subjective probabilities? We do not know. Incorporating utility normally involves the incorporation of a quadratic component into an objective function (since variance is a function of squared parameters) and the development of a stochastic model. Such models are well within the bounds of established techniques. Kennedy & Fransisco[51] have recently been looking at the formulation of risk constraints within linear programming. As will be shown in a later section, the estimate of variance due to season, which is one segment of risk, is a relatively simple task for a simulation model. What is more difficult is the estimate of error due to inadequate system specification. Since the size of this problem is a power function of the number of variables, this in itself is almost sufficient justification for small models.

4.7. EXAMPLES OF OPTIMISATION OF DYNAMIC AGRICULTURAL SYSTEMS

In this section various aspects of optimisation have been explored using two systems familiar to us. Using the first system, linear programming was compared with a more complex procedure to obtain a single optimum gross income from a relatively simple system of utilising a lupin crop. The second system of wool production from sheep grazing a subterranean clover pasture was used to provide an example of using a simulation model to provide response surface data as inputs into an optimisation routine, such that the dynamics of the biological system are retained. It was further used to show the value of using a simulation model to explore the consequences on net income of various management strategies.

4.7.1. System Description

The simulation models used were constructed from sub-models as shown in Figs. 1 and 2. Details of the sub-models are given by Arnold & Campbell[52] (animal growth), Arnold & Galbraith[53] (crop growth), Arnold et al.[54] (pasture growth), Carbon & Galbraith[55] (water balance). The schema shows that these models have many feedbacks and are concerned only with the physical, chemical and biological bases of a ley system and not with the labour or monetary aspects.

The overall model (1500 statements) was developed to describe crop and pasture production and utilisation in a Mediterranean environment. There is a growing season of 4–8 months during the cooler period of the year and inadequate rainfall for growth of most agricultural pastures and crops for the hotter months of the year. Annual crops of cereals and legumes and annual pastures of legumes and grasses are grown.

Lupins (*Lupinus angustifolius* cv. Uniharvest) are a legume grain crop planted in autumn which are harvested in summer and the grain is either sold or used for feeding livestock; the stubble is grazed *in situ*.

Annual pastures based on subterranean clover (*Trifolium subterraneum* cv. Geraldton) can be maintained clover dominant (>80% clover) for several years. These pastures, after an initial year for establishment reseed themselves with the mostly hard seed being buried in a burr. The annual cycle begins with autumn rains which germinate seed that has softened as a consequence of heat exposure in summer. The initial photosynthetic mass is determined by the numbers of seedlings germinating, and if seed supply is low then pasture growth rate is slow. Growth during the season is influenced by environmental conditions but flowering begins in early winter, the time depending on the date of germination. The number

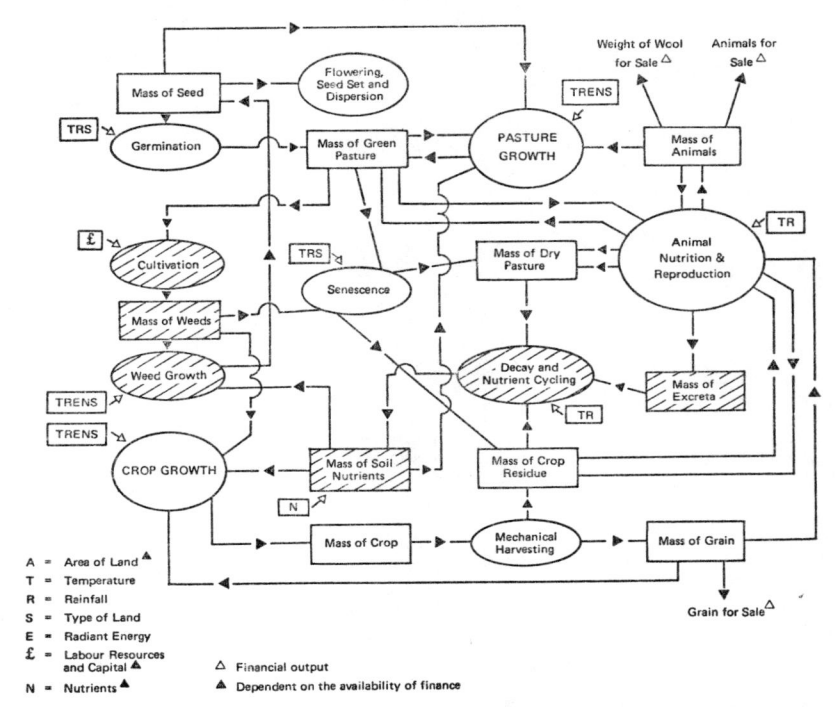

FIG. 1. A descriptive model of the physical inputs, some of the transfer and regulatory mechanisms, the pool of materials within the systems, and the outputs for a ley farm. (Note: The shaded areas have not yet been modelled.)

of seeds to be produced is determined at this time and is influenced by the photosynthetic mass. Thus, heavy grazing prior to flowering will reduce the number of seeds. The quality of the seed is determined by environmental conditions during seed ripening which takes place in late spring. An early cessation of growth due to lack of rain results in a high proportion of soft seed. This is a disadvantage since summer rains will cause germination of much of the soft seed and these seedlings invariably die from drought.

The digestibility and nitrogen content of the pasture vary widely over the year. The potential value over summer is set at the end of spring when the pastures dry off but is changed with leaching by summer rain.

Sheep selectively graze both lupin stubbles and pasture. While much clover seed is buried the sheep can eat that which is on the soil surface and may dig up some buried burr. Burr is not usually eaten until late summer, but at high stocking rates considerable quantities of seed will be eaten. Thus, the system is sensitive to the state of the seed pools.

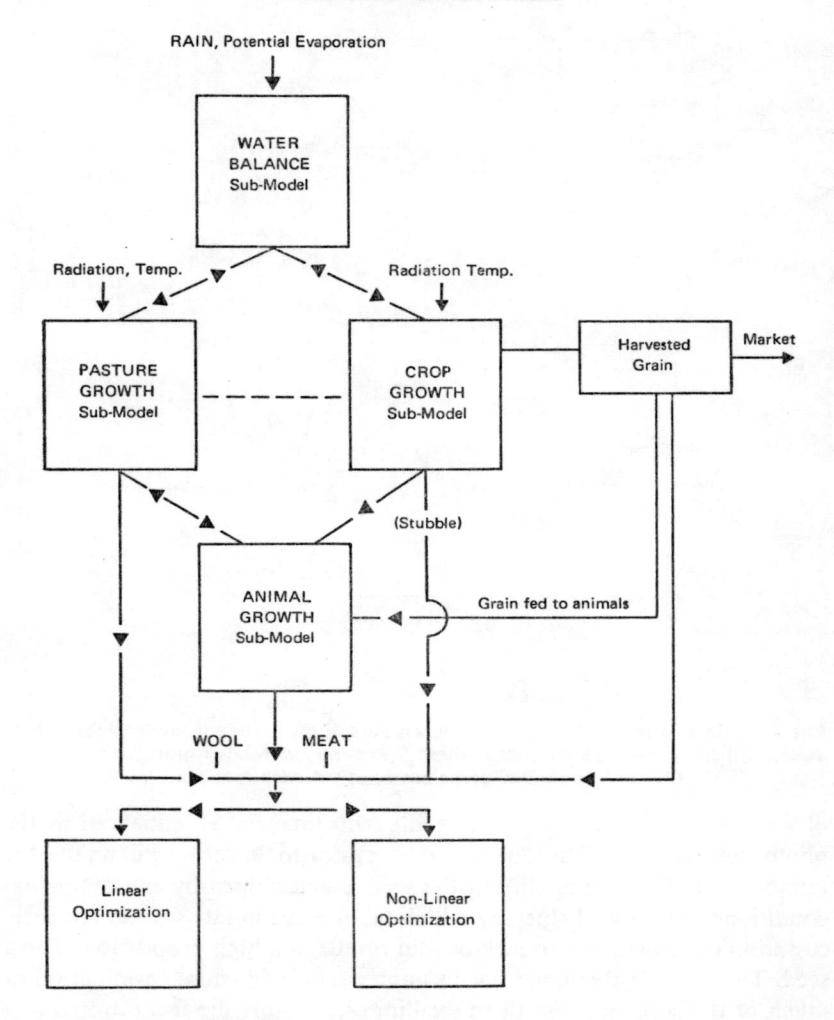

FIG. 2. Sub-models of the ley system showing outputs used in optimisation routines.

4.7.2. Optimum Use of a Lupin Crop

The first simple problem examined was the optimum use of a 50 ha crop of lupins yielding 2000 kg ha^{-1} of grain and 8000 kg ha^{-1} of stubble. The grain could all be sold or part of it used to assist in the fattening of 1000 lambs, already on the farm, on the stubble. The price of the grain when sold was varied at 3·2, 6·4 or 9·5 p kg^{-1}. The lambs could be sold anytime over a six-month period but the lamb price varied from month to month and with the weight of the lambs. The objective function to maximise was return from sale of sheep plus return from sale of residual grain not fed to sheep.

The lupin model of Arnold & Galbraith[53] was used to simulate the monthly liveweights of lambs grazing the stubble at different stocking rates and at three rates of grain feeding, viz. 0, 250 g and 500 g day^{-1}. A linear programme optimisation was compared with a non-linear simplex optimisation (Nelder & Mead[56]).

For the non-linear simplex method a table of the prices of lambs each month at the high and low stocking rates was entered for a level of feeding. The program calculated and interpolated a price for any stocking rate chosen by the simplex regime. Separate runs were produced for each level of feeding and each price of grain. (Three runs at each grain price were executed to get an exact comparison with the linear program since the non-linear simplex method could have been used to interpolate and optimise at any feeding level between 0 and 500 g/day.)

The linear program was a modified simplex package available on the Western Australian Regional Computing Centre's CYBER 72. To describe the system required a matrix of 146 activities with 75 constraints. Seventy-two of the activities were monthly values of grain fed (at three rates) at four different stocking rates (20, 30, 50 and 70 sheep ha^{-1}), with transfer of activity at the end of each month (for six months), which was either continuation of the same feed rate and stocking rate, or sale of sheep at the price at that time. Values of lamb carcasses at each feed rate for each stocking rate for each month were another 72 activities. The other two activities were number of lambs and weight of lupin grain sold. The constraints were total number of lambs, total weight of grain, and total weight of stubble.

In the non-linear simplex programme, starting values and an initial step length had to be entered for each variable. To test whether a global optimum was being found, both starting values and step lengths were varied.

The two techniques gave very similar optima (Table 4).

The non-linear simplex method gave the same optimum for different starting values and increments except when the lowest stocking rate and lowest number of days fed and small stop lengths were used (Table 5), where it found a local optimum at a different stocking rate to the global optimum.

The cost on the computer system we used (CSIRONET) was the same

TABLE 4

Optimum solutions for linear and non-linear optimisation methods

Selling price of grain (pence kg^{-1})	Linear				Non-linear			
	Grain fed (g day^{-1})	Stocking rate (sheep ha^{-1})	No of days fed before selling	Gross profit (£ × 10^3)	Grain fed (g day^{-1})	Stocking rate (sheep ha^{-1})	No. of days fed before selling	Gross profit (£ × 10^3)
3·2	500	20	180	12·5	500	20·0	160	12·8
6·4	250	20	150	14·1	250	20·4	152	14·1
9·5	250	20	120	16·2	250	20·0	120	16·3

TABLE 5

Variation in the optimum solution of non-linear method with different starting values (grain 6·4 pence kg^{-1})

Initial stocking rate (sheep ha $^{-1}$)	Initial step length for stock rate	Initial no. of days fed	Initial step length for days fed	Optimum stocking rate	Optimum days fed	No. function evaluation	Gross profit (£ × 10^3)
37	1	130	1	20·1	146	29	14·2
35	3	90	10	21·0	150	38	14·2
35	5	90	10	20·3	152	36	14·2
60	6	150	15	20·0	150	43	14·3
20	2	30	3	20·0	30	11	12·1
20	2	30	10	20·0	30	11	12·1

for the two methods, assuming two runs of the non-linear simplex to ensure that a global optimum had been reached. However, we could explore further only with the non-linear simplex method to find the optimum level of grain feeding. Linear interpolation of animal responses between levels of feeding was assumed. The optimum solution with grain at 6·4p kg^{-1} was 147 days of feeding at 225 g day^{-1} at a stocking rate of 20·4 sheep ha^{-1} yielding £14 100 gross income. In fact the solution was no different, but the non-linear method allowed this to be checked and was much easier to use.

4.7.3. Optimisation of a Dynamic Pasture System

4.7.3.1. Seeking a Single Optimal Management Strategy

The second problem involved a completely dynamic system of wether sheep producing wool, stocked year-long on an annual pasture of subterranean clover.

The stability of the system is influenced by weather conditions and also by management strategies, i.e. stocking rate, supplementary feeding to spell the pasture from grazing, and fertiliser rate. It has been demonstrated experimentally that spelling the pasture at the beginning of the growing season may improve pasture productivity and subsequent sheep liveweight and wool production under certain conditions (Smith *et al.*[57]).

There are therefore many combinations of management and prices that could be explored and each will have an effect on the system as a whole. The optimisation techniques available to us were not capable of dealing with all possible variables at once, so specific choices had to be made. The first question we examined was whether spelling could be an advantage as a standard practice.

The simulation model was run for periods of nine years using random sequences of climatic data for the Bakers Hill area of Western Australia. Stocking rate was varied from 7·0 to 15·0 sheep ha^{-1} and pastures were either grazed continuously or spelled for three weeks after germinating rains each season. Two sequences of nine years (called MET 1 and MET 2 in Figs. 3(a) and 3(b) were simulated for stocking rates of 7·0, 9·0, 11·0 and 12·0 sheep ha^{-1}, and three sequences at stocking rates of 13·0, 13·5, 14·0 and 15·0 sheep ha^{-1}.

The most important characteristics of the system are shown in Figs. 3(a) and 3(b) for three of the stocking rates in which yearly values of pasture available in early winter, maximum yield, seed pool at the end of each season and sheep liveweights and wool production in mid-November are plotted for the system stocked continuously with sheep.

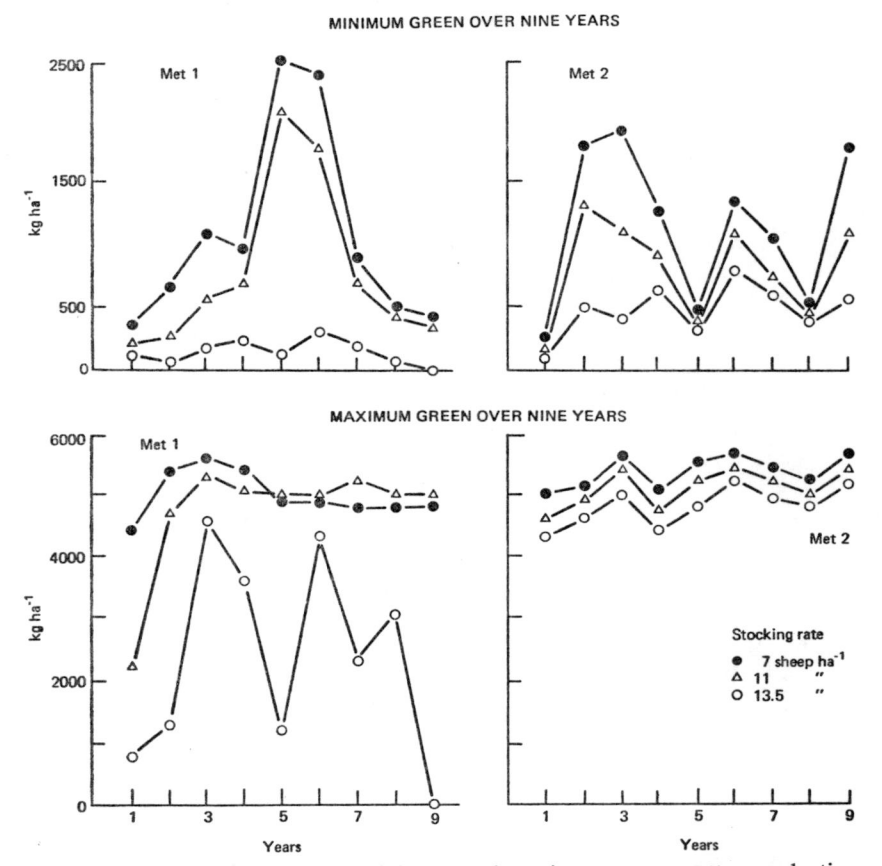

FIG. 3(a). Annual fluctuation in minimum and maximum green matter production over a period of nine years. (Note: Results from only three of the eight stocking rates are plotted. Met 1 = nine years with frequent poor seasons. Met 2 = nine years with frequent good seasons.)

At low stocking rates the amount of pasture available in early winter varies greatly from year to year, depending on seasonal conditions; whereas at a high stocking rate it remains small in all years, despite being spelled to keep the sheep alive. (The system is managed to keep sheep alive by removing them from the pasture and feeding them when their wool-free liveweights are less than 40 kg. Feeding is for the period that the pasture will not sustain this liveweight plus 14 days.) Without such a strategy the system would be much less stable. The maximum pasture yield is similar from year to year at low stocking rates but fluctuates a lot

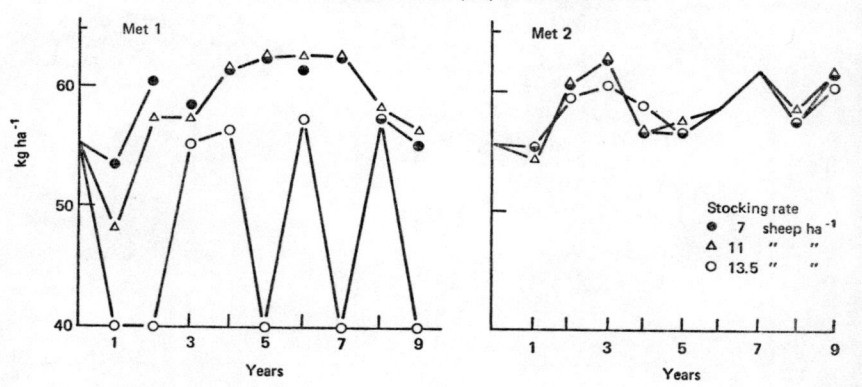

FIG. 3(b). Annual fluctuation in hard seed numbers on 14 November and maximum liveweight over periods of nine years. (Note: Results from only three of the eight stocking rates are plotted. Met 1 = nine years with frequent poor seasons. Met 2 = nine years with frequent good seasons.)

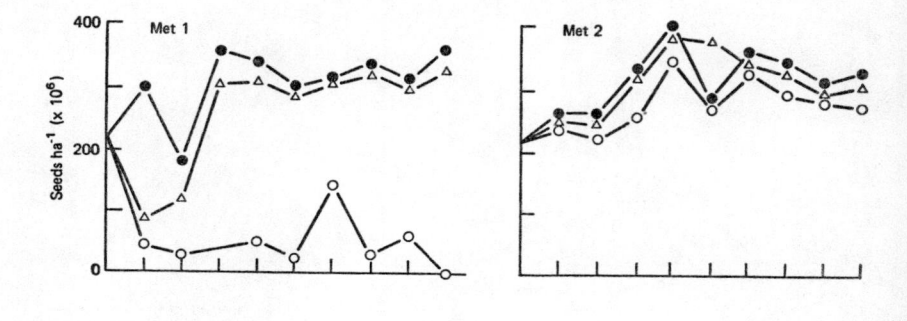

at a high stocking rate. A major determinant of the level of production is the size of the seed pool. That which does not germinate and is not eaten carries over from one year to the next and acts as a buffer to poor years, as seen at low stocking rates, but a sequence of poor years and a high stocking rate (MET 1) causes the pool to drop to a continuous low level.

Sheep liveweight in November and fleece weights reflect the fluctuations in pasture available during each year.

The effect of spelling pastures at the break of season might be expected to reduce the variability in pasture productivity at higher stocking rates and hence the variability in liveweight and wool production. At low stocking rates pasture availability is less likely to limit animal performance.

The differences under the two alternative management systems are shown in Figs. 4(a) and (4b), in which mean curves have been fitted over the simulated periods and standard deviations derived. Expectation of the effects of spelling is substantiated. The question is, what is the optimum stocking rate and management for a 1000 ha property producing wool from wethers which have a productive life of five years?

The above output curves from the simulation model were used in the non-linear simplex method to find the optimum strategy in the given random selection of years with different wool prices and risk levels in net income. The values of liveweight and wool were obtained within the programme as the product of the liveweights and wool yields obtained from the curves and the appropriate prices. The value of the objective function is calculated from linear or curvilinear equations or by using a table function which produces interpolated values from a table of X and Y data.

The objective function used was to maximise the value of liveweight sold in November plus the value of wool but minus the running costs and the cost of supplementary feeding. Running costs were £0·70 wether^{-1} year^{-1}, and superphosphate cost £3·2 ha^{-1} year^{-1}. Supplementary feed cost 1·3p wether^{-1} day^{-1}, cost of replacement wethers purchased in November was £9·5 each and the wool price was either £0·64 or £1·27 kg^{-1}. Value of sheep sold was 13p kg^{-1}.

Constraints applied (using the curves in Figs. 4(a) and 4(b)) were that the seed pool had to be at least 200 million seeds ha^{-1} in November to maintain a viable system, and to prevent soil erosion the amount of green dry matter on 5 June had to exceed 400 kg ha^{-1}. The risk, which was defined as the standard deviation (£) of the objective function, was entered as a constraint and varied to examine its effect on the optimum solution (Fig. 5) by running the program with risk set at £5000, £6000, etc. Clearly, the spelling system of management is preferable, achieving a higher income at all levels of risk. Profitability increases substantially as risk is increased, but only to a certain level.

The above exercise showed that the dynamics of the biological system do not have to be discarded during optimisation of the system. However, the cost of maintaining them may be high as will be discussed later.

4.7.3.2. Would Decision Rules Give an Adequate Guide to Management?

It is conceivable that a single strategy applied every year may not give

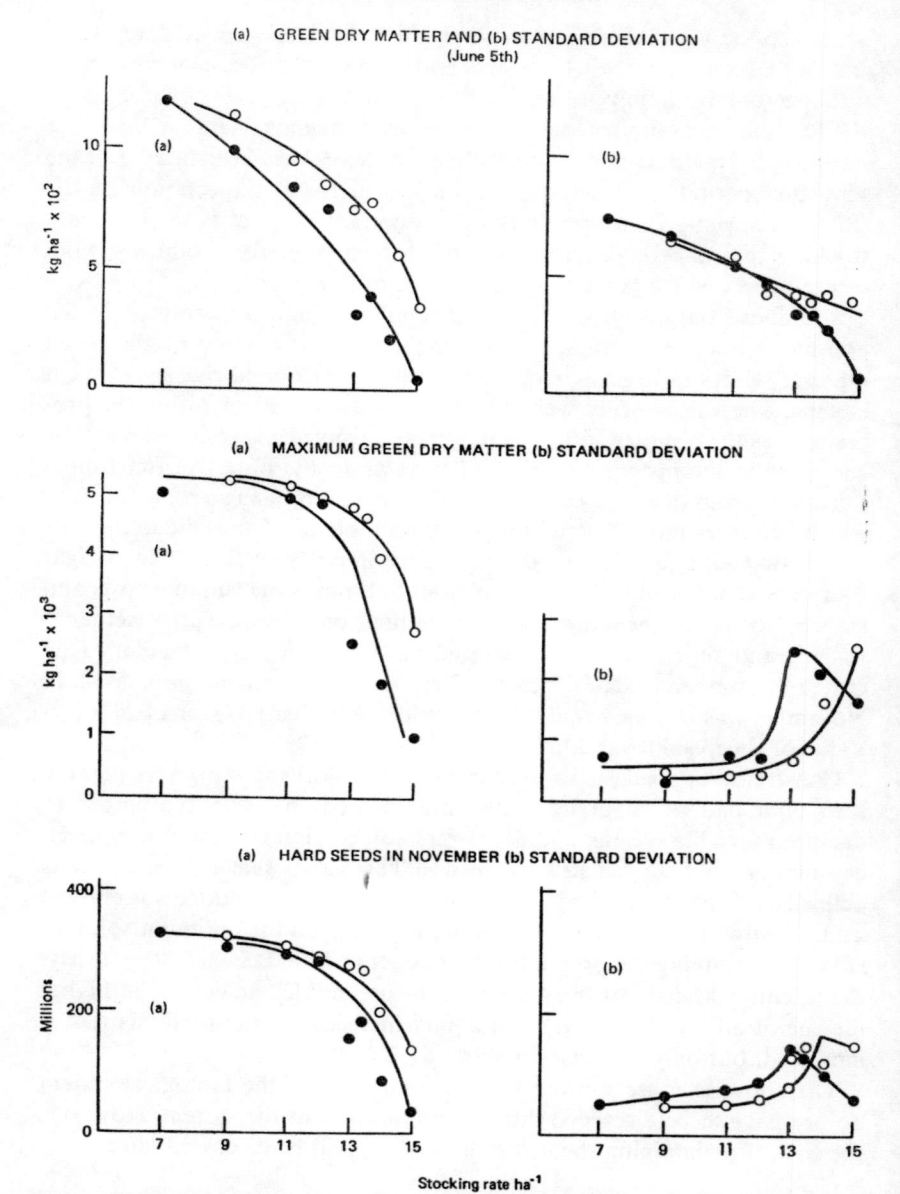

FIG. 4(a). Mean responses (a) and standard deviations (b) for a number of plant parameters with varying stocking rate. ●—● = no spelling used; ○—○ = spelling used.

The Problem of Finding an Optimum Solution 157

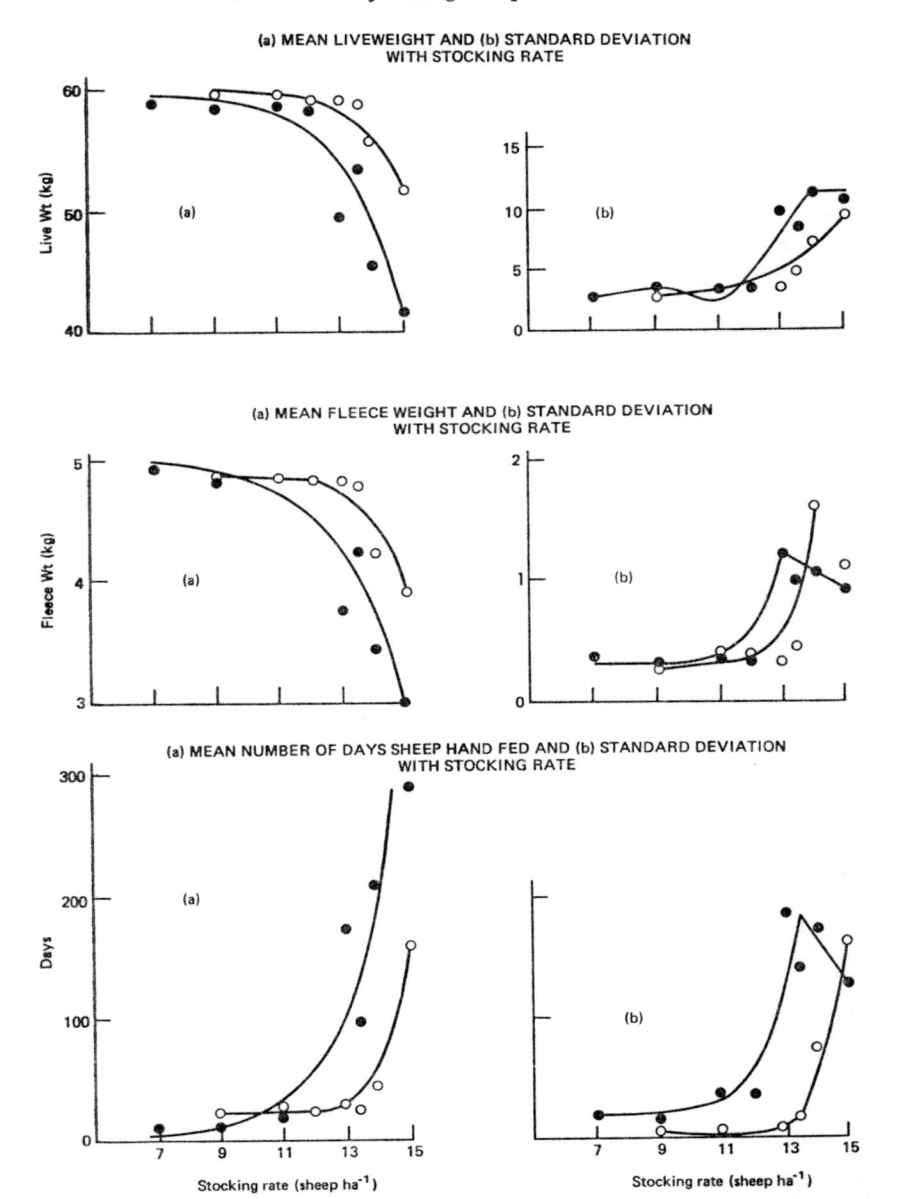

FIG. 4(b). Mean responses (a) and standard deviations (b) for a number of animal parameters with varying stocking rate. ●—● = no spelling used; ○—○ = spelling used.

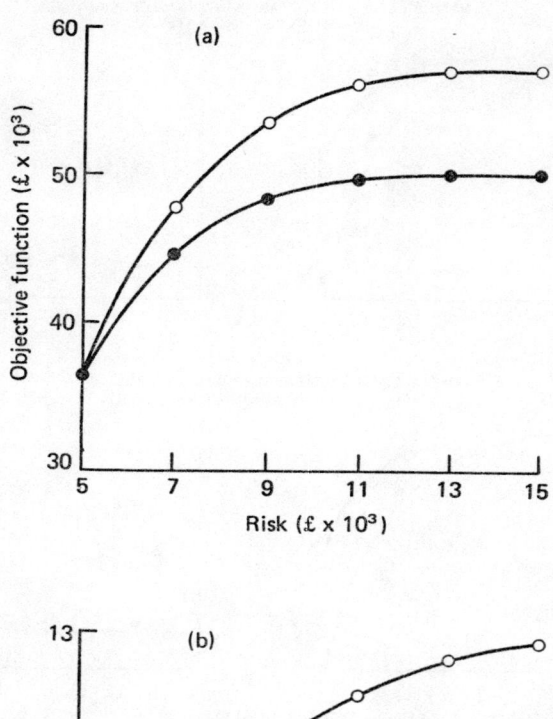

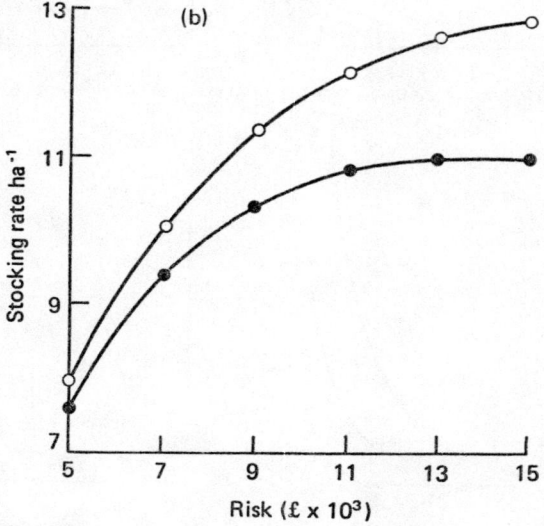

FIG. 5. Changes in value of objective functions (a) and in optimum stocking rate (b) with change in risk. ●—● = no spelling used; ○—○ = spelling used.

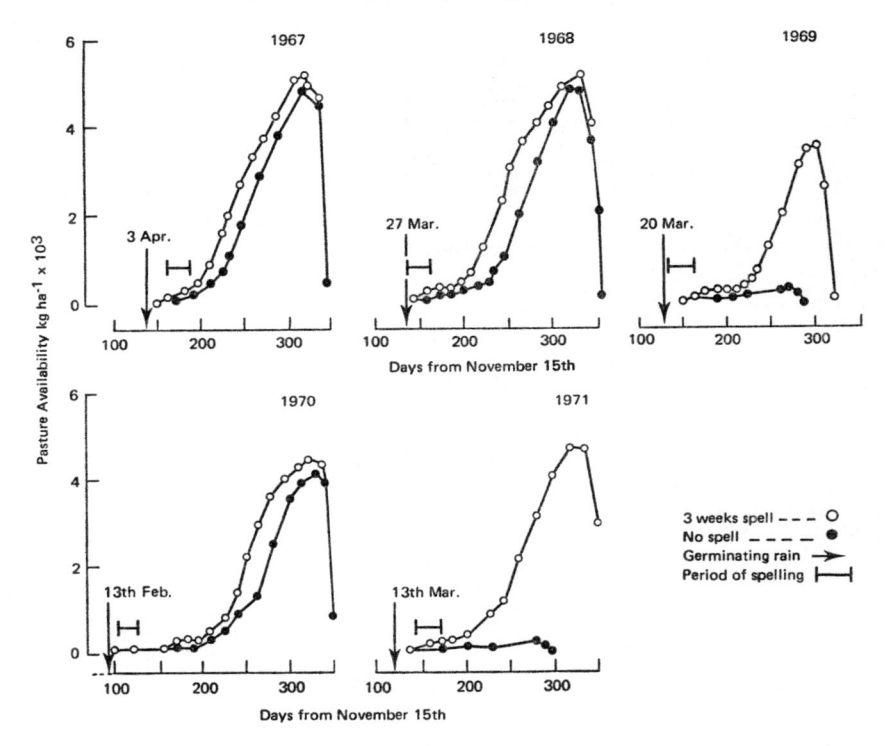

FIG. 6. Effect of spelling for three weeks after germinating rains on pasture available at 12·5 sheep ha^{-1}.

the most profit. As can be seen from Fig. 6, at a quite high stocking rate, spelling the pasture at the beginning of the year had benefit in only half the years, in terms of pasture yield. A set of decision rules for the farmer might be set up which are based on his stocking rate and the conditions at the start of each growing season. From experimental evidence, the rule would be 'in seasons with late or light opening rains and at higher stocking rate, there will be a benefit from pasture spelling'. But this rule will not allow the farmer to judge what stocking rate to use to optimise his profit.

4.7.3.3. Is it Better to Have a Single Strategy Every Year or Take into Account Seasonal Conditions?

The farmer, faced with the decision of what stocking rate to adopt, has three choices. He can choose a 'safe' rate which will minimise the risk of having to feed sheep for long periods during dry years, or he can choose to take this risk, or he can adjust his stocking rate from year to year according to seasonal conditions by buying or selling sheep. To assess the effects on income of choosing either the second or third policies the simulation model was altered to allow buying and selling of stock as shown in Fig. 7. The prices of buying and selling were the same and the decision was made twice each year. In poor years sheep prices were taken to be much lower than prices in good years, but wool prices were held constant at 64p kg^{-1}. This strategy was tested using two sequences of nine

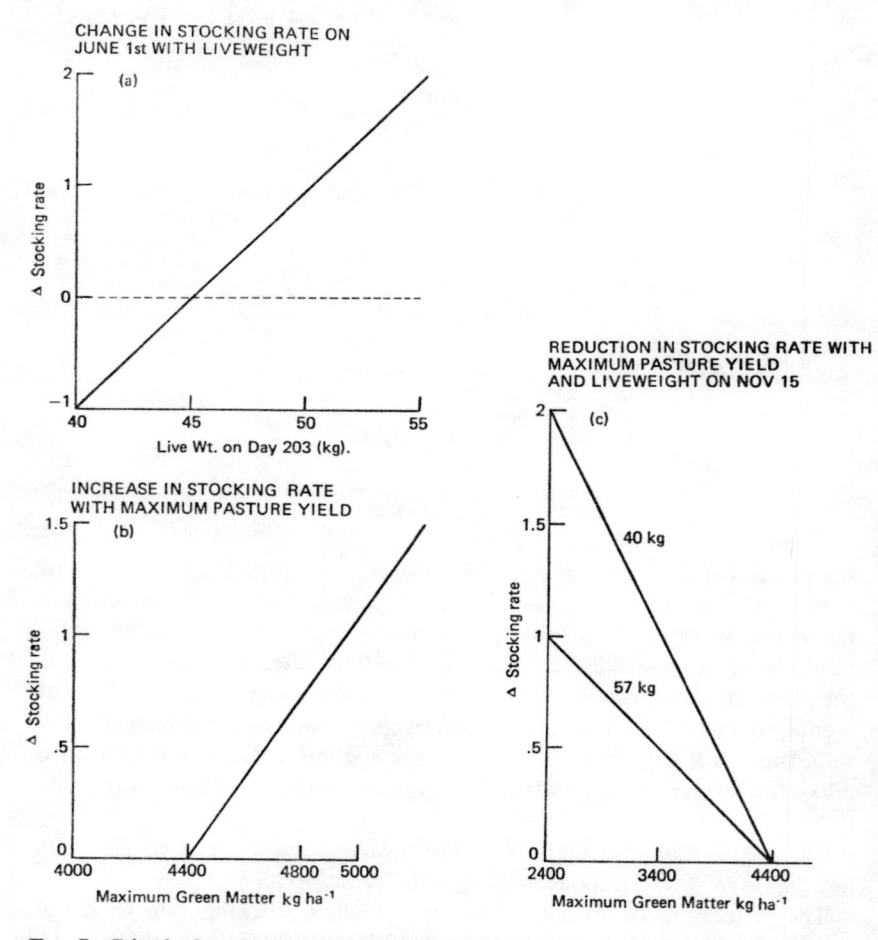

FIG. 7. Criteria for altering stocking rate. Note: In Fig. 7(c), the actual value was derived by interpolating between the two lines as a function of body weight.

years climatic data with starting stocking rates of 11·0, 13·0 and 15·0 sheep ha^{-1}. In this annual pasture ecosystem productivity fails completely if seed pools are run down. In the previous examples when this 'crash' occurred production and costs were calculated as that of feeding for survival for the remaining years in a nine-year run. However, in this example we allowed re-seeding of pastures at 5 kg ha^{-1} and complete spelling of pastures from grazing for a year, during which time sheep were hand-fed at a survival rate. We estimated that the resown pastures produced, when ungrazed, 200 million seeds, sufficient to restart the next year.

The results of this approach are shown in Figs. 8 and 9, and in Table 6. What happens is that after several years the stocking rates are similar in all systems at starting values of 11·0 and 13·0 but in the process the farmers' incomes fluctuated drastically. The fluctuations in income were far greater than if, under the same price regimes, the farmer had chosen to maintain a constant stocking rate of either 11·0 or 13·0 sheep ha^{-1} when his average net income would have been higher also. With the starting rate of 15·0 the system remains different throughout but the income and the standard deviations are similar to the system starting at 11·0 sheep ha^{-1}.

The adjustments in stocking rate allowed were not precise enough to prevent 'crashes' in some years, which means that the decision rules were inadequate. This being so, we studied a further variation of the model in which fluctuations were allowed but 'conservative', 'some risk', and 'very risky' maximum stocking rates were set, i.e. starting at 11·0 the maximum was 13·0, starting at 13·0 it was 15·0 and starting at 15·0 it was 17·0 sheep ha^{-1}. These changes had very little effect on net income (Fig. 9)

TABLE 6

Average net income and standard deviations for two stocking rate policies

Policy	Income (£ × 10^3)	Standard Deviation (£ × 10^3)
Adjusted stocking rate		
Start at 11·0	54·4	30·9
13·0	57·3	38·2
15·0	53·5	27·2
Fixed stocking rate		
11·0	59·2	9·1
13·0	64·0	27·2
15·0	26·4	48·0

but slightly reduced the standard deviation of income. At 15·0 sheep ha^{-1} starting rate there was no effect since stocking rate never reached 17·0.

4.7.4. Conclusion

The above analyses indicate the sorts of problems that can be explored using simulation and/or optimisation procedures. We have used output

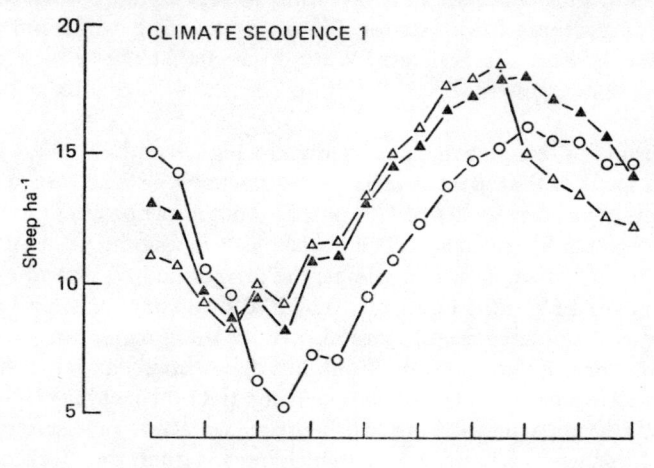

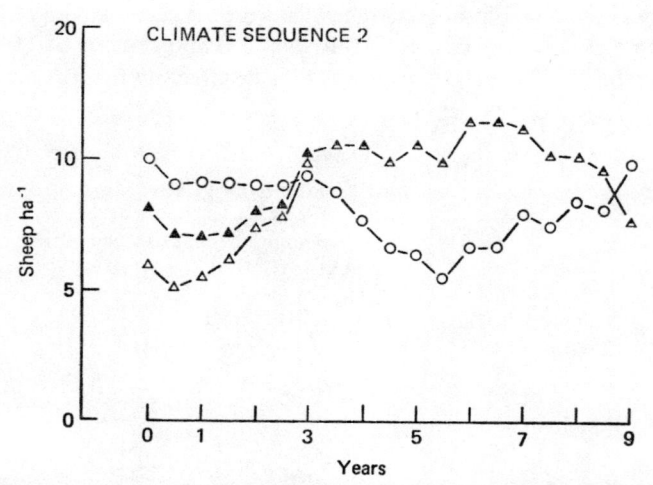

FIG. 8. Stocking rate adjustments through time with two climate sequences, when decisions are made twice yearly (see Fig. 7 for basis of decision). Initial stocking rates were △ = 11·0, ▲ = 13·0 and ○ = 15·0 sheep ha^{-1}.

The Problem of Finding an Optimum Solution

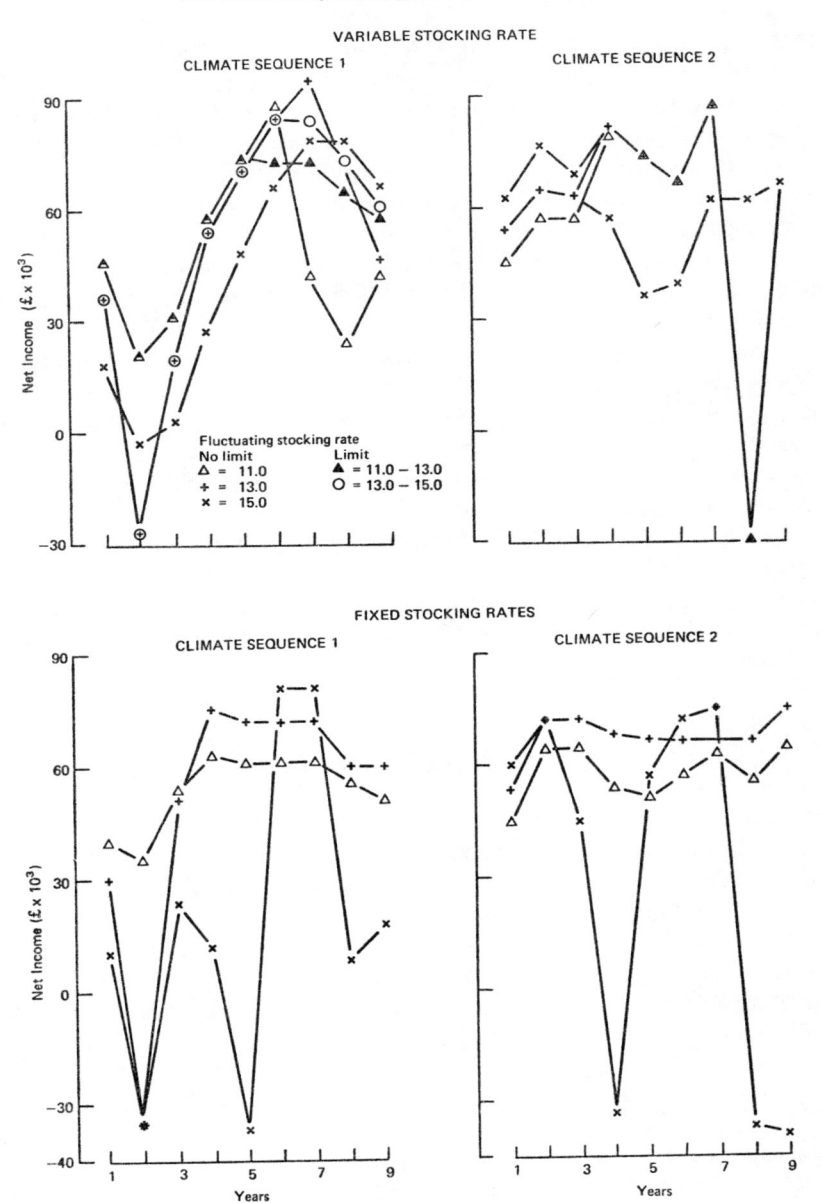

FIG. 9. Comparison of net incomes given two nine-year periods and two climate sequences with both fixed and variable stocking rates.

from simulation models as input for optimisation routines and believe that the non-linear simplex method allows the dynamics of the system to be retained. Simulation can be used to examine the consequences of a variety of decisions (or tactics) in managing a complex system with many feedbacks. Optimisation gives a single strategy for a given set of conditions. Both simulation and optimisation have a role to play in helping the decision-maker. Possibly optimisation is of value in deciding short-term issues, such as the use of the area of lupin crop, and in helping decide long-term strategy. Simulation allows the decision-maker to see the consequences of strategies other than the single optimum one.

Another point that comes from these analyses is that there is really no one optimum solution to a problem, but that the optimum depends, to a large extent, on the variability in income that the farmer is willing to accept, as well as on the constraints applied. For the examples given the constraint (apart from biological ones) used was risk, but others such as labour could have been imposed.

Finally, it must be re-emphasised that the important criterion in optimisation is not so much the method but the objectives of the user of the solution. There are methods available that can optimise most dynamic systems but the choice of which to use depends on the system and the type of answer required.

4.8. THE COSTS INVOLVED

Within our limited experience (we have been involved in problems of simulation of agricultural systems for five years and their optimisation for only three years) it may be of interest to document the developments and costs of this type of work.

The simulation of the crop and animal production from ley systems began in 1970 with a detailed word picture of crop, pasture and animal production in such a system. Sub-models were then begun and each was developed and tested separately. Colleagues in various disciplines have been involved in these at various times and to different extents and the whole would not have been possible without their contributions.

A wether sheep sub-model (Arnold & Campbell[52]) was first produced. Then followed a soil moisture sub-model basic to the plant production sub-models (Carbon & Galbraith[55]). These two were coupled with a pasture growth sub-model (Arnold et al.[54]) and a lupin crop sub-model with revised animal consumption and conversion routines to incorporate weaner sheep and cattle (Arnold & Galbraith[53]). Development of a more detailed sub-model of sheep production began in 1972 and is now complete (Graham et al.[58]).

This framework is far from comprehensive since it includes only one crop and one type of pasture and does not consider nutrient recycling, labour, machinery and capital use. Yet it has taken the equivalent of two people full time for five years (approximately £100 000) and cost £3 200 in computing time.

The other major project which has involved the equivalent of two people for three years is the fertiliser model DECIDE (Bennett & Ozanne[25]) which was conceptualised in 1971, tested against field fertiliser response trials in conjunction with the Western Australian Department of Agriculture and Agribusiness Councillors Pty Ltd in 1972, further developed by scientists in both CSIRO and the Western Australian Department of Agriculture in 1973 and is being applied by both private and governmental farm management advisers in 1974. The cost of computing time in developing this model has been £650 and salaries and support (6 man-years) at £60 000.

Allowing that a simulation model exists and so do optimisation routines, the relative costs of using the two separately or in combination are considerable. If in the grazing management problem the curves used in the optimisation routine had been derived empirically from knowledge of the shape of responses to be expected, then the cost in computer time would have been £30. But it cost £275 to run the simulation model to obtain these curves. Thus careful consideration must be given to whether precise (so far as knowledge of the system allows) responses are more important than are generalised ones in seeking an optimum solution. In many situations the prediction errors of the simulation model will be large and generalised curves will be just as useful. In fact, if optimisation is required for specific solutions on many farms then the use of the large simulation model must be restricted to providing the shape of response curves. The scaling on the axes would then be adjusted for different climatic zones, for soil types within zones, etc. Such an approach has been used in the DECIDE model discussed in the next section.

4.9. IMPLEMENTATION OF THE OPTIMUM SOLUTION

4.9.1. Belief

The degree of difference between the proposed optimal solution and the decision-maker's present actions is likely to be the main stumbling block.

If there is no difference then the farmer will increase his satisfaction through reinforcing his feeling that he is doing the right thing. Slight changes (e.g. ratios of crop to pasture, age at culling, etc.) might be acceptable, but as the change becomes greater, so does the doubt that the objectives, the constraints, or the abstraction of the physical, biological and economic processes have been correctly specified. One has to resort to terms such as counter-intuitivity (Forrester[59]) to explain why thought, accompanied by trial and error, has not lead to a closer proximity of optimum to practice, unless prices of resources and/or products have shown a sudden change.

4.9.2 Short-cuts

While models remain an inexact analogue of reality, and while objective functions remain inexactly specified, there seems to be a considerable market for short-cut techniques to overcome these problems. One such is Powell & Hardaker's[60] 'sub-optimal programming'. In this approach the objective function is exchanged with a constraint to determine new, perhaps more acceptable solutions. Within the example cited, sacrificing 15% of income would allow either labour use to be reduced to one-third of that required for the optimum, or for 40% of the farm to be in lucerne, which is quite a desirable objective where the history of crop rotations is insufficient to determine whether the optimal solution (67% of the farm in wheat) is biologically feasible.

4.9.3. Implementation of DECIDE

The fertiliser decision model (Bennett & Ozanne[25]) has given us considerable experience in the problems of implementation. To a large extent the progress we have made is due to our adoption of a very simple model. We have discovered that nearly all the farmers we have consulted, before accepting advice from any model, whether hand calculated or computerised, demand an understanding of the assumptions and parameter values. With anything more complex than a very simple model the task of explanation could prove too difficult. Associated with but separate from this problem is the selection of functional form. If one can identify the parameters of the response curve and associate their value with some environmental parameter, then explanation and extrapolation are considerably eased. For DECIDE, using a Mitscherlich representation:

$$Y = A(1 - B \exp(-CX)) \qquad (6)$$

we can associate A the yield maximum, with soil and climatic variables, and use local knowledge to provide an estimate. The intercept when no further fertiliser is applied, B, is a function of the residual value of previous applications for which various discounting procedures have been suggested (Barrow[61]), while C is a function of the soil's absorbing properties (Barrow[62]), the method of application (Rudd & Barrow[63]) and the plant species grown (Ozanne et al.[64]). Associating environmental variables with polynomial forms of yield parameters is likely to be far more difficult.

But there is still a large component of husbandry left in farming, if one defines husbandry as the degree to which local knowledge about parameter values and the probability of outcomes has yet to be scientifically documented. It must therefore be a tenet of present agricultural modelling that where local information is superior to that incorporated into the averages normally used by a model, then the local information is used. It is this tenet that is one of the bases of using the decision-maker's own information and assessments of subjective probability. The analyst can argue about parameter values and probabilities but only to influence the decision-maker's final selection. We have incorporated this tenet into the extension of our fertiliser decision model but it is too early to comment further on the problems and advantages involved.

So far all we have done is to provide an optimum, assuming adequate finance to purchase the fertiliser and (in the case of applications to pasture) adequate stock to consume the feed. Often either of these can act as real or imagined constraints within the decision. We have started to solve the allocation with financial constraints using Lagrange multipliers (Anderson[12]) and do not regard the problem as severe. To cope with the second problem we have started work to incorporate the grazing animal model of Arnold & Campbell[52] mentioned earlier, for in many cases stocking rates are insufficient to justify further additions of fertiliser. The most interesting scientific aspect of this addition will be the model's response to risk.

Normally risk increases with investment and avoidance of risk is associated with a reduction in input (e.g. McArthur & Dillon[65]). In the case of fertilisers the nature of the response function is such that losses from under-fertilisation rise much more steeply than losses from excessive fertiliser use with deviations from the optimum rate. Hence the risk avoider should apply more fertiliser and less stock. This interesting aspect has not been explored by surveys of farmers' risk attitudes. We suspect, however, that farmers tend to avoid investment rather than risk, and that the risk-avoider in fact applies lower than average fertiliser rates.

One of our mistakes was to develop an automated computer program as soon as the project began. This was on the assumption that with 16 000 farmers in Western Australia alone, each with approximately 20 paddocks, there was a need for an automated prediction service. But just as one learns what addition means in primary school by performing many sums, so there appears to be a need for many hand calculations in the initial stages of model use. What has sold in this case is a hand calculation sheet and nomographs. No doubt the computer program will be required when demand increases as a consequence of mounting confidence, but that is not yet.

In the extension programme which we commenced in March 1974 with the Western Australian Department of Agriculture and private advisers we have started to work closely with four fairly diverse farmer groups ranging from wholemilk farmers in a 1000 mm rainfall environment to wheat farmers in a 400 mm environment. Since that time, removal of the Australian government's superphosphate bounty, increases in the prices of raw materials, notably phosphate rock, and increases in manufacturing costs have effectively trebled the price of superphosphate to the farmer. This has resulted in a situation of uncertainty in fertiliser decision-making and a demand for extension advice. Part of the cost of this advice will be the need for farmers to learn more about the factors which determine that rate of fertiliser which maximises profit, and how to calculate their own requirements using the model. It will involve the junior author with explanation and communication and provide an adequate excuse for absenting himself from this symposium. But more than anything else, we hope that future model development will act as a communication mechanism between the farmers of Western Australia and hopefully elsewhere who need better advice and the scientists in CSIRO and the Departments of Agriculture, who are concerned with the model's construction and further data collection and validation.

4.10. APOLOGY

This paper has been one of the most difficult we have attempted. Experts in mathematics, computing science, economics, biology, zoology and sociology must cringe when they read both the inadequate summaries and the errors that we have presented. We attempted this paper to give a better perspective of the overall problem. Our better perspective is that we need help from experts in other disciplines. More than ever before we are convinced of the necessity for interdisciplinary teams.

ACKNOWLEDGEMENTS

Ken Galbraith spent many hours on the simulation models used for optimisation studies and adapting the optimisation routines. Without his valuable contribution this study would not have been possible. We would also like to thank Robin Barron for his attempts to improve our manuscript.

REFERENCES

1. Charlton, P. J. & Thompson, S. C. (1970). Simulation of agricultural systems, *Journal of Agricultural Economics*, **21**, 373–89.
2. Dillon, J. L. (1971). Interpreting systems simulation output for managerial decision-making, in: *Systems Analysis in Agricultural Management*, (ed. J. B. Dent and J. R. Anderson), Wiley, Sydney.
3. Charlton, P. J. & Street, P. R. (1975). The practical application of bioeconomic models, in: *Study of Agricultural Systems* (ed. G. E. Dalton), Applied Science Publishers, London.
4. Dillon, J. L. & Heady, E. O. (1961). Free competition, uncertainty and farmer's decision, *Journal of Farm Economics*, **43**, 643–51.
5. Cyert, R. M. & March, J. G. (1963). *A Behavioural Theory of the Firm*, Prentice-Hall, Englewood Cliffs, New Jersey.
6. Officer, R. R., Halter, A. N. & Dillon, J. L. (1967). Risk, utility and the palatability of extension advice to farmer groups, *Australian Journal of Agricultural Economics*, **11**, 171–83.
7. Officer, R. R. & Dillon, J. L. (1968). Probability and statistics in agricultural research and extension; a pro-Bayesian view of modern developments, *Journal of Australian Institute of Agricultural Science*, **34**, 121–9.
8. Dillon, J. L. (1971). An expository review of Bernoullian decision theory in agriculture: is utility futility?, *Review of Marketing and Agricultural Economics*, **39**, 3–80.
9. Officer, R. R. & Halter, A. N. (1968). Utility analysis in a practical setting, *American Journal of Agricultural Economics*, **50**, 257–77.
10. Edwards, W. & Tversky, A. (1967). *Decision Making*, Penguin Books, Harmondsworth.
11. Daw, M. E. (1965). The 'operational' use of linear programming, unpublished discussion paper presented to the Agricultural Economics Society (UK) Summer Conference, cited by Powell & Hardaker (1969) (see ref. 60).
12. Anderson, J. R. (1967). Economic interpretation of fertiliser response data, *Review of Marketing and Agricultural Economics*, **35**, 43–57.

13. Pesek, J. T., Heady, E. O. & Dumenil, L. C., (1960). Influence of residual fertiliser effects and discounting upon optimum fertiliser rates, *7th International Congress of Soil Science Transactions*, Madison, Wisconsin, pp. 220–7.
14. Heady, E. O. & Dillon, J. L. (1961). *Agricultural Production Functions*, Iowa State University Press, Ames.
15. Doll, J. P., Heady, E. O. & Pesek, J. T. (1958). Fertiliser production functions for corn and oats; including an analysis of irrigated and residual response, Agricultural and Home Economics Experiment Station, Iowa State College Research Bulletin 463.
16. Food and Agricultural Organisation of the United Nations (1966). '*Statistics of crop responses to fertilisers*', FAO, Rome.
17. Nicholson, T. A. J. (1971). *Optimisation in industry*, vol. I: *Optimisation Techniques*, Longman, London.
18. Swartzman, G. L., ed. and co-ordinator (1972). Optimisation techniques in ecosystem and land use planning, US/IBP Grassland Biome Technical Report No. 143, Colorado State University, Fort Collins.
19. Kaufmann, A. & Faure, R. (1968). *Introduction to Operations Research*, Academic Press, New York.
20. Wagner, H. M. (1969). *Principles of Operations Research with Applications to Managerial Decisions*, Prentice-Hall, Englewood Cliffs, New Jersey.
21. Tabak, D. & Kuo, B. C. (1971). *Optimal Control by Mathematical Programming*, Prentice-Hall, Englewood Cliffs, New Jersey.
22. Wilde, D. J. & Beightler, C. S. (1967). *Foundations Optimisation*, Prentice-Hall, Englewood Cliffs, New Jersey.
23. Hutton, R. F. (1965). Operations research techniques in farm management, *Journal of Farm Economics*, **47**, 1400–14.
24. Swartzman, G. L., co-ordinator (1970). Some concepts of modelling, US/IBP Grassland Biome Tech. Rep. No. 32, Colorado State University, Fort Collins, 142 pp.
25. Bennett, D. & Ozanne, P. G. (1973). Deciding how much superphosphate to use, Australia, CSIRO, Annual Report Division of Plant Industry 1972, pp. 45–7.
26. Richards, P. A. & McCarthy, W. O (1964). *Linear Programming and Practicable Farm Plans: A Case Study in the Goondiwindi District, Queensland*, University of Queensland Press, Brisbane, Australia.
27. Rae, A. N. (1970). Capital budgeting, intertemporal programming models, with particular reference to agriculture, *Australian Journal of Agricultural Economics*, **14**, 39–52.
28. McFarquhar, A. M. M. (1961). Rational decision-making and risk in farm planning: an application of quadratic programming in British arable farming, *Journal of Agricultural Economics*, **14**, 552–63.
29. Carlsson, M., Hormark, B. & Lindgren, I. (1969). Recent developments in farm planning. I. A Monte Carlo method for the study of farm planning problems, *Review of Marketing and Agricultural Economics*, **37**, 80–103.
30. Rae, A. N. (1971). An empirical application and evaluation of discrete stochastic programming in farm management, *American Journal of Agricultural Economics*, **53**, 625–38.

31. Throsby, C. D. (1964). Theoretical aspects of a dynamic programming model for studying the allocation of land to pasture improvement, *Review of Marketing and Agricultural Economics*, **32**, 149–81.
32. Kennedy, J. O. S., Whan, I. F., Jackson, R. & Dillon, J. L. (1973). Optimal fertiliser carryover and crop recycling policies for a tropical grain crop, *Australian Journal of Agricultural Economics*, **17**, 104–13.
33. Kingma, O. T. (1973). A recursive optimising and simulation approach to farm-firm growth research, *Proceedings of the Australian Society of Operations Research*, **1**, B4,1–11.
34. Bird, J. G. (1968). Alternatives in dairy industry adjustment in the far north coast area of New South Wales, Division of Marketing and Agricultural Economics, New South Wales Department of Agriculture, Miscellaneous Bulletin 6.
35. Jensen. R. C. (1968). Farm development plans including tropical pastures for dairy farms in the Cooroy area of Queensland, *Review of Marketing and Agricultural Economics*, **36**, 139–48.
36. Sinden, J. A. (1970). Poplar growing and farm adjustment on the north coast of New South Wales, *Review of Marketing and Agricultural Economics*, **38**, 121–36.
37. Lindner, R. K. (1969). The economics of increased beef production on dairy farms in Western Australia, *Quarterly Review of Agricultural Economics*, **22**, 147–64.
38. Australia, Bureau of Agricultural Economics (1972). Economic principles for increased beef production, Beef Research Report No. 9.
39. Bracken, T. (1963). Mathematical protein models for selection of diets to minimise weighted radionucleotide intake, United States Public Health Service, Publication 999, R-4, 18 pp.
40. Van De Panne, C. & Popp. W. (1963). Minimum-cost cattle feed under probabilistic protein constraints, *Management Science*, **9**, 405–30.
41. Smith, B. J. (1971). The dairy cow replacement problem: an application of dynamic programming, University of Florida Agricultural Experiment Stations, Bulletin 745 (Technical).
42. Swartzman, G. L. & Van Dyne, G. M. (1972). An ecologically based simulation–optimisation approach to natural resource planning, *Annual Review of Ecology and Systematics*, **3**, 347–98.
43. Trebeck, D. B. & Hardaker, J. B. (1972). The integrated use of simulation and stochastic programming for whole farm planning under risk, *Australian Journal of Agricultural Economics*, **16**, 115–26.
44. Christian, K. R., Armstrong, J. D., Davidson, J. L., Donelly, J. R. & Freer, M. (1974). A model for decision-making in grazing management, *Proceedings XIIth International Grassland Congress*, Moscow.
45. Mann, S. J. (1968). A mathematical theory for the exploitation and control of biological populations, Operations Res. Dept., Case Western Reserve University, Cleveland, Ohio, Tech. Memo No. 114.
46. Heady, E. O. & Whittlesley, N. K. (1965). A programming analysis of inter-regional competition and surplus capacity of American agriculture, Iowa Agricultural Experiment Station, Research Series Bulletin R-538, **44 pp.**

47. Van Dyne, G. M. (1966). Application and integration of multiple linear regression and linear programming in renewable resource analysis, *Journal of Range Management*, **19**, 356–62.
48. Maruyama, Y. & Fuller, E. I. (1965). An inter-regional quadratic programming model for varying degrees of competition, Massachusetts Agricultural Experiment Station Bulletin 555.
49. Bennett, D., Batini, F., Sharpe, R. & Havel, J. J. (1973). An allocation model in catchment land use planning, *Australian Institution of Engineers, Proceedings of the Hydrology Symposium*, Perth, August, pp. 181–3.
50. Townsley, R. (1964). The maximisation of revenue fron New Zealand sales of butter on the United Kingdom market: a dynamic programming problem, *Australian Journal of Agricultural Economics*, **8**, 169–80.
51. Kennedy, J. O. S. & Fransisco, E. M. (1973). On the formulation of risk constraints for linear programming: aggregative programming model for Australian agriculture, Department of Agricultural Economics and Business Management, University of New England, Report No. 2.
52. Arnold, G. W. & Campbell, N. A. (1972). A model of a ley farming system, with particular reference to a sub-model for animal production, *Proceedings of the Australian Society of Animal Production*, **9**, 23–30.
53. Arnold, G. W. & Galbraith, K. A. (1974). Predicting the value of lupins for sheep and cattle in cropping and pastoral farming systems, *Proceedings of the Australian Society of Animal Production*, **10**, 383–6.
54. Arnold, G. W., Carbon, B. A., Galbraith, K. A. & Biddiscombe, E. F. (1974). Use of a simulation model to assess the effects of a grazing management on pasture and animal production, *Proceedings of the XIIth International Grassland Congress*, Moscow.
55. Carbon, B. A. & Galbraith, K. A. (1974). Simulation of the water balance for plants growing on coarse soils, *Australian Journal of Soil Research* (in press).
56. Nelder, J. A. & Mead, R. (1965). A simplex method for function minimisation, *Computer Journal*, **8**, 308–13.
57. Smith, R. C. G., Biddiscombe, E. F. & Stern, W. R. (1973). Effect of spelling newly sown pastures, *Australian Journal of Experimental Agriculture and Animal Husbandry*, **13**, 549–55.
58. Graham, N. McC., Black, J. L., Faichney, G. J., Arnold, G. W. & Campbell, N. A. (1974). Computer model of growth and production in sheep, in: *Proceedings of International Symposium, Animal Requirements and Linear Programs of Animal Diets*, Logan, Utah (in press).
59. Forrester, J. W. (1969). *Urban Dynamics*, MIT Press, Cambridge, Mass.
60. Powell, R. A. & Hardaker, J. B. (1969). Recent development in farm planning: 3. Sub-optimal programming methods for practical farm planning, *Review of Marketing and Agricultural Economics*, **37**, 121–9.
61. Barrow, N. J. (1974). The slow reactions between soil and anions. I. Effects of time, temperature and water content of a soil on the decrease in effectiveness of phosphate for plant growth, *Soil Science* (in press).
62. Barrow, N. J. (1973). Relationship between a soil's ability to absorb phosphate and the residual effectiveness of superphosphate, *Australian Journal of Soil Research*, **11**, 57–63.

63. Rudd, C. L. & Barrow, N. J. (1973). The effectiveness of several methods of applying superphosphate on yield response by wheat, *Australian Journal of Experimental Agriculture and Animal Husbandry*, **13**, 430–3.
64. Ozanne, P. G., Keay, J. & Biddiscombe, E. F. (1969). The comparative applied phosphate requirements of eight annual pasture species, *Australian Journal of Agricultural Research*, **20**, 809–18.
65. McArthur, I. D. & Dillon, J. L. (1971). Risk, utility and stocking rate, *Australian Journal of Agricultural Economics*, **15**, 20–35.

5

Constraints and Limitations of Data Sources for Systems Models

J. N. R. JEFFERS

Institute of Terrestrial Ecology, Merlewood Research Station, Grange-over-Sands, Cumbria, England

5.1. INTRODUCTION

In this paper I shall argue the thesis that the present emphasis on the compilation and construction of general-purpose data banks in environmental research is based on a misconception of the necessary processes of scientific research. As a basis for the modelling of ecological processes and management systems, the data bank has little or no part to play and the primary unit of information retrieval for the scientist is the model itself. I cannot pretend that this will be a popular thesis: it may well offend many of the participants in this symposium and many of those who will read the proceedings subsequently. I expect the argument that I will advance to be vigorously attacked, and I will be happy for the argument to be the target for critical examination. If it can be shown that the doubts I have about the value of the data bank concept are unfounded, or that there are simple solutions to the problems I see in the extension of this concept to practical applications in environmental research, I will feel that this paper has served a useful purpose.

My argument will begin from a consideration of the types of modelling activities which are clearly recognisable, classified in relation to the sources of the data used in the modelling process. Having identified four main categories of modelling activity, I shall attempt to show that ideas about the use of data which are currently in vogue stem more from accounting theory than from the philosophy of the scientific method. In particular, I shall stress the structure imposed upon data by the methods of data collection, and the constraints imposed by the types of variable measured and the relationships between them. Finally, I shall attempt to show the implications of the essential structure of data and variable constraints for data storage and manipulation.

5.2. TYPES OF MODELLING ACTIVITIES RELATED TO DATA SOURCES

Four main types of modelling activity can be readily distinguished and, in this paper these activities are termed:

(i) conceptual models;
(ii) post-analysis, data-based models;
(iii) designed models, and
(iv) model validation.

5.2.1. Conceptual Models

Conceptual models are formulated on the basis of experience or intuition. They are frequently derived with a minimal input of data, or even from no data at all, and represent the modeller's understanding of a particular set of circumstances, and of the simplifications which he feels may be made to inherently complex relationships. Such models are valuable as checking the consistency of theories and hypotheses, and they have the advantage over other forms of conceptualisation that the assumptions made by the modeller are at least explicit. However, for this type of activity to be regarded as part of the main body of reputable science, it is necessary that the models should lead to hypotheses which are capable of verification. Models which are incapable of practical verification, or which are constructed by the multiplication of causal entities, are, at best, speculative. Perhaps the best known examples of conceptual models are the world dynamic models which have led to the discussion on 'limits to growth'. The models represent the ideas of several groups of working scientists on the complex interrelationships between various factors involved in the production of food, changes in population through birth and death, use of nonrenewable resources and industrial and agricultural investment. The results of the manipulation of input factors to a series of relationships with positive and negative feedback are frequently counter-intuitive, and have led to much argument about the value and the meaning of the models. Nevertheless, although these models have highlighted the very different assumptions which are made by thinking, both expert and non-expert, about world problems, they are not strictly capable of verification. Their usefulness lies in prompting further work which may establish relationships between the component sub-systems.

5.2.2. Post-analysis, Data-based Models

Many models are formulated as a result of preliminary analysis of extensive data. Extensive data are already collected through designed experiments and surveys carried out in many different fields of research, and even more data become available through almost haphazard collection by scientists and others working in a relatively uncoordinated fashion. While one may deplore the inefficient use of resources which results in data collection not subjected to the critical design of either experiments or surveys, it would be wasteful to neglect such data entirely, provided they are first subjected to critical analysis.

The distinction between this form of modelling activity and that described under the heading of conceptual models is that the models are based on sometimes extensive data that have already been subjected to analysis. Such models may need verification, leading to the design of further experiments and surveys to collect the necessary data for the critical testing of hypotheses. Frequently, because of difficulties or deficiencies in the design of the data collection, it may not be possible to reject competing theories upon which the models may be based, and this inability will itself lead to the need for further verification and data collection. Furthermore, the models which are constructed as a result of the analysis of previously collected data will sometimes be severely limited by the availability of the data and the form in which they were originally collected. This type of modelling activity places constraints upon the modellers which the conceptual modeller would be unwilling to accept. Nevertheless, those of us who have been educated in the more traditional schools of mathematical philosophy will generally feel more comfortable when engaged in modelling activities which are at least partly data-based, and for which the data have previously been subjected to analysis. Examples of models of this kind exist in the factor models of intelligence, the models of farm structure resulting from statistical and economic analysis of agricultural census data, and the yield models for forests.

5.2.3. Designed Models

It is perhaps only recently that the simultaneous design of models and data collection has come to be regarded as either feasible or necessary. Much of the earlier modelling activity has either taken place independently of, or much later than the collection of data, or has preceded any form of data collection. In the projects of the International Biological Programme, for example, while national and international synthesis of the projects was written firmly into the objectives of the programme, data collection largely preceded the modelling activity, and the final syntheses were attempted, where possible, from such data as had managed to be collected. The resulting models are weakened severely by the lack of co-ordination between the modelling and data collection stages. With greater experience, it is now evident that the data collection and modelling needs to progress in close parallel, the preliminary models being formulated partly on the basis of existing data and then augmented by research to provide the necessary data base for the later stages of the modelling. In the systems modelling which is proposed for the Man and Biosphere projects, for example, five interrelated phases are recognised in the simultaneous design of models and data collection, namely:

(i) The setting of objectives and the construction of a preliminary synthesis. The objectives will need to specify (a) the range of topics the model is expected to cover; (b) the types of manipulation, modification or disturbance to be included in the model; and (c) the variables it is intended to measure and to predict. Following a review of the existing information, and the re-examination of the objectives, a preliminary synthesis of the model will be attempted.

(ii) The experimentation phase, both in the field and the laboratory, made in conjunction with experiments on the model. Part of this experimentation may be regarded as a validation of the model, in which the output of the model is tested against the field or laboratory results, but considerable adjustment and refinement of the model will also be taking place.

(iii) The management phase, during which small-scale, prototype management plans and manipulations will be defined and undertaken, with extensive consultation between the scientists developing the synthesis, experiments, and models.

(iv) The evaluation phase, during which the changes proposed from the results of the earlier phases on the structure, functioning and stability of the system are assessed. The information derived from this phase will usually be valuable in suggesting policies for management, use and allocation of resources.

(v) The final synthesis will summarise and integrate all the information collected in the earlier phases.

In each of these stages, modelling and data collection are so clearly interrelated that one flows from another, and the model begins as a first approximation to a solution which is gradually refined to the final synthesis. This type of designed model has, so far, been attempted only by a few closely integrated organisations, but has been applied to such problems as environmental impact assessment, ecosystem modelling, and water resource engineering.

5.2.4. Model Validation

This activity is often regarded as being distinct from model formulation, but can be incorporated as an integral part of designed models. It is important, however, to distinguish model validation from model exploration, with which it has become largely confused in much of the present-day discussion of the concepts of systems ecology. Model validation is essentially concerned with the testing of the hypotheses represented by the model and by their assumptions. This testing assumes that it is possible to make predictions from the model which are capable of direct verification. Sometimes it will be possible to test that observed parameters predicted by the model are approximated sufficiently well for the model to be regarded as a useful predictor. In other situations, and perhaps more commonly, it will be necessary to derive a series of sub-hypotheses from the overall hypothesis of the general model structure and to test these independently. In the derivation of these sub-hypotheses, the formal logic of mathematics may play an important part, with the attendant difficulty that many intending users of the model may well not understand the formal derivations, or the importance of the mathematical argument in deriving the sub-hypotheses. Such lack of understanding may well lead to uncritical acceptance of models produced as a stage in thinking of systems ecology and systems research. It will certainly be the task of the scientist to ensure that the models he has produced, and which he knows to be imperfect, are not used uncritically and are not accepted as more valuable predictors than they really are.

5.3. THE CONCEPT OF THE DATA BANK

There is a major discrepancy between two philosophies of data collection:

(i) The accounting theory assumes that the subsequent use of data is independent of the methods of collection. An accountant believes that it is possible to collect data in some neutral sense, and that any subsequent manipulation of these data can be justified, if the manipulation contributes to the understanding of the problem.
(ii) The 'statistical' theory insists on the essential interdependence of the way in which data are collected and the methods of analysis which are appropriate for those data. The methods of collection

(including such questions as the population sampled, the sampling units, and the scales of measurement) define the valid methods of analysis which may be employed. Alternatively, if we wish to use particular methods of analysis or estimation, we have to ensure that appropriate methods of data collection have been used.

Much of the discussion on data collection and data banks assumes the acceptance of the 'accounting' theory of data manipulation. In contrast most, if not all, of the available methods of modelling numerical data assume the statistical theory of data collection, management and manipulation. The loss of the essential structure inherent in the collection of data when they are stored in the 'data bank' is, therefore, an insuperable limitation to the valid use of such data. It is important, perhaps, to stress that this loss of structure is not a function of computer-based storage. The loss of information about methods of data collection is as much a part of data storage in files, notebooks, and scraps of paper. While in the early years of a research investigation the methods of data collection may remain within the memories of the working scientists associated with the data, as soon as these scientists disappear through retirement, transfer to other jobs, or death, the necessary details of data collection become lost to those who might subsequently seek to use the data. Working scientists are well aware of the rapidly deteriorating value of data which are held within their research files, often at great expense, as the associated memories of the procedures used in their collection disappear. Regrettably, managers and administrators steeped in the accounting theory of data continue to place increasing reliance upon the concept of the data bank, believing that it will be possible to manipulate data for whatever purpose in an apparently neutral sense. The greater ease of data extraction and manipulation made possible by the electronic digital computer, and the ability to hold large quantities of numerical and alpha-numeric information within computer memories, has given an added impetus to the idea that it is somehow possible to create a bank of valuable data which will become, by definition, important in the construction of management information systems, often totally unrelated to the original purposes of the collection of the data.

5.4. IMPOSITION OF STRUCTURE ON DATA

The structure imposed on data at the time of their collection may derive from the definition of the population to be sampled, the constraints of the sampling frame, the methods of sampling, the relationships between variables, and the constraints imposed by experiment and survey designs. It will be appropriate to explore some of these ideas, at least in outline, although complete discussion of the many problems involved in the recognition of imposed structure is beyond the scope of a short paper.

In theory at least, every set of data collected refers to some defined population. Hopefully the population will have been defined explicitly so that inferences can be made from the sample data set to the defined population. More usually the population sampled is implicit and there may even be a marked bias in the way in which the samples from that population have been selected so that fair inferences cannot be made about the population. There is also the marked danger that individuals other than the research workers concerned may, inadvertently, extend the inferences which may be drawn from the sample data to a wider population than that for which valid inferences can be made. The primary responsibility for defining the population, and ensuring that the inferences are not invalidly made, is essentially that of the research worker, but he may not be in a good position to check on the way in which his data are subsequently used.

The limitations on the population are usually those of time and geographical location, although other limitations will sometimes be imposed. For example, it is quite common for the research worker, in collecting data, to seek to exclude marked heterogeneity by imposing constraints of uniformity on the sample areas that he is prepared to accept. Rigorous application of such exclusion may lead to the situation in which the only valid inferences that can be made are to some sub-population of falsely homogeneous areas. Similar truncation of natural populations may take place if, for example, crops which are markedly diseased or suffering from some abnormality are excluded from survey.

The constraints of the sampling frame used in the collection of data may also impose a structure on the data. For example, the size of the sample unit may itself impose constraints of scale on the data and may prevent examination of variability at entirely different scales being undertaken, particularly when no attempt has been made to group the sample units to test the effect of the size. Similar considerations apply in the shape of sample units, and again little or no work may have been done for a particular data collection to ensure that the shape of the sample unit is not imposing some particular constraint upon data, possibly interacting with size.

The method of sampling used in the collection of data often has a major effect upon the value of the data for subsequent manipulation, and especially in modelling activities. In many fields of activity subjective sampling is traditional during which a skilled, or sometimes less skilled, observer selects those individuals which he regards as being representative. Such practices still continue, despite the fact that statisticians have shown for nearly half a century that subjective sampling is a dangerous practice and that serious bias is almost always introduced into data collection if such sampling is used. Random sampling, although always preferable for the sake of avoiding bias and providing a ready check of the variability of the population, may not always give the maximum precision for future estimates, and the position and location of random samples is frequently lost in data storage, although, for subsequent purposes, such information may be vital. Systematic sampling, although likely to be more precise in the provision of estimates, and also unbiased, can lead to problems of periodicity and contiguity of individual sample units, and again, the information of the contiguity or position of samples from a systematic sampling design may well be lost when data are stored.

The types of relationships which are assumed between the variables to be included in the model may also place important constraints upon the subsequent use of the data. For many statistical purposes, for example, it is important to distinguish whether the model parameter is based upon a variable or a variate, i.e. whether or not a probability distribution may be associated with any particular variable. The type of problem which may occur will be readily familiar to any research worker who has calculated the regression of a dependent variable on one or more regressor variables. Only the dependent variable has, strictly speaking, a probability distribution. The resulting regression equation cannot be regarded as a functional equation linking the several variables, and any attempt to treat a regression equation as a functional equation will lead to biased estimates. Again, the important distinction between the valid use of the techniques of correlation and regression depend upon the method of sampling that was employed, although the arithmetic calculations for the two techniques appear very similar. The distinction is one which is frequently and carelessly ignored by many of those engaged in ecological modelling, and it is certainly not a distinction which is usually included in the information held in the data bank. To add even more to the confusion, it is frequently necessary to replace missing values in complex data sets by values estimated from various procedures. Unfortunately the 'best' missing value for some particular purpose is not necessarily the 'best' missing value for some quite different purpose. The compiler of a data bank, therefore, has the option of showing a value as 'missing' by some special symbol, or of replacing it by a value which may well be appropriate for only a limited range of purposes. When the data bank contains only primary data, these difficulties may not be severe, but as soon as any form of preliminary computation is attempted in order to summarise primary data, the considerations of the dependence or independence and the presence of missing values may place important constraints upon the stored data.

Finally, in this catalogue of possible disasters in the compilation of data banks, it is important to mention some of the problems which will almost certainly arise if data from designed experiments are included in the data bank. First, any designed experiment will have an associated structure designed to reduce the size of the experimental error. Frequently, this structure will involve the grouping of individual treatment plots into blocks, so that the appropriate comparison is between treatments within the same block. Any data derived from comparisons in designed experiments, therefore, must distinguish between those comparisons which can be made within the same block and those comparisons which have to be made between different blocks. In more complex designs, it may even be necessary to make adjustments for the presence of the treatment in different blocks. Primary data derived from experiments cannot, therefore, be given equal status and the underlying structure of the experiment has to be associated as a constraint on the data themselves. Similar considerations arise from repeated measurements on the same and different plots and will certainly influence comparisons which are made over long time-scales as is frequently necessary in both agriculture and forestry.

5.5. IMPLICATIONS FOR DATA STORAGE

Because of the various constraints described above, which may be attached to individual units of data, there is a need to associate the structure imposed by the collection of the data with any form of information storage. Thus, while a data bank may represent the results of a single type of data collection, and may provide valuable information for the creation of a model which bears in mind the methods of data collection which were used, there is a danger that the information may well be used for some quite different model, requiring different methods of data collection for any valid inferences to be drawn.

Nowhere is the implication of the structure imposed upon collected data more easily seen than in the procedures which are developed for the rejection of outliers and for data screening as a preliminary to model construction. In the extensive sets of data collected by the Meteorological Office, and which are frequently required for work associated with the effects of climate upon ecosystems, a minimum amount of screening of assumedly correct values has been carried out. Most research workers, however, will feel that it is necessary to undertake extensive screening of such data, and the form of the screening to be undertaken will vary markedly from model to model and will depend upon the ultimate use of the data. It will be important, therefore, to know what types of rejection of outliers have already been undertaken and where any data screening has previously been done. The Meteorological Office data have certainly been screened for some, but not all, variables. Again, many sets of data have been subjected to interpolation and smoothing, or to various adjustments to remove excessive degrees of variation. Yet this variation may be the very feature that is required for the construction of the model, particularly if the model is to be based upon probability functions or is to test the parameters of stochastic relationships.

Finally, in the interests of reducing the size of the data bank, much of the essential redundancy of the collected data may have been reduced. Yet this redundancy provides the very material necessary to the modeller for testing that a set of relationships compiled on one part of the data is equally applicable to the relationships which may be detected in the data as a whole. In the modelling of ecological processes and systems, redundancy of data is sometimes all-important.

However, the constraints that will have been imposed in the collection of data have to be considered in relation to the use of computers for the storage and transmission of information. Essentially, computers make the problem easier in one way but harder in another. Thus, while computers have transformed the possibilities of handling large quantities of data and making them readily accessible from convenient forms of storage, the quantities that we now attempt to use and store have increased the difficulty of retaining the associated structure and constraints, and, unfortunately, the recording of structures and constraints is more difficult than the recording of numerical values.

So far, the emphasis of computer development has been on the hardware and software problems of the storage, transmission and manipulation of data. Many of these problems have arisen from the rapidly changing types of storage which have become available, during which the emphasis has moved from storage on fixed magnetic drums and discs to magnetic tape, and is now turning to removable discs. The shift from magnetic tape to magnetic disc itself involves a change of emphasis from serial access to random or pseudo-random access of the stored data. These preoccupations have been sufficient to place considerable emphasis on changes in programming and access techniques, and yet have not given the practical user sufficiently large bodies of data to determine the advantages and disadvantages of the various systems. Bearing in mind the rapidity with which changes take place in a quickly developing technical field, there has been little stability in the computer systems available to the working scientist.

In the programming of computers for the storage and transmission of data, as well as for their use in subsequent modelling, much of the argument has rested on the general and relative merits of special-purpose languages and of general languages. Special-purpose languages for data storage and for report generation have usually been favoured, although these languages often make the data inaccessible in curious ways and lead to difficulties if the language is not available on an alternative make or type of computer. There is also a considerable difficulty in the special-purpose language itself imposing additional constraints on the types of model that can be built, or the ways in which the data can be retrieved. The computer language itself may provide a solution to the association of a constraint on the use of data if this language is sufficiently specialised to indicate the constraints on the ways in which data may be used. However the manipulation is to be managed, there needs to be particular protection against the undue removal of redundancy in data, as a result of the natural desire to reduce the data to the smallest possible quantity. Now that electronic computers can store and manipulate very large quantities of data in a relatively small space, there is certainly less reason for early summarisation and loss of the redundancy essential for model building.

5.6. CONCLUSIONS

The main purpose of this paper has been to show that data collection itself imposes constraints upon the ways in which data may subsequently be used. The problem in the construction of a data bank is that these constraints have to be held at the same time as the data themselves. While it may be possible, with more sophisticated use of electronic computers, to construct specialised languages for data storage and manipulation which so structure the data that only certain and valid operations can be performed, it seems more likely that the concept of the data bank, as it is held by most administrators and managers, is firmly based on the accountant's theory of data which places no emphasis on these constraints.

For those of us concerned with the wider application of systems analysis and modelling to problems of ecology and resource management, the future needs are relatively clear. Our ability to model will depend on our ability to design and implement research programmes in which data collection and modelling are closely integrated, as in the designed models described earlier in this paper. It will be for us to resist the pressures from administrators and, perhaps, from our less systems-orientated colleagues, to embark on vast data collection and data banking exercises. It is even more important that we should resist the emergence of that new breed of scientist, whose habitat is frequently to be found in international organisations, and who regards the collection of other people's data (garnered from all possible sources, with little regard to the original objectives and methods of the data collection) as a justification for spending large sums of money on data banks and reference systems.

I would hope that there would be little dissent, at a symposium like this, from the idea that the true unit of the transmission of concepts in the progress of science is the model itself and not the data used to construct or validate the model. There is little difficulty in the exchange of models between scientists and the computer has helped rather than hindered the development of this exchange by its dependence upon exact algorithms for computer programmes. If we can implement our model on a computer, then, by definition, we have described our model sufficiently exactly for its concepts to be understood by a fellow scientist. If our colleagues do not wish to explore the inner working of our models, they can at least test that the output from our models is consistent with the data they have collected or can collect, with due attention to the requirements of data collection for a valid test.

Systems analysis is not, therefore, an extraneous activity in which we may indulge if we wish but which may be avoided by banking the data we collect. It is an essential chain in the logic of investigation of complex systems, and the models of systems analysis are the units of conceptual advance in science. Still less can we embark upon systems analysis and modelling as an activity divorced from the collection and storage of data. Rather, it is an activity demanding the close integration of modelling and data collection.

Discussion Report

R. N. CURNOW

Dean of Faculty of Agriculture and Food, and Professor of Applied Statistics, University of Reading, England

M. UPTON

Department of Agricultural Economics and Management, University of Reading, England

An impression gained from some of the papers was that they were excessively introspective, concerned with the taxonomy and classification of models rather than their practical application in problem solving. Nonetheless, some speakers considered that the study and definition of methods is useful. Some discussion centred on the question of whether there is a distinction to be drawn between the 'systems approach' and 'operational research'. One speaker suggested that the systems approach is a philosophy or paradigm while operational research is the application of quantitative methods in problem solving. Alternatively, it was argued that there are two schools of systems methodology. On the one hand there is the systems engineering, control theory and communications theory school and on the other hand the operational research school.

Various general problems associated with modelling were raised. For instance, there are the dangers of extrapolation. There is a clear distinction to be drawn between interpolation and extrapolation, many models being used for the latter purpose. Considerable uncertainty must be associated with extrapolations. Then the question was raised of the number of variables to include in any experiment or model. It was suggested that we frequently try to incorporate too many variables, particularly in searching for an optimum. Fixing the values of some of the variables may reduce the attainable optimum but make it easier to get near the optimum in the restricted dimensions.[1] Concern was expressed over the large size of some models. Large models are not easily explained to decision-makers and are costly in terms of both money and time. If a model takes ten years to develop the results may well be out of date and the decision which the model was designed to guide will already have been made on some less formal basis. Against this it was argued that a model may have multiple uses, such as aiding decision-making, extending public awareness, aiding experimental analyses of data or policy-making, research and extension, and the cost should therefore be spread over all these uses. This may be true, for instance, of fertiliser response models. Furthermore, the size of the model must to some extent depend upon the size of the problem. Also, many biological factors do not change over time so that biological results, at least, do not become out of date. Nevertheless, some thought that simple, less costly methods could be used to produce results that were just as useful as those from larger models.

Much time was devoted to the discussion of the special problems of applying the systems approach in farm management and agricultural economics as compared with natural, biological/ecological systems. Whereas the latter are plainly the result of either random or mechanically determined processes, agricultural systems are much influenced by human decisions. It is very difficult to model these human decisions which are complex, subjective and each one unique in place and time. Attempts to involve sociologists in modelling agricultural systems apparently have not proved very useful. One reason put forward was that sociology is a body of research methodology rather than a theory of human behaviour.

There is a conflict between generality and realism or precision. To justify their costs, farm planning models must be applicable to more than one farm at one point in time. They must be generalised. This has been achieved either (a) by building models based entirely on technical relationships and used for forward accounting (the skeleton and autonomous models proposed by Professor Dent are thought to fall into this category) or (b) by heroically simplifying assumptions such as that all farmers are optimisers. However, there is some debate about what is optimised. To argue that the objective is to maximise welfare begs the question since we do not have a generally acceptable measure of welfare. Important considerations in relation to decision-making in agriculture are (a) differing time patterns of costs and benefits for different activities, and (b) costs and benefits which are unpredictable or subject to risk. It was suggested that farmers and other human agents prefer their incomes to remain stable or constant from year to year, whereas ecological systems are variable by nature. Perhaps we must learn to accept and adapt to this natural variability. From this point of view agricultural systems need to be both robust and flexible, two requirements which may be incompatible to some extent. The robustness is necessary for the system to be insensitive to unpredictable, chance variations in the environment. The flexibility is needed so that, in future, options remain open and adjustments can be made in the system in response to changes in costs, prices or the environment.

This raises the question as to whether it is meaningful to evaluate alternative farm plans or systems in terms of a single measure such as profit or even utility. Three alternative suggestions were made regarding decision-makers' objectives.

The first is that decision-makers may have a hierarchy of objectives, that they rank in some way (known in the literature as the lexicographic approach). For instance a model is being developed for studying the control of yellow fever. This involves a hierarchy of objectives, namely (a) total elimination of the disease, (b) preventing outbreaks and (c) curing outbreaks. The ranking of these objectives is clearly a political decision depending on value judgements. Indeed decision-makers may prefer not to make their choice of objectives explicit.

The second, alternative suggestion is that decision-makers may not seek to optimise but rather to 'satisfice', that is, to search until a satisfactory solution is found, rather than an optimum. If indeed this is the objective of most farmers, simulation would seem to be a more appropriate planning tool than the various optimising models discussed in the papers.

The third alternative suggestion is that optimisation may be a long-term process of progressive improvement rather than a once-and-for-all choice. Farmers may be concerned with finding the direction of greatest improvement rather than an overall most profitable solution at any one point in time. In this connection the technique of 'evolutionary operation' which is used in industry may be relevant.[2] This involves continual deliberate modification of the system in operation, searching for improvements and following what may be a moving optimum. It was suggested that Dent's skeletal models might be used in this gradual continuing search for an optimum.

Professor Dent's concept of a skeleton model is not a new idea. Such models are widely used in the USA for agricultural extension work. The building of a complete simulation model for the individual farm would require a high degree of skill on the part of the extension worker. The skeletal model on the other hand may be handled by an unskilled man. In fact it was suggested that the linear programming model is a widely used form of skeleton model particularly when used in planning least cost-rations. However, some ascribe the growth and development of new planning models to dissatisfaction with the unreality of some of the assumptions involved in linear programming.

Others argued that linear programming is essentially a long-term planning tool, the solution of which may be relevant 10 or 20 years hence. However, farmers have many short-term decisions to make and other models may be useful for this purpose. Indeed there is difficulty for the adviser to maintain contact with the farmer over the long period if he does not offer advice on short-term problems; this is important because the farmer can visualise the outcome more easily than with longer-term plans.

In order to apply models generally to large numbers of farms it is necessary to describe and classify farm types. However, it is more logical to base the classification on the farmer's resource base, that is, the amounts of land, labour, buildings, machines and other capital at his disposal, rather than the current farming system. After all the farmer may be producing beef when by all the criteria available it appears that he should be keeping dairy cows.

At various points in the discussion, the relevance of our techniques to less developed countries was mentioned. The argument that our models should be simple enough for the decision-makers to understand cannot be applied to illiterate farmers in such countries. However, it may nonetheless be relevant when applied to government policy-makers. The study and understanding of tropical farming systems is very important, perhaps more important than building optimising models at this stage. At the same time our existing farm planning techniques are inadequate to deal with the problems of farmers in less developed countries. Their special problems include (a) greater susceptibility to risk, (b) fewer resources and hence less ability to cope with risk, (c) complex timing of labour inputs and (d) greater interdependency between farm and household.

REFERENCES

1. Curnow, R. N. (1972). The number of variables when searching for an optimum, *Journal of the Royal Statistical Society*, **B34**(3) 461
2. For listed references *see* Hahn, G. J. & Dershowitz, A. F. (1974). Evolutionary operation today: some survey results and observations, *Applied Statistics*, **23**, 214; and Lowe, C. W. (1974). Evolutionary operation in action, *Applied Statistics*, **23**, 4218.

PART 2

Application of a Systems Approach in Practice

6

A Systems Approach to the Control of the Sugar Cane Froghopper

G. R. CONWAY, G. A. NORTON, N. J. SMALL

Environmental Resource Management Research Unit, Department of Zoology and Applied Entomology, Imperial College of Science and Technology, Silwood Park, Ascot, Berkshire, England

A. B. S. KING*

Caroni Research Station, Waterloo Estate, Carapichaima, Trinidad

6.1. INTRODUCTION

A 'pest' is a living organism; but the term has no biological meaning. It implies a value judgment and hence is only meaningful in a social or economic context. An organism becomes a pest when it causes damage or illness to man or his possessions or is otherwise, in some sense, 'not wanted'. For the purposes of economic analysis it is assumed that the damage or illness caused can be quantified in monetary terms, so that the benefits associated with control of the pest can be compared with the costs of control.

In a recent paper Southwood & Norton[1] applied to pest control the classical economists' tool or marginal analysis. A control action is rational when its cost is less than or equal to the net increase in revenue it produces and is economically most efficient when the marginal cost of control equals the marginal revenue produced (Fig. 1). If we take for our argument an agricultural crop, where the pest reduces the quantity and quality of the product, then in mathematical symbols we aim at maximising:

$$Y(A(S))P(A(S)) - C(S) \qquad (1)$$

where $A(S)$ is the level of pest attack associated with control strategy S, $Y(A(S))$ is the yield at level of attack $A(S)$, $P(A(S))$ is the price at level of attack $A(S)$, and $C(S)$ is the cost of control using strategy S.

---

* Present address: Centre for Overseas Pest Research, College House, Wrights Lane, London W.8, England.

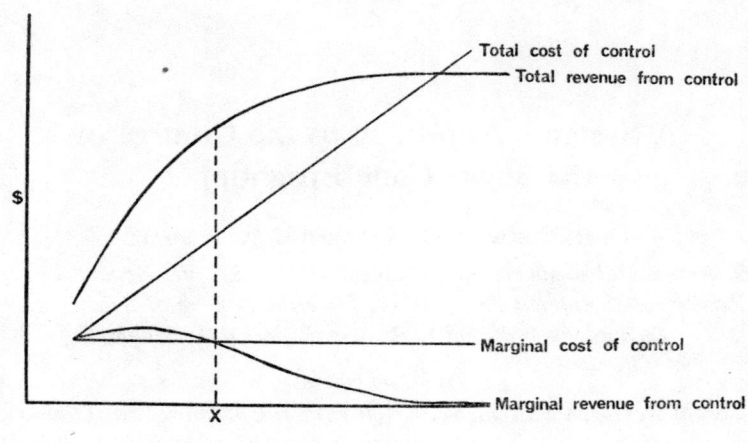

FIG. 1. Cost and revenue curves for pest control. X indicates the optimal level of control.

Two important functions are implicit in this expression: the damage function, $Y(A)P(A)$, which relates pest numbers to change in crop revenue, and the control function, $A(S)$, which relates control measures to the change in pest numbers. From this point onwards biological, and in particular ecological, considerations play an increasingly important role. Thus the damage function depends on the feeding habits of the pest and on the dynamic interrelationship between the crop plant and the pest. Rarely is the function of a simple linear form; there may be thresholds and further complications due to the plant's ability to compensate for damage. Similar effects determine the form of the control function. The efficacy of an insecticide, for example, will depend on the physiology of the pest and on the degree to which the pest is protected from the insecticide by its habitat or behaviour.

As we expand the temporal and spatial limits of the system, ecological considerations become critical. Climatic factors, the population dynamics of the pest and its migratory patterns affect when and where control is taken. When insecticides are used there are often pervasive effects, including the development of resistance, interference with natural enemies and injurious effects on wildlife and man, which have to be taken into account in the decision-making process.

It is clear, then, that rational pest control must rest on both economic and ecological analysis, carried out within a broad holistic framework.

Unfortunately, as we have learned from personal experience and as any survey of the literature will reveal, past and present pest control practice falls far short of this ideal. Economic analysis has been largely theoretical[2,3] and it is rare to find more than the sketchiest information on the damage function. Indeed, one suspects that pest control is frequently irrational, the cost of the control measures being greater than the value of the damage they are aimed at eliminating.

Recent ecological studies have done much to improve our understanding of the dynamics of insect populations[4] but the results have been only indirectly related to the end objective of control. There has been little attempt to bring ecological and economic parameters together in a meaningful fashion. We now have considerable information on the environmental effects of pesticides and other control techniques but again the data are frequently not available in a form which is of use to decision-makers.

6.2. BACKGROUND TO THE PRESENT STUDY

In the early 1960s a number of workers began to recognise the potential value of systems analysis and mathematical modelling as guides to pest management. The pioneer in the field is K. E. F. Watt who has developed a number of theoretical population models, one of which was used in the Canadian Spruce Budworm study. He was also the first to use the optimising techniques of simulation and dynamic programming in assessing the costs and benefits of pest control.[5] Recently one of us (G. R. C.) reviewed progress since Watt's work and concluded that while there have been a number of perceptive theoretical models, particularly in the area of biological control, and one or two directly useful models, mostly related to the sterile mating control technique, the impact on pest control theory and practice has been small.[6] For the most part the models have been focused on particular aspects of the problem and have not attempted the task of integrating the ecological and economic components.

In 1970 the Environmental Resource Management Research Unit was established at Imperial College under a grant from the Ford Foundation to develop systems techniques for the management of resource problems. As a first task the Unit concentrated on problems in pest control and attempted to bring together ecological and economic analyses using the framework of the operations research approach and the methods of systems analysis.

The study reported here is a collaborative project between the Unit and the Caroni Research Station in Trinidad and has the general objective of developing analytical methods which will be of practical value for a specific problem. A preliminary study of the problem was made in 1972 by one of us (G. A. N.) and the then entomologist at Caroni, Dr. D. E. Evans;[7] but the major part of the work has been carried out in 1973 and 1974 by the Unit's staff in collaboration with the fourth author of this paper (A. B. S. K.) who was appointed entomologist at Caroni in 1972.

At the time of writing the study has still to be completed. This paper is intended as a presentation of the approach and methods being used. More detailed accounts are being published elsewhere.

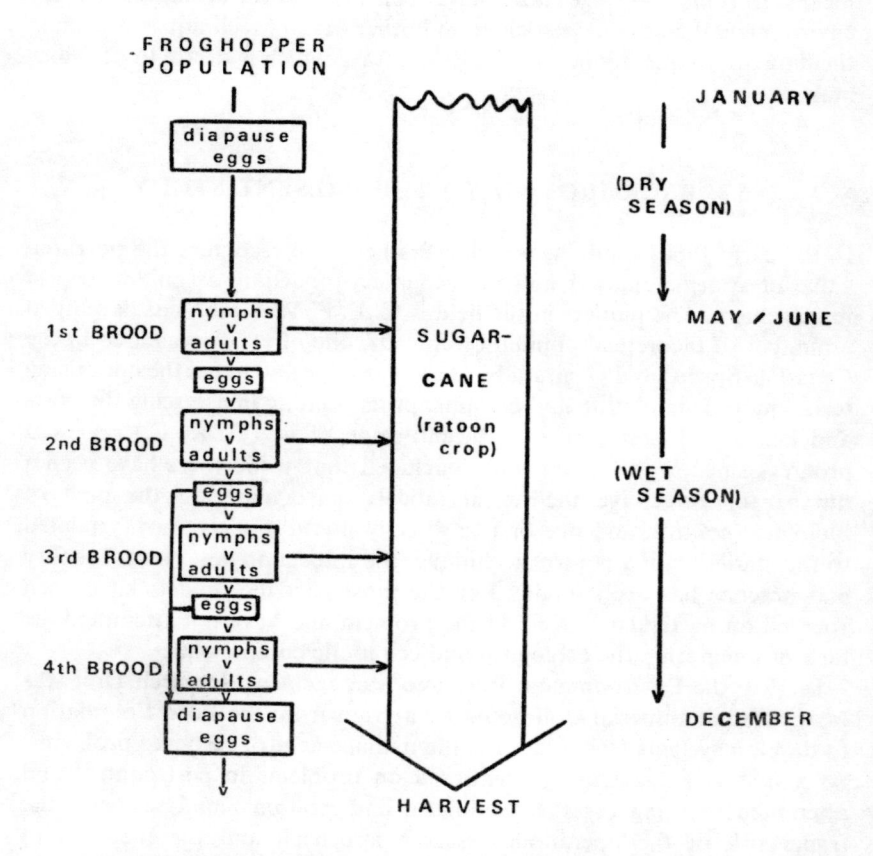

FIG. 2. The seasonal cycle of the froghopper population on sugar cane in Trinidad.

6.3. OUTLINE OF THE PROBLEM

The froghopper is a small bug, with sucking mouthparts, which feeds on the sap of the cane plant. Young froghoppers, termed nymphs, live in cracks in the soil and feed on the roots. However, the important damage is caused by the adults which feed on the uppermost leaves.[8] When severe, this feeding causes 'blight', a necrotic condition of the leaves which reduces photosynthetic capacity. Damage results in a loss in total cane weight per acre and in juice purity.[9,10]

There are four generations or broods of froghopper each year, the first arising after the onset of the wet season which occurs in May and June. Later broods lay diapause eggs which survive the dry season from December to May and give rise to the first brood nymphs of the succeeding season (Fig. 2).

Caroni Ltd. grows approximately 50 000 acres (20 200 ha) of cane in Trinidad and current control practice is to spray infested fields from the air with an insecticide that kills the adult froghoppers. The fields are regularly monitored to determine the level of the froghopper population and spraying is carried out when the density exceeds 5 adults per 100 cane stems in the first brood or 50 adults per 100 stems in later broods. In 1971 the cost of spraying was around £200 000, representing an average of three sprays per acre. The value of the sugar cane crop in that year was approximately £6 million.

6.4. OBJECTIVES

Our initial aim was to investigate ways of improving the economic efficiency of the current control practice of aerial spraying. As we have already seen in expression (1), analysis of efficiency in this context depends on knowledge of two key relationships, the damage function and the control function.

On the basis of the limited evidence then available to them, Norton & Evans[7] suggested that damage was probably a linear function of the froghopper population measured in terms of the total number of adult-days. More recent work by one of us (A. B. S. K.) indicates that the relationship may be non-linear, but our initial analysis assumed a linear function of the form:

$$Y = Y_0 - dA(S) \qquad (2)$$

where Y is the yield in tons of sugar, Y_0 is the yield in the absence of pest attack, $A(S)$ is the number of adult-days per 100 stools in a froghopper population subject to control strategy S, and d is the damage coefficient in terms of the loss of yield per adult-day per 100 stools.

The parameter values derived from the Norton & Evans study were as follows:

$$Y_0 = 3 \cdot 5 \text{ tons per acre}; \quad d = 1 \cdot 307 \times 10^{-5} \text{ tons}$$

The objective of improving control efficiency can now be redefined in terms of maximising:

$$P(Y_0 - dA(S)) - C(S) \tag{3}$$

where P is the price per ton of sugar (\$TT 200 in 1973) and $C(S)$ is the cost of control strategy S.

We approached the control function, $A(S)$, from two directions. First we were concerned with developing, through data analysis and computer simulation, an understanding of the ecological characteristics of the pest which are relevant to control. Secondly, we examined the theoretical basis for insecticidal control using simulation and programming techniques. These two phases of work are nearing completion and we are now embarking on the construction of a practical management scheme for aerial spraying. Finally, as a fourth phase, we are exploring the feasibility of control methods that will complement or replace aerial insecticide spraying.

In this paper we summarise the work of the first two phases and refer briefly to current progress on the management scheme and on alternative methods of control.

6.5. POPULATION DYNAMICS OF THE FROGHOPPER

Froghopper populations vary in size from year to year and from place to place; but because of the scarcity of detailed field records we concentrated initially on explaining the seasonal pattern of population change which occurs in a single field. Even for this situation there are no records for large unsprayed populations. However, a detailed sampling programme was maintained for a single acre field (Orange Field 3) over the 1973-74 season where a large population developed and spraying was delayed until late in the second brood (Fig. 3). We have tried to model the population pattern exhibited by this set of records using information obtained from the literature or from laboratory or field experiments.

We began by attempting to build a computer model which included all the information on froghopper populations currently available in the literature. This approach was not fruitful, since much of the data was incomplete or measured in such a way as to be incompatible with other

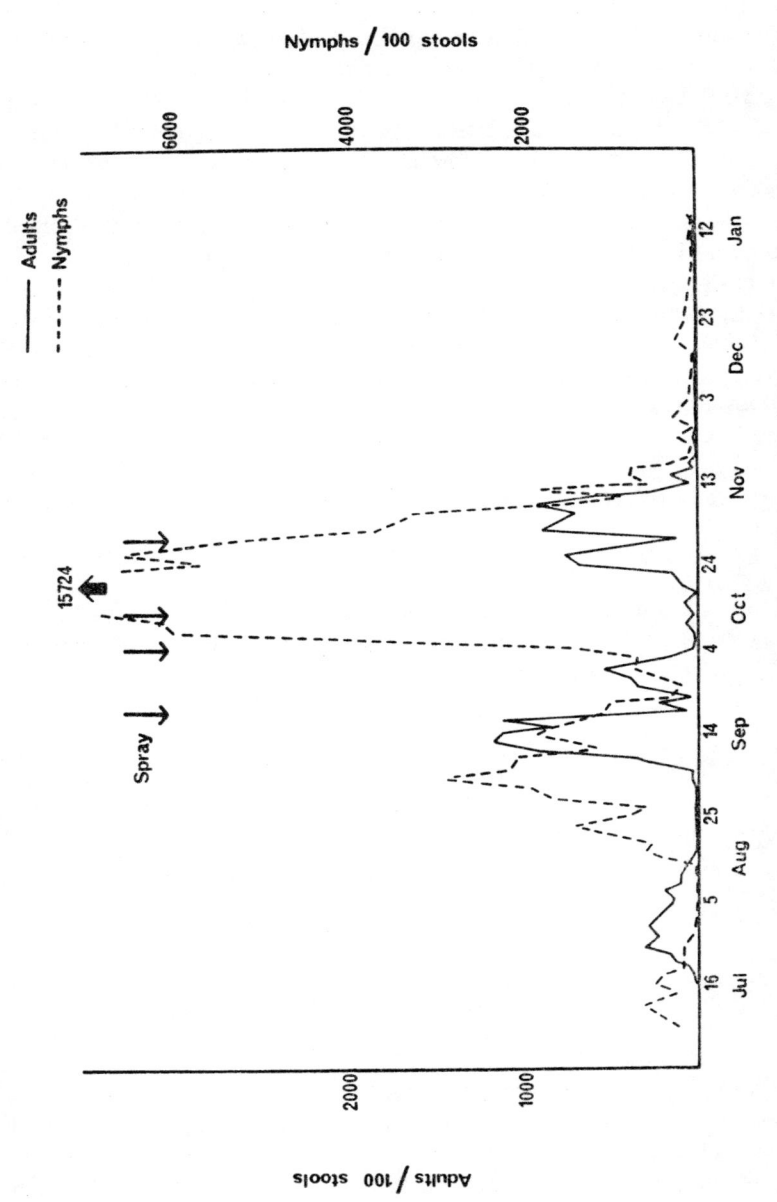

FIG. 3. Record of the numbers of adult and nymphal froghoppers during the 1973–74 season for a single field (Orange Field 3) at Caroni Ltd. in Trinidad.

data. In addition the model was clumsy and we had difficulty in distinguishing the relative contributions made by different components of the model to the pattern of population change predicted by the model. To overcome this we adopted a building-block approach, starting with a very simple model based on the Leslie Matrix[11] and progressively seeking more data and adding more complicated and realistic components. At each step in the model-building process the output was compared with the Orange Field 3 data set, the aim being to assess how much explanation of events could be accounted for by each added component.

The Leslie Matrix model has the merit of describing population change in a neat and conceptually clear manner and we found that, coupled with the building-block approach, it enables a highly productive relationship to develop between model building and field and laboratory experimentation. Developments in the use of the Leslie Matrix have recently been reviewed by Usher.[12] Applied to the froghopper population the model is as follows.

If the age distribution of the population on day d is taken to be a column vector

$$\mathbf{n}_d = \{n_d(0), n_d(1), \cdots, n_d(k)\}$$

where $n_d(x)$ = number of female froghoppers alive on day d in the xth age interval (x to $x + 1$ day-olds). Then the age distribution of the survivor and descendents on the next day, $d + 1$,

$$\mathbf{n}_{d+1} = \{n_{d+1}(0), n_{d+1}(1), \cdots, n_{d+1}(k)\}$$

will be given by

$$\mathbf{n}_{d+1} = \mathbf{A}\mathbf{n}_d \quad (4)$$

where

$$\mathbf{A} = \begin{pmatrix} f(0) & f(1) & \cdots & f(k-1) & f(k) \\ p(0) & 0 & \cdots & 0 & 0 \\ 0 & p(1) & \cdots & 0 & 0 \\ 0 & 0 & \cdots & 0 & 0 \\ 0 & 0 & \cdots & p(k-1) & 0 \end{pmatrix}$$

and where $p(x)$ = probability that a female alive in the xth age interval on day d will survive to be in the $x + 1$th age interval on day $d + 1$, and $f(x)$ = number of female eggs produced in the interval d to $d + 1$ by each female alive in the xth age interval on day d, which will be alive in the 0'th age interval on day $d + 1$.

The age intervals are taken to be days (0–1 day-olds, 1–2 day-olds, 2–3 day-olds and so on). We assumed that the values of $p(x)$ and $f(x)$ changed only with each new brood, a brood lasting from new-laid eggs to the death of the adults to which they give rise. Each brood is thus described by a characterstic **A** matrix.

Clearly the construction of the matrix requires information under three headings: (a) development time, which determines the size of the matrix; (b) fecundity, which gives values for $f(x)$; and (c) survival and migration, which give $p(x)$. Because of the paucity of existing data at the beginning of the study the necessary information was obtained from a series of laboratory and field cage observations carried out during 1973. Table 1 summarises this information on a brood-by-brood basis.

The estimates of fecundity and development so obtained can be reasonably expected to approximate the values occurring naturally in the field; whereas the survival estimates obtained from the laboratory or field cages are more likely to be overestimates. Hence in building the model we began by including the fecundity and development data and then adjusted the survival rates to provide a fit, comparing them subsequently with our estimates.

The model was initiated using a field estimate of the number of viable eggs present in Orange Field 3 in May (19 600 eggs per 100 stools* of which a sample gave a 40% hatch) and it was assumed that all the eggs hatched on 19 June when heavy rains first fell. Figure 4 shows the best fit to the female adult population, obtained with brood survival rates of 0·91, 0·96, 0·93 and 0·91 for broods 1 to 4 respectively.

We then progressively added components to improve the fit and obtained a reasonably realistic outcome with five additional components:

(*i*) *Hatching as a function of rainfall.* The initial assumption of a single egg hatch was clearly wrong. When we plotted field records for the number of newly emerging adults for Orange Field 3 against time 28 days previously (the modal nymph development period for both sexes combined) we obtained a close correlation with the daily rainfall (Fig. 5(a)). We therefore adjusted the model so that the initial population of eggs ($7840 = 19\ 600 \times 0\cdot4$) hatched on the dates and approximately in the proportions indicated by the adult emergence curve.

(*ii*) *Fecundity as a function of female age.* In an attempt to reduce the spacing between brood peaks we examined in the laboratory the effect of

---

* A stool, representing a single cane plant, contains an average of 17 stems.

TABLE 1

Estimates of development, fecundity and survival for the sugar cane froghopper obtained from laboratory and field cage experiments.

Brood	Development times (days)				Fecundity		Survival (daily)		
	Non-diapause[a] eggs (mean)	Nymphs (mode[b])			Eggs per ♀	Proportion non-diapause eggs laid	Non-diapause eggs	Nymphs	Adults (♀)
		All	♂	♀					
1	19	28	28	32	13·38	0·932	—	?	0·894
2	20	28	28	32	12·93	0·703	0·981	09·35	0·838
3	24	28	28	32	10·77	0·253	0·983	?	0·803
4	27	28	28	32	13·26	0·321	0·990	?	?

[a] Non-diapause eggs are defined as eggs hatching in less than 35 days.
[b] Nymphal development times observed for the first brood are assumed to be similar for later broods.

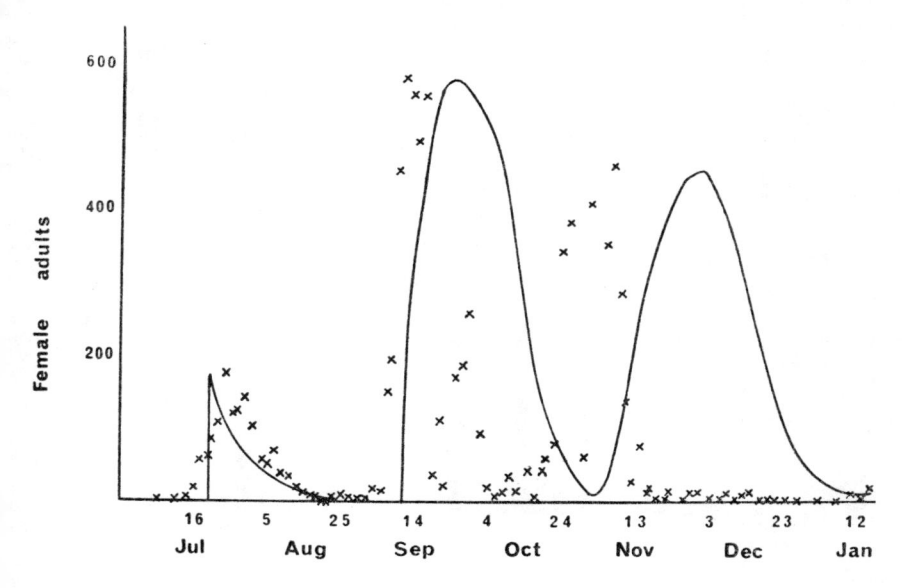

FIG. 4. Preliminary fit of the model to female adult froghoppers recorded in Orange Field 3 in 1973–74.

age of adult froghopper on the fecundity rate. We found that the rate rises rapidly to a maximum on about day 5 and then declines (Fig. 5(b)). However, including this relationship in the model only resulted in a slight reduction in the spacing.

(*iii*) *Inclusion of the male froghopper population in the simulation.* In the initial model only the female population was simulated, the assumption being that there was at all times a 1:1 sex ratio in the field. However, field observations showed that male development is several days faster than that of females (see Table 1). The adult sex ratio thus changes, at least in the first brood, males predominating at the beginning of the brood and females at the end (Fig. 5(c)). We felt that inclusion of the shorter male development time would further reduce the spacing between broods and so ran a male model in parallel with the female one, employing a matrix in every respect similar to the female matrix but without the $f(x)$ row vector.

(*iv*) *Simulation of the effect of insecticide spraying.* Four sprays of the carbamate insecticide Unden (arprocarb) were applied during the season to Orange Field 3 and their effects clearly have to be included if the model is

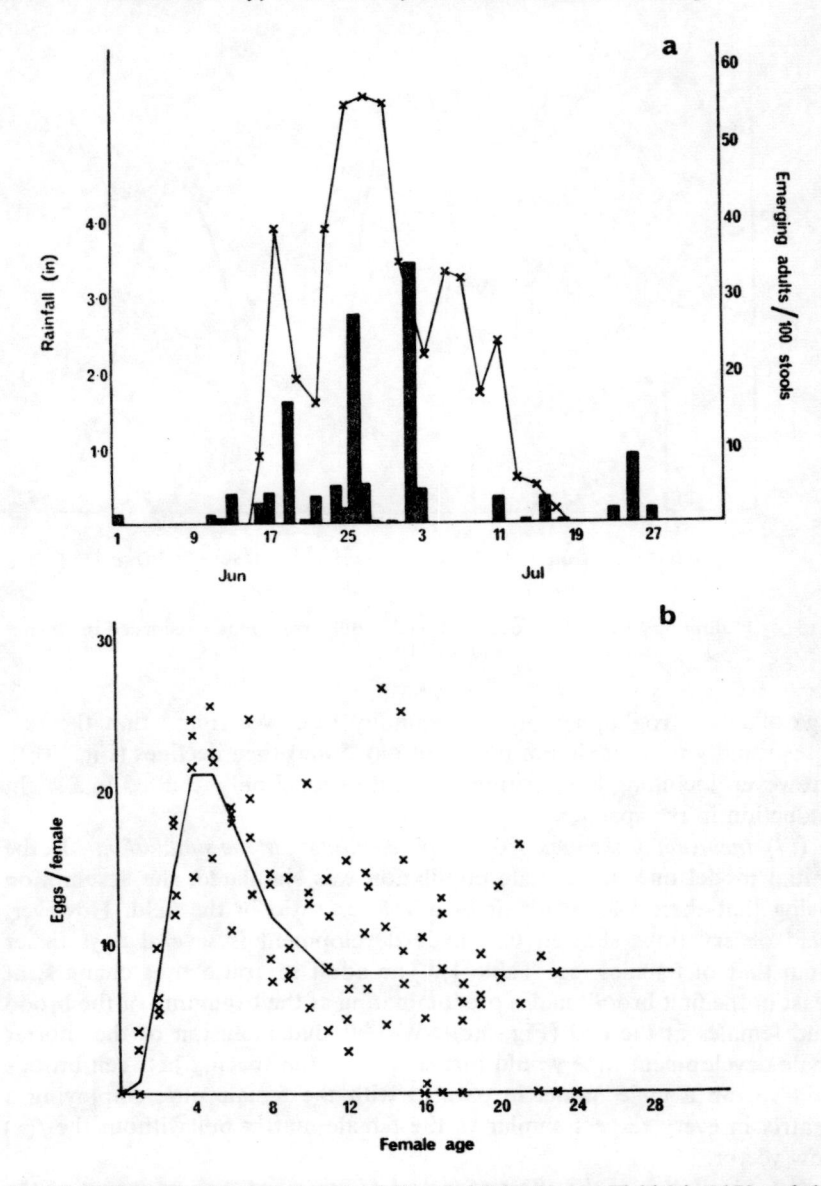

FIG. 5. (a) Relationship between the daily rainfall in Orange Field 3 in 1973 and the numbers of adults emerging 28 days later. (b) Results of cage trials on the effect of adult female age fecundity.

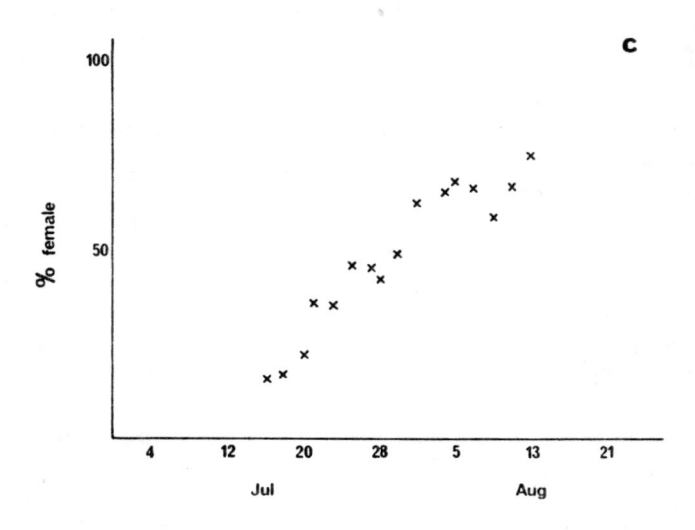

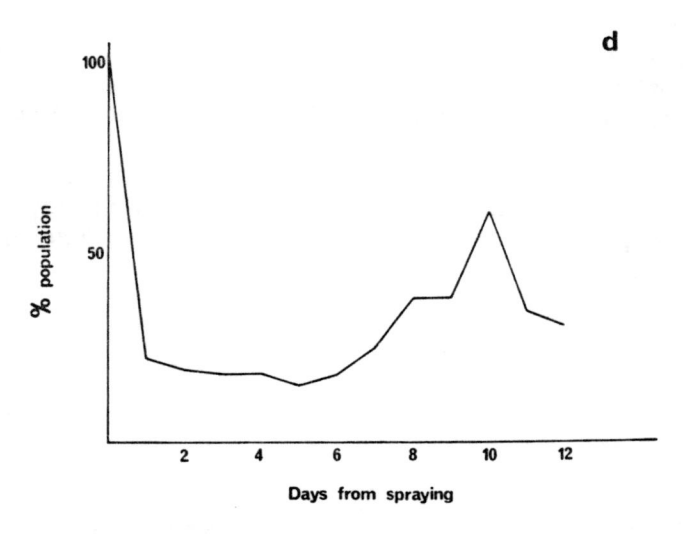

FIG. 5. (c) Adult sex ratio changes during the first brood of the Orange Field 3 population in 1973. (d) Effect on adult froghopper numbers of a field sprayed with Unden.

to be realistic. Unden is a residual insecticide and field trials suggest that it gives an initial kill of 80% of the adult population which is repeated on each of five subsequent days (Fig. 5(d)). (The fourth spray in the season received by Orange Field 3 was not directly applied but resulted from drift from adjoining fields; it was simulated by assuming a non-residual action.)

(v) *A shorter development time.* Inclusion of age-dependent fecundity and male development time still left the broods with too great a spacing between them. We re-examined our data and felt that the source of error probably lay in the egg development times which had been measured in our experiments on a weekly rather than daily basis. Sensitivity analysis suggested that the estimates were approximately too large by seven days.

Figures 6 and 7 show the fit of the model, with these five components included, to the field population data. The final fit was determined by altering the survival rates to obtain a close fit to the adult population curve and subsequently, through minor adjustments, to give approximate fits to the other stages. Table 2 shows the survival rates for the four broods.

The model could be improved by adding further components. For example, we have treated development time as fixed for all individuals whereas in the field populations show a large variance in this respect. Some of the laboratory and field data which were used in building the model are of poor quality. We require, for example, confirmation of the revised estimates of development time, more detailed information on the residual effects of insecticides and independent data on the relationship between rainfall and egg hatching.

Nevertheless a number of conclusions can be drawn at this stage. First, the large discrepancy which occurs in the first two broods between the model prediction and the field records for the fifth instar nymphs

TABLE 2

Computed daily survival rates obtained by fitting the model to the recorded froghopper population in Orange Field 3

	Brood			
	1	2	3	4
Eggs	—	0·98	0·97	0·94
1st–3rd nymphs	0·95	0·97	0·96	0·92
4th nymphs	0·95	0·95	0·95	0·90
5th nymphs	0·95	0·95	0·90	0·90
Adults	0·67	0·84	0·55	0·70

suggests that the sampling technique considerably underestimates this stage. The discrepancy is, in fact, even greater for the earlier stage nymphs. Field observation of the sampling procedures suggests that the model is more likely to be correct. Adult froghoppers, because they are larger and live on the upper leaves, are more easily sampled and probably provide the only accurate basis for monitoring the population.

Secondly, the daily survival rates obtained by fitting the model tend to be higher than expected. The second and third brood egg survival rates and the second brood nymphal survival rates are as high as or higher than the laboratory estimates. It was expected that they would be lower because of mortalities due to parasites and predators which were excluded from the laboratory experiments. If the model is correct it suggests that these stages in the second and third broods are little affected by natural enemies. Survival rates were lower, however, for the first, third and fourth brood adults, reflecting the presence in the field of either adult migration or mortalities due to predators or disease. The drop in survival rate which occurs from the third to fourth brood is interesting since it suggests that the smaller size of the fourth brood cannot be attributed solely to the increased production of diapause eggs but is partly due to increased migration or mortality (possibly a fungus disease).

Although these findings are of relevance for control, more important implications stem from our assumptions about the form of the fecundity and survival functions. We have assumed in the model that the survival and fecundity rates change for each brood but in a manner which is independent of the size of the brood. Thus the total number of adult days produced in a season is a simple multiple of the starting population, that is, of the number of diapause eggs which hatch following the first rains. The critical question is, how realistic is this? If it is true then the most effective control measures are those taken against the first brood; the greater the kill in the first brood, the lower the population in later broods. But if the assumption is incorrect, that is, if a density-dependent relationship exists between the broods, then a more complicated control strategy is required.

6.6. DENSITY DEPENDENCE

The traditional test for density dependence is to plot the log of population size at time $t + 1$ against log population size at some previous time t. The slope of the line is then a measure of the density dependence that has

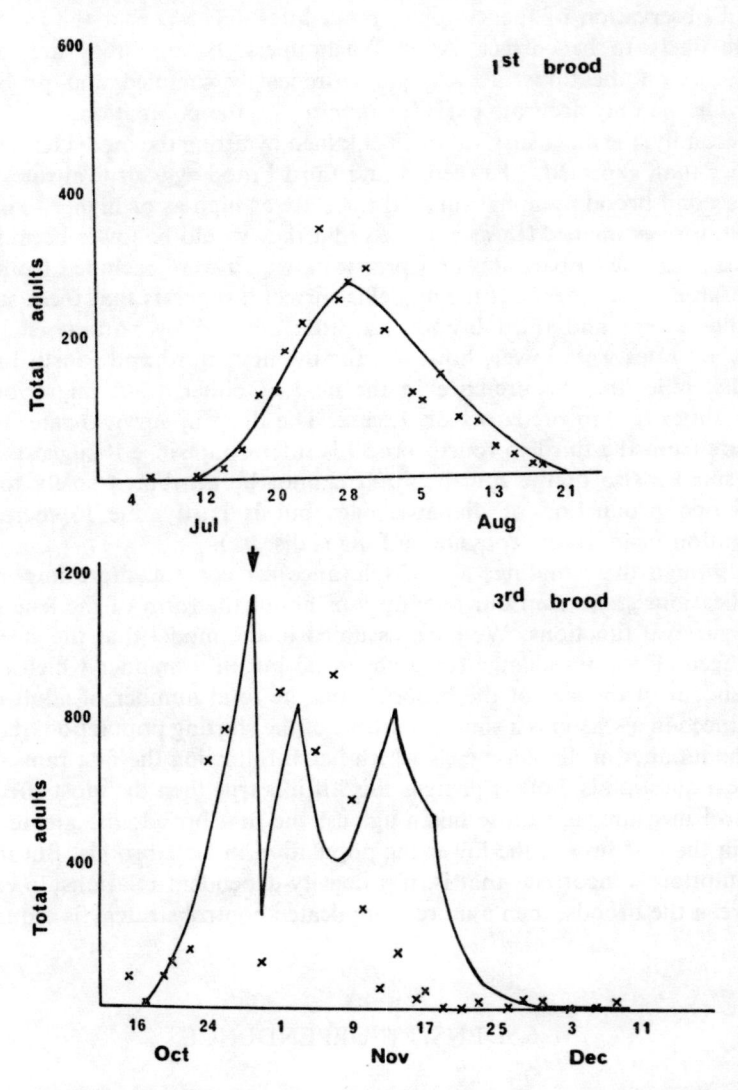

FIG. 6. Final fit of the model to adults in the four broods recorded in Orange Field 3 in 1973–74. Arrows indicate days of spraying with Unden.

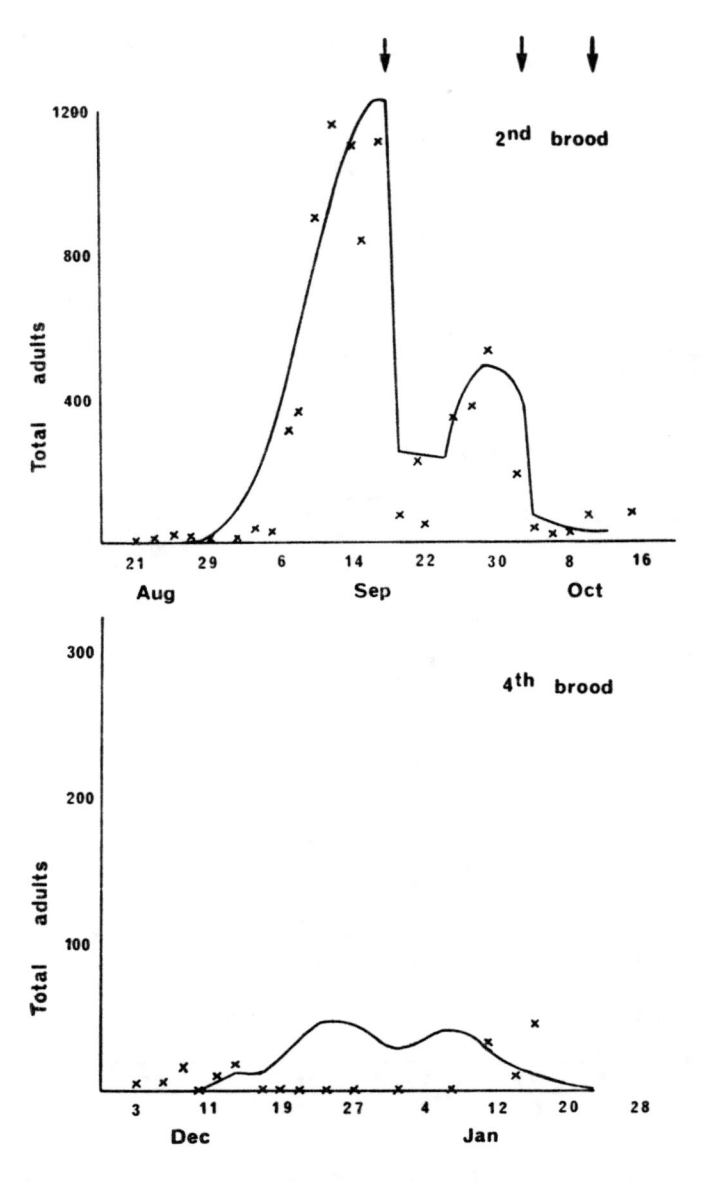

FIG. 6—contd.

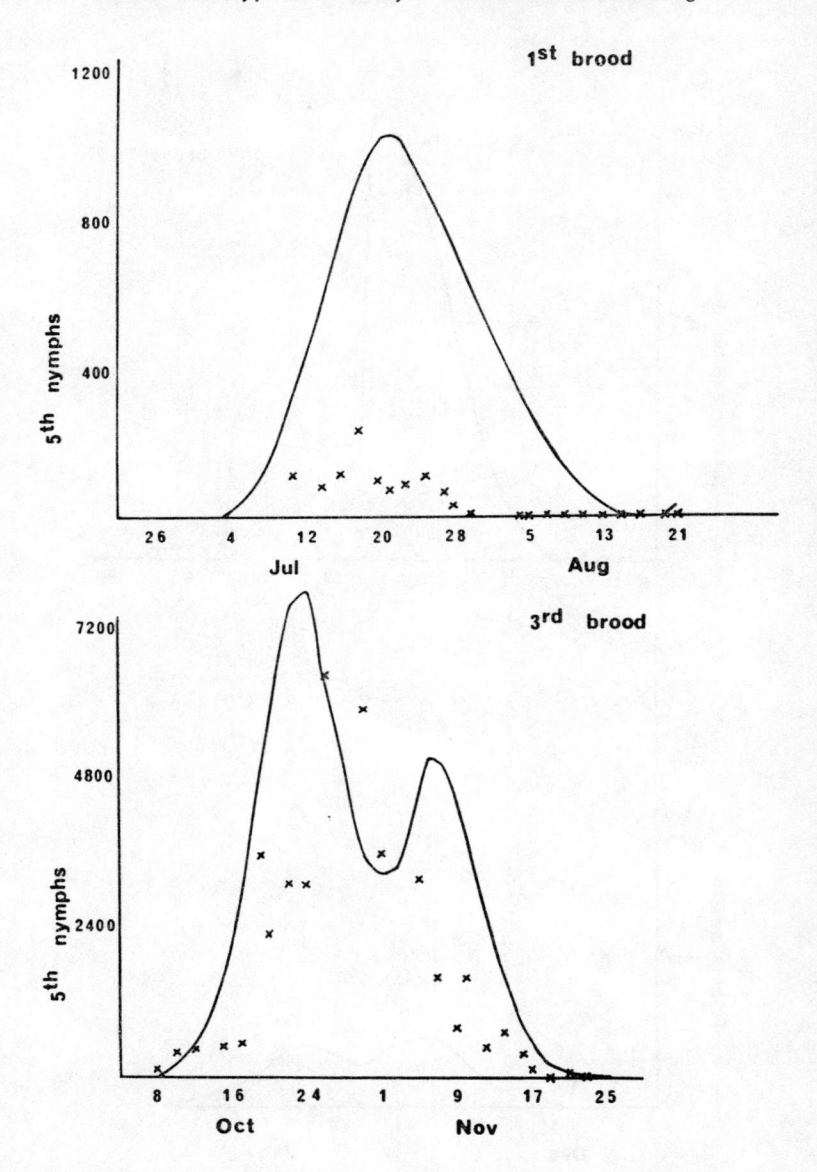

FIG. 7. Final fit of the model to fifth instar nymphs in the four broods recorded in Orange Field 3 in 1973–74.

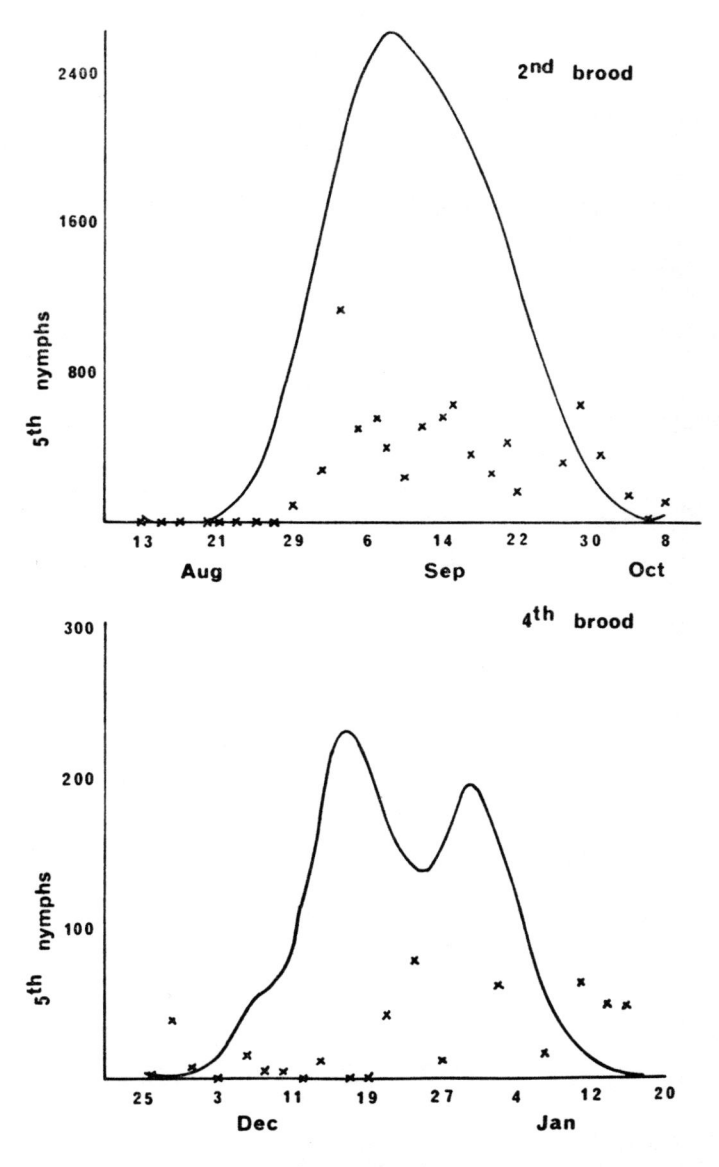

FIG. 7—contd.

occurred. For the froghopper the appropriate plot is of adult days in brood $g + 1$ against adult days in brood g:

$$\ln A_{g+1} = \ln \lambda + (1 - b) \ln A_g \tag{5}$$

or
$$A_{g+1} = \lambda(A_g)^{-b} A_g \tag{6}$$

where λ is the net rate of increase and b, known as the characteristic return rate, measures the degree of density dependence (Fig. 8).[13]

We can regard each brood as having its own equilibrium size which is a function of the carrying capacity of the sugar cane crop at the time. When $b = 1$ there is perfect density dependence in the relationship between the broods; a brood will attain its equilibrium value irrespective of the size of the previous brood. However, when $0 < b < 1$ any perturbation from equilibrium in one brood results in an undercompensation or 'undercorrection' and the equilibrium value of the next brood is approached but not reached. In contrast when $1 < b < 2$ perturbation results in overcompensation and the population overshoots the equilibrium value in the next brood (Fig. 9). A value of $b > 2$ or $b < 0$ indicates an unstable situation with a high probability of extinction. Finally, when $b = 0$ there

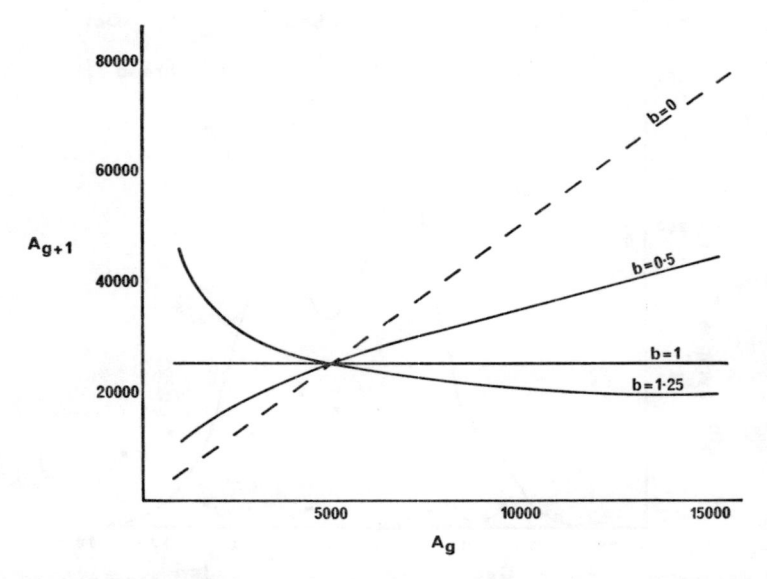

FIG. 8. Effect on the interbrood relationship of different degrees of density dependence. A_g, A_{g+1} are the adult days in succeeding broods and b is a measure of density dependence as before.

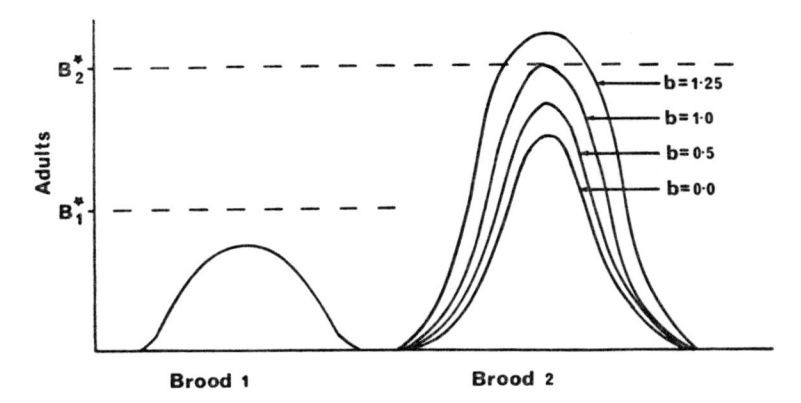

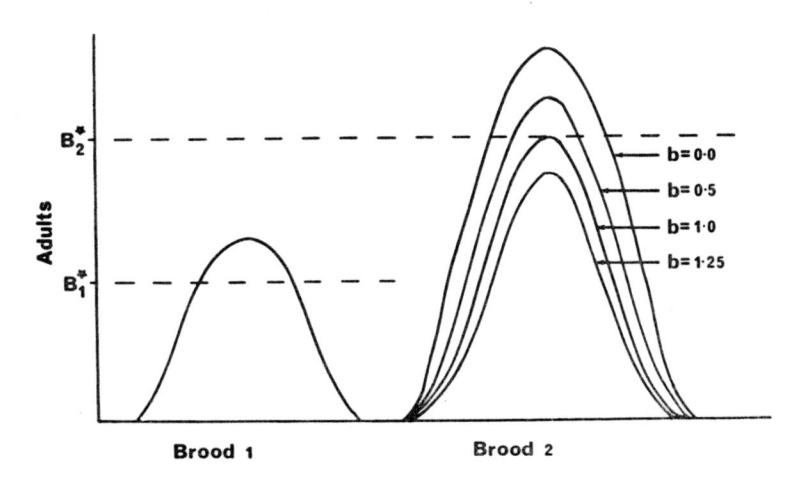

FIG. 9. Effect of density dependence on the relationship between froghopper broods. B_g^* and B_{g+1}^* are the equilibrium levels for brood g and brood $g + 1$. The characteristic return rate, b, measures the degree of density dependence (see text).

is no modifying effect of density on the relationship and the change in numbers is simply given by λA_g.

If present in the froghopper life cycle, density dependence could affect one or more of several processes relating one brood to the next (Fig. 10). Unfortunately, estimation of the b values for these processes requires extensive data sets, preferably with replication over time and space. Moreover, nearly all field records relevant to such analysis were for

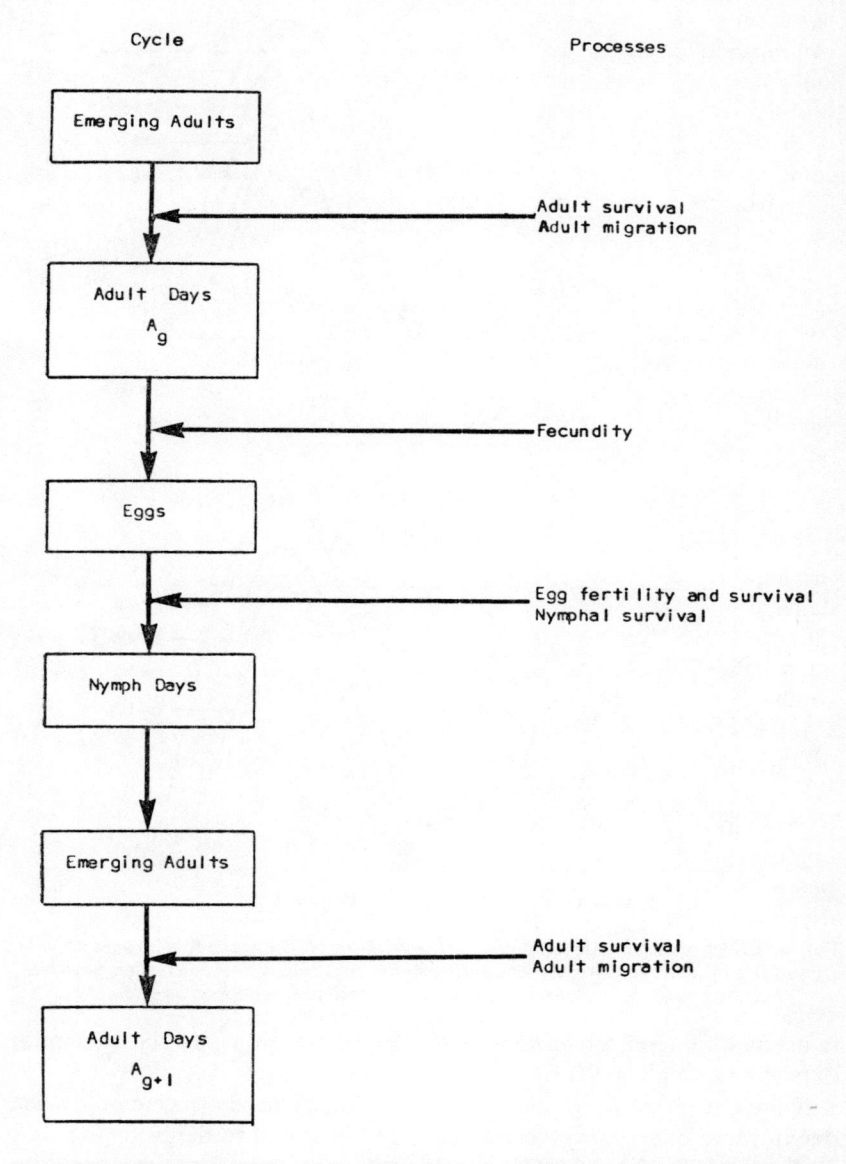

FIG. 10. Processes in the life cycle of the sugar cane froghopper liable to be affected by density dependence.

populations which had been sprayed. Nevertheless, by a process of combination and comparison of several different data sets from the field and from unsprayed cage experiments, we tentatively identified two interbrood processes as subject to density dependence: nymphal survival, which is undercompensating ($b = 0.6$), and adult migration, which is overcompensating ($b = 1.4$). As might be expected, mortality due to spraying is strongly overcompensating ($b = 3.4$), presumably due to the policy of spraying when a fixed population threshold level is exceeded. For the other processes no detectable density dependence was apparent.

Clearly many more observations and experiments are required before the true pattern of regulation in the froghopper population is uncovered. But the evidence suggests that density-dependent processes may be present and should therefore be taken into account in exploring the theoretical basis for control.

6.7. CONTROL BY SPRAYING

The purpose of applying insecticide to a froghopper population is to reduce the number of froghopper adult-days and so bring about a reduction in damage and loss of sugar. In this second phase of the study we investigated the relationship between insecticide application and the ensuing pattern of adult-days and considered the principles determining optimal control strategies.

Insecticide control strategies can be defined here in terms of the number of applications, the time of application and the types of insecticide used. We assumed that a residual insecticide (e.g. Unden) and a non-residual insecticide (e.g. Malathion) are available and that these can be applied in any combination up to a total of three sprays per brood. As in the first phase of the study (see Fig. 5(d)) we took the initial kill of the residual insecticide to be 0.8 which is then repeated against newly emerging adults for a total of five further days. The non-residual insecticide was assumed to give a single kill of 0.8. Costs (calculated for Unden and Malathion respectively) were set at $TT 7.40 for a single application of the residual spray and $TT 2.85 for the non-residual.

For ease of calculation we chose to represent the pattern of an adult population in a brood by an empirical equation. We assumed that a typical brood was represented by the data for newly emerging and total adults in the first brood in Orange Field 3, and obtained an acceptable fit with a beta function (Fig. 11(a)):

$$e_t = E(1.857 \times 10^{-7})(t - 4)^{1.4}(44 - t)^{2.7} \qquad (7)$$

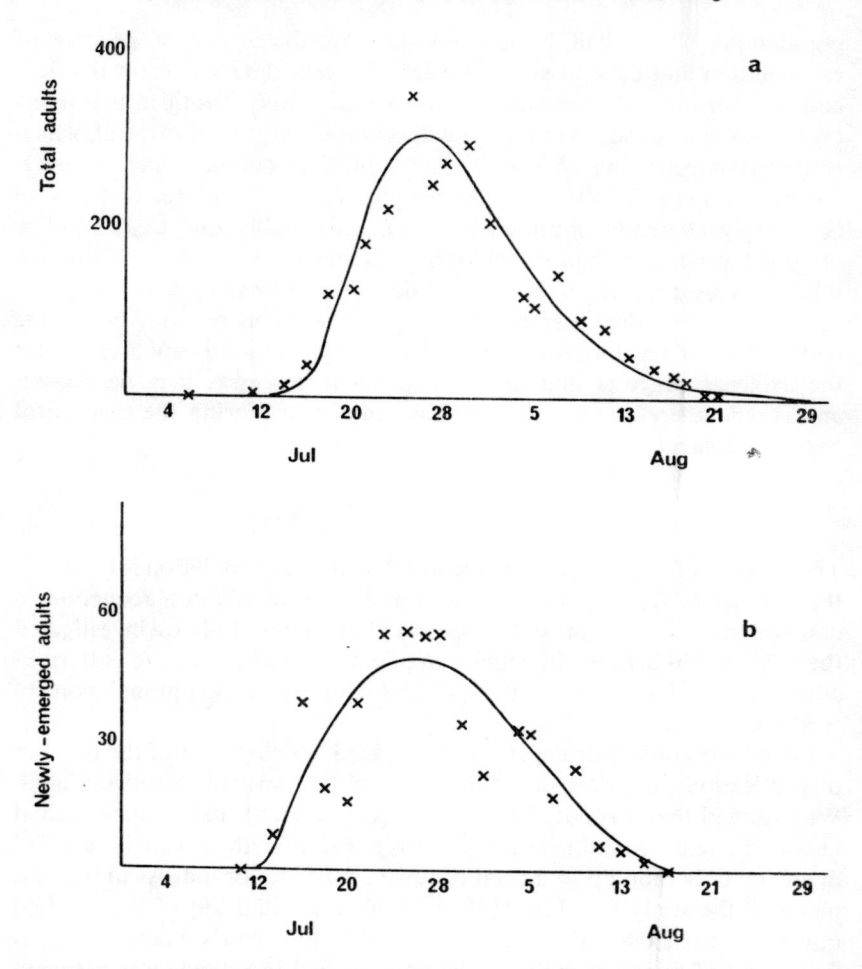

FIG. 11. (a) Fit of the equation $e_t = E(1\cdot857 \times 10^{-7})(t - 4)^{1\cdot4} (44 - t)^{2\cdot7}$ to the numbers of first brood emerging froghopper adults in Orange Field 3 in 1973. (b) Fit of the equation $a_t = A(3\cdot553 \times 10^{-7})t^2(55 - t)^8$ to the numbers of first brood total adults in Orange Field 3 in 1973.

where e_t is the number of adults emerging on day t with t lying between the first day of emergence (4) and the last day (44), and E is the total number of adults emerging during the brood.

Similarly, for total adults (Fig. 11(b)) we obtained

$$a_t = A(3\cdot553 \times 10^{-7})t^2(55 - t)^8 \tag{8}$$

where a_t is the total number of adults present on day t, and A is the total number of adult days in the brood.

We assumed that daily adult survival rate (s) is constant for the brood. Hence the total number of adult days produced by the emerging adults on a single day, $e_t s$, is

$$e_t s + e_t s^2 + e_t s^3 + \cdots \simeq \frac{e_t s}{1-s}$$

It follows that the total number of emerging adults and the total number of adult-days in a brood are related by $A = \dfrac{Es}{1-s}$. For the purpose of this phase of the study we used a value of $s = 0{\cdot}75$ for all broods, computed from the values of A and E in the first brood in Orange Field 3.

By similar reasoning the number of adult-days removed by an insecticide application can be calculated. For a non-residual insecticide a single application will remove

$$\frac{a_t k s}{1-s} \text{ adult-days}$$

where k is the proportionate kill and t is the day of spraying. A single application of a residual insecticide will remove

$$\frac{(a_t + \sum_{i=t+1}^{r} e_i)\, k s}{1-s} \text{ adult-days}$$

where r is the number of days of residual action.

On the basis of these relatively simple expressions we attempted to define optimal spraying strategies for a single brood and then for the progressively more complicated but more realistic cases of two and four broods.

6.7.1. Control of a Single Brood

We examined first the timing of application and then its profitability. Where a brood is sprayed once with a non-residual insecticide the best time for application, in terms of the maximum number of adult-days removed, is when the number of adults, a, is at a maximum. But if a residual insecticide is used, the best time is when

$$a_t + \sum_{i=t+1}^{r} e_i$$

is a maximum.

TABLE 3

Optimal application times for ten spraying strategies directed against a single froghopper brood

Strategy	Time from first adult emergence (days)																		Percentage of adult-days removed
	1	2	3	4	5	6	7	8	9	10	11	12	13	14	15	16	17	18	
1																			0·0
2											NR								19·3
3									NR										32·7
4										R									37·6
5							NR				NR								42·8
6								NR		NR						NR			49·0
7							NR					R							57·2
8							R	R					R						60·9
9						NR								R	R				67·4
10				R													R		74·9

R = residual spray; NR = non-residual spray.

A Systems Approach to the Control of the Sugar Cane Froghopper

Where two or more sprays are applied in a brood the algebra becomes more complicated since one spray will affect the number of adults present at the time of the next spray. For a non-residual insecticide applied on day t and day $t + x$, the number of adults present on day $t + x$ is given by eqn. (8) less those killed on day t who would have otherwise survived to the later date. That is:

$$a_{t+x} - a_t k s^x$$

Even with this small degree of complexity, searching for optimal times of spraying soon becomes laborious and a systematic computer search is required. We have done this for a maximum of three applications using any combination of residual and non-residual insecticide. A total of 15 strategies; defined in terms of number, sequence and type of insecticide, are possible. However, five of these are inadmissible since other permutations of the same combination of insecticides eliminate more adult-days. The nine dominant strategies for a single brood and the optimal times of application are shown in Table 3.

Whereas optimal spraying times for a particular strategy are solely a function of brood pattern, its profitability is affected by brood size. A strategy S is profitable when the revenue from a strategy is greater than its cost, that is, when

$$Pd\alpha(S) > C(S) \tag{9}$$

where $\alpha(S)$ is the number of adult days eliminated by strategy S. This condition is essentially equivalent to the economic threshold criterion of Stern et al.[14]

For a non-residual insecticide the condition is expressed as

$$\frac{Pd\, a_t k s}{1 - s} \geq C(NR) \tag{10}$$

or

$$a_t \geq C(NR)\frac{1 - s}{Pdks} \tag{11}$$

and for a residual insecticide:

$$a_t + \sum_{i=t+1}^{r} e_i \geq C(R)\frac{1 - s}{Pdks} \tag{12}$$

Both of these criteria are practicable where a_t and e_t are being monitored and only a single spray is contemplated. But where more than one spray can be applied, an estimate of total adult days A is required to find the most profitable strategy. Again a systematic computer search procedure has to be followed in which net revenues of the admissible set of strategies are compared.

Figure 12 shows the results of a profitability search involving seven of the ten control strategies described in Table 3. (Three strategies, 4, 5 and 8, are nearly always inferior to at least one of the others in terms of net revenue.) Since the damage function is linear the net revenue curves in Fig. 12 are also linear. A strategy becomes profitable when its net revenue line intersects the non-sprayed revenue line (strategy 1) and for each brood size the

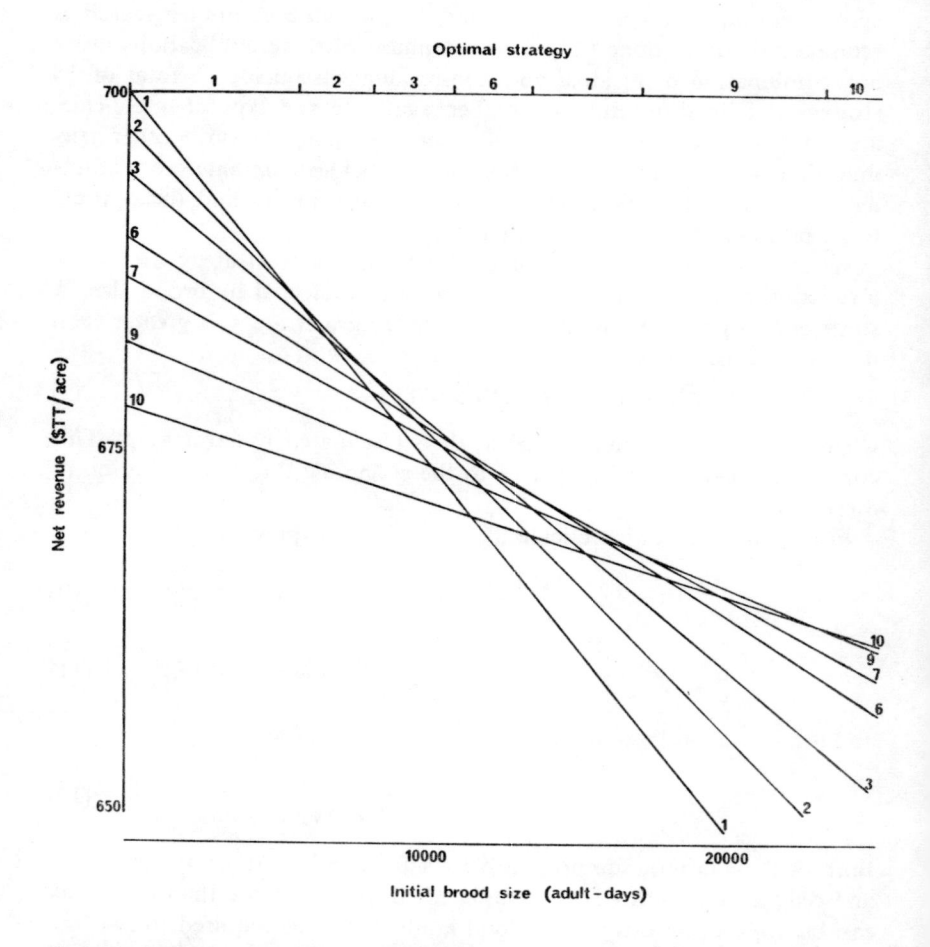

FIG. 12. Optimal spraying strategies (see Table 3) against a single froghopper brood.

uppermost revenue line indicates the most profitable strategy. The arc subtended by these upper lines describes the optimal spraying policy for a range of brood sizes.

6.7.2. Two Broods

When we consider a second brood arising from the first, the optimal spraying policy against the two broods depends on the form of the interbrood relationship. If a density-dependent relationship exists between the broods of the form described in eqn (6) then the net revenue function for each strategy is no longer linear and the optimal spraying policy is more complicated. To illustrate the effect of including a density-dependent relationship we took two broods with equilibrium values of 5000 and 25 000 adult-days respectively as illustrated in Fig. 8.

In studying the implications for control strategies in the two brood situation we assumed that both broods have the same population pattern, as described by eqns (7) and (8), and that for each brood there is a choice of ten strategies as before. We also continued to assume a constant adult survival so that in effect we looked at the consequences of density dependence operating between rather than on the adult stages. Table 4 gives the optimal spraying strategies under these conditions. With no density dependence most of the spraying effort is directed against the first brood, but with increasing density dependence ($b \to 1$) heavier spraying is applied to the second brood. When $b > 1$ the best policy can be largely determined on the basis of the best strategies against each separate brood.

6.7.3. Four Broods

Since in a season the froghopper population produces four broods, we examined finally the optimal spraying policy for this situation. Computationally we assumed this to be a logical extension of the two brood case. However, since the total number of possible strategies is now 10^4, we turned to dynamic programming as a more efficient search procedure. The technique of dynamic programming is well described in standard texts on operations research[15] and was first used in the solution of a pest control problem by Watt.[16] The most thorough treatment of the potential value of the technique in pest control is due to Shoemaker.[17] Applying this technique to froghopper control, we take the last decision stage, which here concerns the control strategy for the fourth brood, as the starting point and the analysis then proceeds backwards, finishing with the first decision (control of the first brood). The advantage of the technique is

TABLE 4

Optimal spraying strategies (see Table 3) for two froghopper broods with different degrees of density dependent relationship between the broods.

Initial size of 1st brood	Density dependence (value of b)							
	b = 0			b = 0·5			b = 1·25		
	Brood strategy		Net revenue ($TT/acre)	Brood strategy		Net revenue ($TT/acre)	Brood strategy		Net revenue ($TT/acre)
	1st	2nd		1st	2nd		1st	2nd	
0	1	1	700	1	1	700	1	1	700
1 000	2	1	684	1	3	672	1	10	650
2 000	3	2	673	1	7	664	1	10	652
3 000	7	2	667	2	7	658	1	10	651
4 000	9	2	662	3	9	654	1	10	650
5 000	10	2	658	3	9	650	1	10	648
6 000	10	2	655	6	9	647	1	10	646
7 000	10	3	652	7	9	644	1	9	644
8 000	10	3	649	7	9	642	2	10	642
9 000	10	3	646	9	9	640	2	9	641
10 000	10	3	643	9	9	639	2	9	639

that less profitable strategies can be removed progressively, so reducing the total number of comparisons that have to be made.

For the purpose of analysis we again assumed that successive broods are related as in eqn (6). In the absence of pertinent field observations we also assumed equilibrium brood sizes of 5 000, 25 000, 20 000 and 1 000 adult-days respectively. Table 5 gives the optimal spraying policies under three different conditions of density dependence. The results further illustrate the importance of density dependence on control policy. It is clear from this analysis that the currently recommended policy for froghopper control,[18] which places heavy reliance on first brood spraying, will be optimal only where there is little or no density dependence, at least between the first and second broods. If further research confirms the preliminary findings that the broods are related in an overcompensating fashion ($b = 1.15$) the optimal policy will include a greater amount of second and third brood spraying.

6.8. THE MANAGEMENT SCHEME

There are three broad strategies that management can follow in spraying adult froghoppers. First, sprays can be applied according to a strictly pre-arranged schedule, possibly utilising predictions based on sampling the diapause eggs or the nymphal stages of the first brood. Secondly, sprays can be applied on the basis of frequent monitoring of the population throughout the season, either employing a simple threshold criterion, as at present, or using a complex threshold which contains elements of prediction of the final size of each brood. Thirdly, some combination of monitoring and schedule spraying can be pursued.

At this stage in our study it is clear that whichever scheme proves to be most suitable, its success will depend heavily on information on two aspects of the seasonal cycle: the development of the first brood from the diapausing eggs and the density-dependent relationships between the broods. So far we have only been able to provide a very incomplete picture of these phenomena and more experimentation and observation is required before any definite control recommendations can be made.

Furthermore, if monitoring is to form the basis of the management scheme then a more accurate sampling technique is required. As our simulation models have demonstrated, sampling of the older nymphs appears to give considerable underestimates. We also have some doubts on the accuracy of adult sampling.

TABLE 5

Optimal spraying strategies (see Table 3) for four froghopper broods with different degrees of density dependence between the broods.

Initial size of 1st brood	Density dependence (value of b)													
	b = 0					b = 0.5					b = 1.25				
	Brood strategy				Net revenue ($TT/acre)	Brood strategy				Net revenue ($TT/acre)	Brood strategy				Net revenue ($TT/acre)
	1st	2nd	3rd	4th		1st	2nd	3rd	4th		1st	2nd	3rd	4th	
0	1	1	1	1	700	1	1	1	1	700	1	1	1	1	700
1 000	2	2	1	1	675	1	9	2	1	649	1	10	10	1	608
2 000	7	3	1	1	664	1	10	3	1	641	1	10	10	1	608
3 000	10	2	1	1	657	2	10	3	1	635	1	9	10	1	607
4 000	9	6	1	1	652	3	10	3	1	630	1	9	10	1	606
5 000	10	5	1	1	648	6	10	3	1	627	1	9	10	1	604
6 000	10	7	1	1	644	7	10	3	1	624	1	9	10	1	602
7 000	10	7	1	1	640	9	10	3	1	621	1	9	10	1	500
8 000	10	9	1	1	638	9	10	3	1	619	2	9	10	1	598
9 000	10	9	1	1	635	9	10	3	1	617	2	9	10	1	597
10 000	10	10	1	1	633	10	10	3	1	615	2	9	10	1	595

6.9. ALTERNATIVE FORMS OF CONTROL

As a final phase of the study we plan to consider the feasibility of developing alternative methods of controlling froghopper populations. The need for such alternatives arises from two important problems associated with insecticide spraying. The first concerns the development of insecticide resistance in the froghopper population. In the past 20 years this has occurred twice on a significant scale; on the first occasion against organochlorine insecticides [19,20] and on the second, against the carbamate insecticide, Sevin, which had to be discarded in 1970.[21] Models would be of value here in predicting the pattern and rate of evolution of resistance and in determing an optimal rotation of different insecticides.

A second problem concerns the effect of insecticides used for froghopper control on the pest status of other sugar cane pests, in particular the moth borers (*Diatraea* spp.). In recent years borer populations have increased to such levels that they are of comparable economic importance to the froghopper. It has been suggested that the upsurge of borers may be due to the adverse effects on its natural enemies of the insecticides used in froghopper control. If this is indeed so it will be important to explore a number of non-insecticidal techniques of control for the froghopper.

One such technique is control through the release of sterilised insects. As first developed by Knipling,[22] the technique relied on mass breeding of insects which were sterilised and released against a natural population, with the aim of bringing about its eradication. In one or two instances the technique has been applied successfully and pest populations eradicated in this way.[6] More recently though, with the development of chemosterilants and the discovery of attractant compounds, interest has focused on the capture, sterilisation and release of individuals from the natural population itself.[23] There has been a concomitant shift of objective from eradication to control, the technique being repeated as required.

In the case of the froghopper, rearing in the laboratory has so far proved extremely difficult and the mass rearing that would be required for an eradication programme does not appear feasible. Furthermore, the presence of a reservoir population of froghoppers in wild vegetation near the cane fields makes eradication unlikely. However, trapping of male froghoppers is now possible using either paraffin lamps or a recently discovered sex pheromone.[24] Hence control through sterilisation is feasible provided a suitable chemosterilant can be found.

To explore the consequences of sterilising a proportion of males in the

TABLE 6

Simulation results of continuous sterilisation of first brood males compared with two spraying strategies, using the model of the froghopper population in Orange Field 3. (Under the decision rule, spraying is carried out when first brood populations exceed 5 adults per 100 stems and when later broods exceed 50 adults per 100 stems.)

Proportion of fertile males sterilised each day	Total adult-days	Total diapause eggs laid
0	80 088	201 123
0·1	60 058	146 945
0·2	46 225	109 723
0·3	36 139	82 685
0·4	28 480	62 204
0·5	22·474	46 174
0·6	17 643	33 298
0·7	13 674	22 732
0·8	10 357	13 906
0·9	7 543	6 424
1·0	5 126	0
Spraying according to decision rule	27 030	68 820
Spraying as carried out in field	43 332	90 925

field population we used the model of the Orange Field 3 data. Assuming random mating and complete mixing of the population, we examined the effects of the proportion sterilised on the subsequent number of adult-days and diapause eggs produced. In Table 6 these results are compared with those obtained when insecticides are applied according to the standard decision rule. The simulations show, at least for the 1973 population in Orange Field 3, that trapping and sterilising first brood males at a rate of 40% per day will give control as good as current spraying practice. However, a significant amount of damage is still likely to occur, partly due to the fact that the number of adult-days in the first brood is unaffected. A more efficient strategy may be to spray the first brood prior to sterilisation.

6.10. CONCLUSION

Our study of sugar cane froghopper control is not yet complete. However, we consider that our adoption of a systems approach and our use of various modelling and simulation techniques has already been of demonstrable value in creating an understanding of the problem, in identifying some of the key elements in the system, and in directing research effort.

In particular, we feel we are able to demonstrate the value for pest control problems of an interdisciplinary team composed of individuals with expertise in field and laboratory work, in economics, in theoretical ecology and in mathematical modelling and systems techniques. Unlike some other systems groups there has been no set or imposed work-programme. Different members have come at the problem from different directions and the programme has evolved out of the discussions which their findings have stimulated. Progress was slow at first but has quickened as different pieces of the 'jigsaw puzzle' have been identified and seen to fit. There has been much discussion about the overall system and about our objectives, as well as much, apparently unconnected, tinkering with small components. We have found this approach to be the most effective way of stimulating the feedback of ideas and information.

Greater understanding and more purposeful research are widely accepted benefits of the systems approach in a great variety of problem areas. Although in the field of pest control the use of systems techniques is still rarely encountered, it is apparent that guidelines to decision-making can be readily produced in this way. However, it remains to be seen whether this approach can also be of value in the practicalities of day-to-day management in pest control; it will take considerable skill and new techniques to produce specific management models that can cope with the inherent variability of such natural systems, without being unduly expensive in money and time.

ACKNOWLEDGMENTS

We are grateful to Professor T. R. E. Southwood who was instrumental in initiating the project and has given valued advice and encouragement. We are also indebted to Dr. Christine Shoemaker who introduced us to the technique of dynamic programming and worked with us on the theory of control by spraying.

The work was carried out jointly under the auspices of Caroni Ltd., Trinidad, and the Environmental Resource Management Research-Unit at Imperial College which is funded by a grant from the office of Resources and Environment in the Ford Foundation. We are very grateful to both these sponsors.

Computing was carried out on the University of London CDC 6600 computer and the CDC 6400 computer of Imperial College.

REFERENCES

1. Southwood, T. R. E. & Norton, G. A. (1973). Economic aspects of pest management strategies and decisions, in: *Insects: Studies in Population Management* (ed. P. W. Geier, L. R. Clark, D. J. Anderson and H. A. Nix), Ecological Society of Australia (Memoirs 1), Canberra, pp. 168–84.
2. Headley, J. C. (1971). Defining the economic threshold, in: *Pest Control Strategies for the Future*, National Academy of Science, Washington, DC, pp. 100–8.
3. Hall, D. C. & Norgaard, R. B. (1973). On the timing and application of pesticides, *American Journal of Agricultural Economics*, 55, 198–201.
4. Southwood, T. R. E. (1975). Dynamics of insect populations, in: *Insects, Science and Society*, Cornell University Press, Ithaca, NY (in press).
5. Watt, K. E. F. (1968). *Ecology and Resource Management: A Quantitative Approach*, McGraw-Hill, New York.
6. Conway, G. R. (1973). Experience in insect pest modelling: a review of models, uses and future directions, in: *Insects: Studies in Population Management* (ed. P. W. Geier, L. R. Clark, D. J. Anderson and H. A. Nix), Ecological Society of Australia (Memoirs 1), Canberra, pp. 103–30.
7. Norton, G. A. & Evans, D. E. (1974). The economics of controlling froghopper (*Aeneolamia varia saccharina* (Dist.)) on sugar-cane in Trinidad, *Bulletin of Entomological Research*, 63, 619–27.
8. Fewkes, D. W. (1969). The biology of sugar cane froghoppers, in: *Pests of Sugar Cane* (ed. J. R. Williams, J. R. Metcalfe, R. W. Mungomery and R. Mathes), Elsevier, Amsterdam, pp. 283–307.
9. Merry, C. A. F., Fewkes, D. W. & Vlitos, A. J. (1963). Chemical control of the sugar cane froghopper *Aenolamia varia saccharina* (Distant), *Proceedings of the International Society of Sugar Cane Technologists*, 11, 642–50.
10. Fewkes, D. W. & Buxo, D. A. (1966). Yield losses in sugar cane due to froghopper infestations, Annual Report of the Tate & Lyle Central Agricultural Research Station, Trinidad, 1965, pp. 364–72.
11. Leslie, P. H. (1945). On the use of matrices in certain population mathematics, *Biometrika*, 35, 183–212.
12. Usher, M. B. (1972). Developments in the Leslie Matrix Model, in: *Mathematical Models in Ecology* (ed. J. N. R. Jeffers), Symposium of the British Ecological Society, Blackwell, Oxford, pp. 29–60.
13. May, R. M., Conway, G. R., Hassell, M. P. & Southwood, T. R. E. (1974). Time delays, density dependence and single-species oscillation, *Journal of Animal Ecology*, 43 (in press).
14. Stern, V. M., Smith, R. F., van den Bosch, R. & Hagen, K. S. (1959). The integrated control concept, *Hilgardia*, 29, 81–101.
15. Hillier, F. S. & Lieberman, G. J. (1969). *Introduction to Operations Research*, Holden-Day, San Francisco.
16. Watt, K. E. F. (1963). Dynamic programming, 'look ahead programming', and the strategy of insect pest control, *Canadian Entomologist*, 95, 525–636.

17. Shoemaker, C. (1973). Optimisation of agricultural pest management. II. Formulation of a control model, *Mathematical Biosciences*, **17**, 357–65.
18. Evans, D. E. & Buxo, D. A. (1972). Insecticide control of the sugar cane froghopper, *Proceedings of the 14th Congress of the International Society of Sugar Cane Technologists*, pp. 507–15.
19. Little, F. B. (1958). Annual Report Experiment Department, Crop Year 1957–58, Ste. Madeleine Sugar Co., Trinidad (private circulation).
20. Blackburn, F. M. B. (1954). Further developments in froghopper control, *Proceedings British West Indies Sugar Technologists*, pp. 137–40.
21. Evans, D. E. (1973). Resistance to carbamate insecticides in *Aeneolamia varia saccharina* (Distant), *Tropical Agriculture (Trinidad)*, **50**, 153–63.
22. Knipling, E. F. (1959). Sterile male method of population control, *Science*, **130**, 902–4.
23. Campion, D. G. (1972). Insect chemosterilants: a review, *Bulletin of Entomological Research*, **61**, 577–635.
24. Fewkes, D. W. & Buxo, D. A. (1965). Report of the Tate & Lyle Central Agricultural Research Station, Trinidad, 1964, pp. 12–120.

Discussion Report

J. PHILLIPSON

Animal Ecology Research Group, Zoology Department, University of Oxford, England

Systems analysis is the formulation and manipulation of a set of mathematical relationships which represent the ways in which the physical and/or biological components of the selected system are likely to interact. However, the approximation of a set of mathematical relationships (= mathematical model) to reality is largely dependent on two factors: (a) clear recognition of the primary objective underlying the study, and (b) subjective considerations used in constructing the initial model. In the case of present-day ecological/agricultural systems models it is generally recognised that the models are imperfect and abstract representations of the real world.

Implicit in the present paper is the view that a systems approach, despite its imperfections, aids decision-making in a more rational manner than empirical/subjective-based methods of analysis. With regard to pest control it is suggested that measures not based on a systems approach are frequently irrational. Nevertheless, for this contributor, the interest of this paper lies in the interdependence of the two 'disparate' approaches.

The primary objective of this study was the control of the sugar cane froghopper. Formally, the objective was summarised as maximisation of the expression $P[Y_0 - dA(S)] - C(S)$; this can be expressed in words as:

$$\text{Price of sugar} \left(\begin{array}{c} \text{Yield of sugar in} \\ \text{absence of pest} \end{array} - \begin{array}{c} \text{Damage coefficient} \\ \times \text{ Number of adult days} \\ \text{under strategy } S \end{array} \right) - \begin{array}{c} \text{Cost of} \\ \text{strategy} \end{array}$$

Clearly, the objective is an outcome of economic considerations; however, maximisation of the expression requires quantification of the ecological characteristics of the pest when subjected to particular control strategies. Insofar as the present model has been developed, the control strategies relate to the application of insecticides. One immediately recognises that subjective considerations have imposed limits on the system being studied.

The cost-effectiveness of insecticide application is explored without reference to other possible control measures. This procedure, although mathematically valid, does not necessarily lead to maximisation of the general expression where strategy S remains undefined. It is feasible that maximisation might be achieved by reference to control measures other than insecticide application.

The authors acknowledged their model as representing a sub-system of the real world of the sugar cane field and stressed the point that it was presented as an illustration of a systems approach to a pest control problem. The model explores possible interactions between pest and pesticide but does not, as yet, take into consideration certain biological factors such as the role of density dependence in regulating pest numbers. Other factors which should be incorporated are interactions with other sub-systems, for example moth borers and syrphid predators. Spatial heterogeneity as shown by the data presented in Table 1 is clearly of importance; without taking its occurrence into account it could be argued that insecticide treatment increases froghopper density. It should be recognised also that the preliminary findings about control measures arising from this ongoing work are based in part on model validation using data from a single field (Orange Field 3) over one season. The lack of precise information about spatial heterogeneity (pest aggregations) and sampling accuracy could mean that the validation simply reflects the 'noise' of the field situation. Note also that model validation was carried out against a background of high management standards; it would be of interest to learn what contribution the present exercise could make to less well-managed situations.

The present model is an imperfect abstraction of the real world but the authors themselves recognise this. Clearly, a systems approach is valuable in that it aids reasoning by focusing thought processes and channelling the information acquired into a general relationship. It allows evaluation of the model in terms of realism, precision and generality; but only if and when sufficient field data are available to test the predictions of dynamic change over time. Model manipulation frequently suggests the type of additional field data required but one must not lose sight of the possibility that a modeller might concentrate on those system variables, transfer and forcing functions, and parameters deemed by him to be relevant to the objective. Further, he might seek, to the detriment of the reality of the model, simplicity and mathematical perfection. Mathematical manipulation alone is no substitute for the real world. It is at this point that the interdependence of the systems analyst and empiricist schools should be acknowledged. The empiricist analyses information on the basis of subjective considerations unrelated to mathematical formulations and, as demonstrated by the points made in this discussion, intuitively formulates views which enable him to challenge the validity of certain model predictions. The contribution made by the systems analyst is such that the empiricist is obliged to think more rationally about the problem at hand. It is a pleasure to record therefore that Dr. Conway and his co-authors constitute an interdisciplinary team, the activities of which have made us all think very carefully about the pest status of the sugar cane froghopper.

In conclusion one can only hope that decision-makers will take note that however convenient a mathematical prediction may be, prediction may only be considered to reflect the real world when the most rigorous field validation procedures have been followed.

7

The Practical Application of Bio-economic Models

P. J. CHARLTON

Department of Agriculture and Horticulture, University of Reading, England

and P. R. STREET

Department of Agriculture and Horticulture, University of Nottingham, England

7.1. INTRODUCTION

Computer simulation methods have now been used for some years to study the time-dependent behaviour of agricultural systems.[1] The models that have been constructed have varied considerably in their structure and in the ways in which they have been applied. As with most newly developed techniques, claims have been made for the potential benefits which have subsequently proved to be considerably exaggerated. The use of these modelling methods has in fact received a considerable amount of criticism and the whole worthwhileness of modelling studies of any kind has been called into question.

Much of the criticism of the approach has arisen because of a failure of many studies to produce any clear practical benefits, either at the research or extension level. This paper will attempt to provide some guidelines on ways in which future studies can be of greater relevance to actual research and extension problems. The usefulness of the 'systems approach' and its associated methodologies will be discussed and the problems of building effective systems models will be illustrated by reference to two models, constructed by the authors, which have been used for research and extension purposes.

The underlying theme of the paper will be to emphasise that the objective of all systems modelling exercises should be their practicality and usefulness to research workers or farmers. Model building is an expensive and time-consuming occupation. The construction of complex models for their own sake, without clearly defined and achievable objectives, can no longer be justified. A strong argument will be made for a greater simplicity of approach and for the wider use, in the future, of relatively simple 'calculator' models of the type that will be described in this paper.

7.2. THE SYSTEMS APPROACH

In order to be able to discuss the ways in which a 'systems approach' can be of practical benefit to agriculture, it is necessary to consider the reasons for the current interest in this approach. The recognition of the need to consider whole systems of agricultural production is, after all, not new. Most farmers have always adopted what amounts to a systems approach in running their businesses. They have clearly recognised that no part of the business or any one aspect of husbandry can receive attention to the complete exclusion of all others.

It is principally only by non-practitioners, that is, by academics and advisers, that the need to consider complete systems, rather than just their component parts, has often been neglected.

Traditionally, agricultural science was developed by researchers with a good understanding of all the components of agriculture and its related disciplines. More recently, however, there has been a very significant increase in specialisation which has inevitably channelled research into progressively more restricted fields. This greater specialisation has brought an accompanying decrease in the ease of communication both between and within the traditional disciplines.[2]

This reduction in communication between specialists in different disciplines has resulted in significant problems. Research carried out in isolation in this way has often been found to be irrelevant or meaningless when put into a practical context.

Similarly, in advisory and extension work, advice has often been given referring only to an isolated part of the system without regard for its significance within the whole farm situation.

It was the recognition of the need to try once again to take an overview of a situation, simultaneously considering a wide range of factors, which produced the so-called 'systems approach'. This has been widely heralded by its protagonists as a totally original concept capable of giving insight into the behaviour of highly complex situations. In fact, it is important to recognise that it is really only an acknowledgement of the inadequacy, in a practical situation, of much of the highly specialised academic research which ignores important reactions between system components. It is simply a reversion by academics to the more general or 'holistic' approach that farmers and other practitioners have themselves always adopted. This is a very pragmatic view of the concept of the 'systems approach'. However, we believe it to be a very necessary one if the study and analysis of whole systems is to be restored to a realistic perspective.

It is unfortunate that this rather simple, naïve concept of needing to look at whole systems has been developed by theoreticians far beyond the level at which the approach can still be of any practical relevance. The development of cybernetics is one aspect of this rather academic interest in the theory of whole systems. Cybernetics is founded upon the mathematical techniques developed for designing effective electrical and mechanical control systems. These techniques certainly apply to all systems which obey relatively well-defined physical laws. It does not follow, however, that the same detailed techniques can be meaningfully applied to other systems, particularly economic and social ones, which do not conform to any such rigid behavioural rules. Close analogies can certainly be drawn between the *gross* behaviour of such systems and their electrical or mechanical equivalents. The gross behaviour, however, is usually relatively well known, even if not fully understood, by anyone with practical experience of the system.

The major contributions that systems theory, in the form of cybernetics, industrial dynamics or system dynamics philosophies, can make is in an educational role—that is, to communicate, to those unfamiliar with a particular system, an understanding of its gross behaviour under a variety of circumstances. A good example of this type of study is that of Meadows' 'commodity production cycles'.[3] The system dynamics methodology used in this exercise enables one to construct a conceptually realistic model of the feedback mechanisms which can produce cycles in the production of commodities.

Such relatively simple models can be extremely useful as educational tools in communicating general principles. To anyone familiar at the practical level, however, with all the factors which can distort the real behaviour of such systems, such modelling efforts are extremely simplistic and naïve. They cannot be used in any way as predictive models capable of giving guidance on specific issues, or of producing information directly useable by decision-makers.

It has been argued by many researchers that all efforts in the systems modelling area are only concerned with general responses and obtaining a 'feel' for a system's behaviour. This is a valid viewpoint provided the purpose is educational in communicating this 'feel' from one group to another. However, it can easily be used as an excuse for looking at, and modelling, everything without actually achieving anything.

Systems research, in common with other forms of research, must have clearly defined objectives. It must normally be possible to anticipate some direct use for the studies carried out and the models that are created. Some greater insight, not obtainable in any other way, or some practical applications, must be visible before embarking on them.

Systems research, because it attempts to draw together many components, is necessarily time-consuming and expensive. Considerable resources have already been applied to general research in this area, in many different countries, with remarkably limited effect in terms of practical and useable results. If the subject is not to lose all credibility, it is essential that future efforts have far more pragmatic and specific objectives. The output of academic-looking papers might well decrease, but this would be adequately compensated for by the increased relevance of the research results.

The next section will consider the adjustments that have to be made to the theoretical ideals of the systems approach to produce models of practical use.

7.3. THE DEVELOPMENT OF PRACTICAL SYSTEMS MODELS

It is deceptively easy to draw up a flow diagram of a real-life system which appears both conceptually correct and visually complex. The particular flow-charting convention adopted is not in itself of any great importance. Forrester-type notation is very convenient for systems involving continuous flows of materials and involving neither decisions nor discontinuities. Other forms of representation, whether conventional logical flow charts or circular or relational diagrams, similarly have their advantages for other types of systems. No one form is suitable for all circumstances. They are only convenient tools which can, if properly used, help in communicating the structure of a system to others who are also familiar with the same set of conventions.

The problems start as soon as one attempts to translate such a diagrammatic model into one which can produce quantitative results of some tactical or strategic use. It must be recognised, in fact, that in many circumstances it will *never* be possible to progress beyond a purely conceptual or diagrammatic model. The system may well be so interrelated, or data so lacking, or subjective or political factors so important, that the components can never be rigidly defined.

To recognise this basic truth is extremely important. There is no advantage to be gained in attempting to define rigidly or to quantify in mathematical terms the essentially unquantifiable, as no confidence could ever be placed in a model developed on such a basis. Conversely, there is no justification for dismissing the exercise of developing the conceptual model as useless, solely because the model can never be quantified.

One of the greatest benefits to be gained from any systems study arises from the discipline of developing some form of initial diagrammatic model. Such an exercise forces one to consider the system objectively and helps to consolidate one's ideas on the nature of the relationships. It may well also be important as a catalyst, helping to focus attention on critical areas. It also helps, as has been mentioned, in providing a basis for discussion and communication with others less familiar with the system. The recognition of one's inability to define a system completely should therefore be regarded as a realistic appraisal of the situation and not as an admission of having wasted time and effort.

Even if it is believed that individual components of a system *can* be adequately quantified, it is easy to underestimate the problems of actually developing a practical systems model.

The general difficulties are of the following kinds:

(i) The difficulty of deciding on an adequate level of complexity for the specific use to which the model is to be put.
(ii) The need to have a truly effective interdisciplinary approach to ensure that individual components are adequately modelled.
(iii) The logical difficulties of incorporating sufficient flexibility into the model's structure so that it can be applied to a range of situations.
(iv) The problem of obtaining data in such a form that it does apply to a range of circumstances.
(v) The technical problems of constructing, testing, validating and 'selling' such a model.
(vi) Finally, the problem of achieving compromise solutions to the above difficulties within a limited time and budget.

The following two case studies based on research and development carried out by the authors will be used to illustrate the considerable difficulties involved in achieving adequate compromise solutions to these problems and in developing models of some practical usefulness. Both the studies will illustrate how it was not possible or desirable to use any highly theoretical concepts or methods. In both cases a very much simpler, more naïve, less academically satisfying, but more practical, approach had to be adopted.

7.4. THE DEVELOPMENT OF A SIMULATION MODEL OF FIRM GROWTH

The objective of this study was to develop a means of studying firm growth which could be used both as a research tool and also as the basis for advisory and extension work. In particular, this work by Charlton[4] set out to study the physical and financial factors affecting the growth of relatively large-scale intensive pig units.

The growth of firms is not a new area of research. Studies have ranged from the purely descriptive work of Penrose[5] to completely quantified multi-period linear programming (LP) studies.[6,7]

It was considered that both these extremes were deficient in terms of being able to satisfy the objectives of this research. The descriptive, qualitative accounts of Penrose convey an understanding of the general factors affecting the growth of firms. They could not be applied, however, to a specific business to give any quantitative indications as to the relative importance of different factors affecting its growth.

The multi-period LP approach in turn suffered from the deficiency of being too rigid. It is a time-consuming and expensive process to adapt such a matrix to the details of a particular farm situation. It is also highly questionable whether the criterion of maximising some form of objective function over a rigidly fixed time span is realistic in a farm situation. Although this approach has some conceptual attractions as a research method, it would certainly be extremely difficult to convince a farmer of its validity.

7.4.1. Theoretical Aspects of Firm Growth

The theoretical aspects of firm growth are concerned partly with the sociology of business development, but mainly with the economic and technical factors which make it possible. Much of the previous work on firm growth had used very sophisticated and complicated methods to achieve relatively naïve conclusions on the relationships between these factors. In particular, most were concerned with the interactions between different levels of profitability, interest rates and debt/equity ratios, or 'gearing' levels.

General conclusions on the effect of these factors on growth could, however, have been far more readily obtained from an extremely simple model of the following form.

A_t, L_t, I_t, P_t and E_t are respectively the total assets, loans, interest payments, profitability (after tax and depreciation) and the equity of a business at end of year t. r_t is the interest rate, g_t the gearing level and k_t the profitability rate expressed as a proportion of the assets available at the beginning of the year.
Then

$$E_t = A_t - L_t$$
$$I_t = r_t L_t$$
$$P_t = k_t A_{t-1}$$
$$L_t = g_t E_t$$
$$A_t = A_{t-1} + L_t - L_{t-1} + P_t - I_t$$

assuming that all profits are reinvested.

These equations imply that

$$E_t = \left(\frac{1 + k_t(1 + g_{t-1})}{1 + r_t g_t}\right) E_{t-1}$$

For growth in value, or equity, of the business it is necessary for the factor

$$F = \left(\frac{1 + k_t(1 + g_{t-1})}{1 + r_t g_t}\right)$$

to be > 1.

A very simple analysis of the conditions under which this relationship is greater than unity enables almost all the general conclusions concerning business growth to be drawn.[4]

7.4.2. Progressing Beyond a Theoretical Model

The difficulty is in progressing beyond this stage of drawing very general conclusions and being able to apply a model to a specific farm situation. Fig. 1 illustrates the range of factors which are important in practice in affecting growth of the individual business. Such a diagram only illustrates the system in general terms and not the way in which it can or should be modelled.

In order to choose a suitable methodology for modelling the growth of a pig enterprise it was necessary to refer again to the original objectives. It was essential that the study could be used as the basis of an advisory tool. It therefore had to be acceptable both to advisers and to farmers.

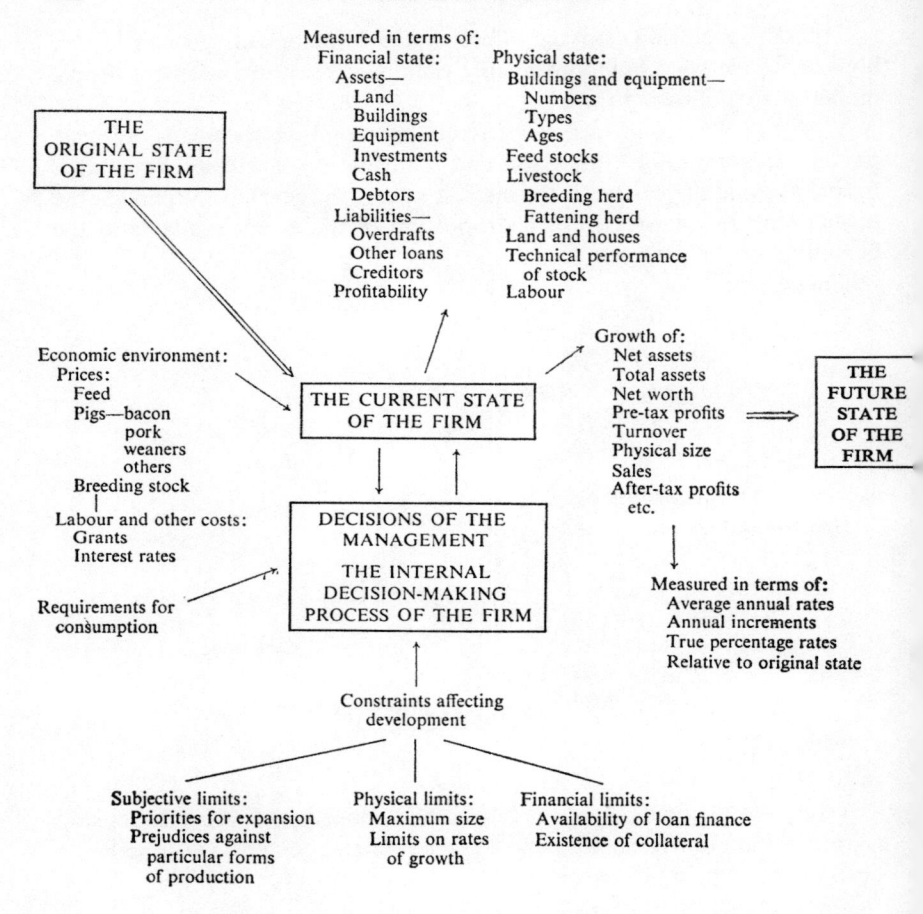

```
                    Measured in terms of:
                    Financial state:      Physical state:
                    Assets—               Buildings and equipment—
                      Land                  Numbers
   THE                Buildings             Types
ORIGINAL STATE        Equipment             Ages
OF THE FIRM           Investments           Feed stocks
                      Cash                  Livestock
                      Debtors                 Breeding herd
                    Liabilities—              Fattening herd
                      Overdrafts          Land and houses
                      Other loans         Technical performance
                      Creditors             of stock
                      Profitability       Labour

                                              Growth of:
Economic environment:                           Net assets
  Prices:                                       Total assets             THE
    Feed                 THE CURRENT STATE      Net worth               FUTURE
    Pigs—bacon              OF THE FIRM         Pre-tax profits  ⟹    STATE
         pork                                   Turnover                OF THE
         weaners                                Physical size            FIRM
         others                                 Sales
    Breeding stock                              After-tax profits
                         DECISIONS OF THE         etc.
  Labour and other costs:   MANAGEMENT
    Grants
    Interest rates         THE INTERNAL
                         DECISION-MAKING       Measured in terms of:
  Requirements for       PROCESS OF THE FIRM     Average annual rates
    consumption                                  Annual increments
                                                 True percentage rates
                                                 Relative to original state

                         Constraints affecting
                             development

        Subjective limits:     Physical limits:    Financial limits:
          Priorities for expansion  Maximum size    Availability of loan finance
          Prejudices against        Limits on rates  Existence of collateral
            particular forms          of growth
            of production
```

FIG. 1. A conceptual model of the principal factors affecting firm growth.

This implied that it should:

(i) Not involve obscure mechanisms which could not readily be explained;
(ii) require only readily available data on the farm situation;
(iii) Contain the minimum amount of built-in data;
(iv) Be capable of producing readily understandable output; and
(v) Be capable of answering the most frequent and important questions likely to be posed by a farmer about the growth of his farm.

This last point is the most important of all and yet perhaps the most frequently neglected in the enthusiasm for the development of a new tool or technique. It is all too easy to produce a model of any system which is not capable of answering the key questions.

It was concluded, after discussion with farmers, that the most common forms of question likely to be raised in this particular context were:

(i) The effect of possible future price changes on the way in which growth could develop.
(ii) The effect of improvements in the technical performance of the stock and its management.
(iii) The effect of the farmer making major changes to his present production system—for example, continuing to fatten weaners rather than selling them.
(iv) The effect, when considering the setting-up of a new unit, of having a highly capital-intensive system as opposed to a more extensive one.
(v) The effect of different financing arrangements and interest levels.
(vi) The levels to which prices and performance indices could fall before there was a danger of being unable to meet any loan commitments.
(vii) The effect on growth of an increased level of consumption by the farmer and his family.
(viii) The extent of the advantages of having conservative growth policies and maintaining adequate cash reserves in periods of fluctuating prices.

A simple mathematical model of the type outlined above would be quite incapable of giving detailed answers to these questions. Similarly, a very much more complex model based on an inappropriate technique would have been equally unsuitable. For example, the industrial dynamics methodology was immediately rejected, since the detailed structure of a business was too subject to discontinuities and required the representation of interrelated decision processes. There were no continuous flow processes which could be identified for the development of an industrial dynamics structure based on differential equations.[8] Some form of modelling approach midway between the extremes of the highly simple, generalised model and the sophisticated mathematical approach was therefore required. It was recognised that all the questions listed above would normally be answered by the farmer or adviser through the development of forward budgets. It therefore appeared that the most logical approach for modelling such a system was the automatic generation of these budgets.

Conceptually and mathematically such an approach had several attractions. First, it conformed with what the farmer and his bankers or advisers already did by hand. The approach was therefore likely to be readily accepted and understood. Secondly, the mechanics and logic of the approach were relatively well defined and so model development would not pose major problems of data availability. Thirdly, the simulation of the implications of certain initial assumptions, as made by the farmer or adviser, was sufficiently flexible to allow a wide variety of factors to be considered simultaneously.

The basic logical structure of the model that was constructed is shown in Figs. 2 and 3. This is an essentially simple structure which belies the difficulty of achieving a model which could actually be used. The difficulty

A. The setting-up of the model

1. INPUT PHASE
↓
2. PRELIMINARY CALCULATION PHASE
↓

B. Cycling to simulate the behaviour of the system over time

3. CYCLING PHASE

Repeated for each year of simulated period

↓

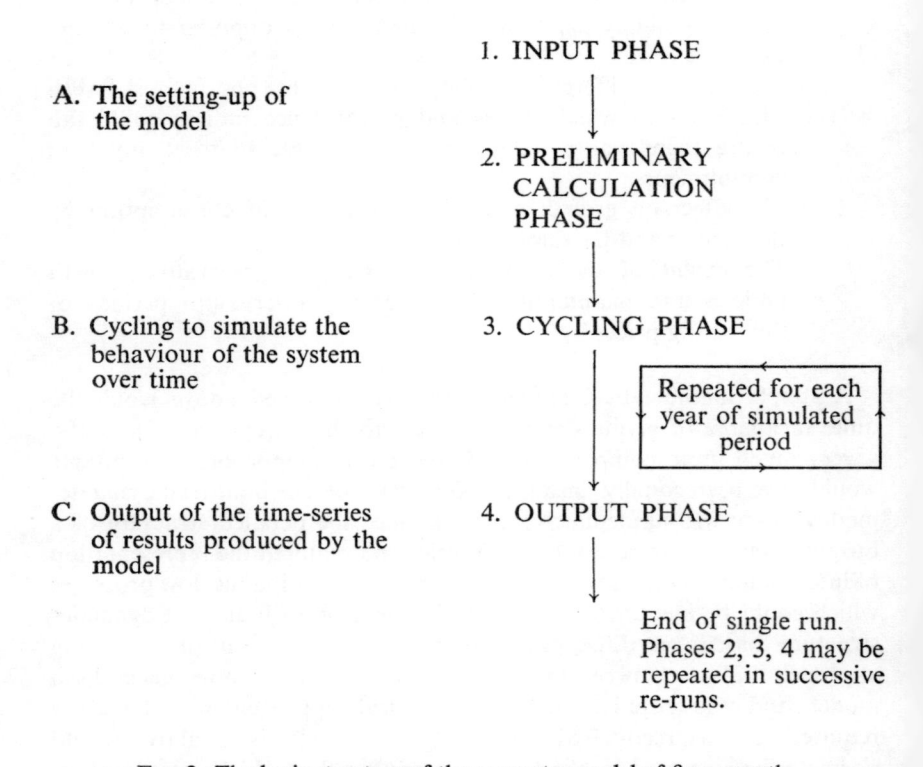

C. Output of the time-series of results produced by the model

4. OUTPUT PHASE
↓

End of single run. Phases 2, 3, 4 may be repeated in successive re-runs.

Fig. 2. The basic structure of the computer model of firm growth.

START OF CYCLING PHASE

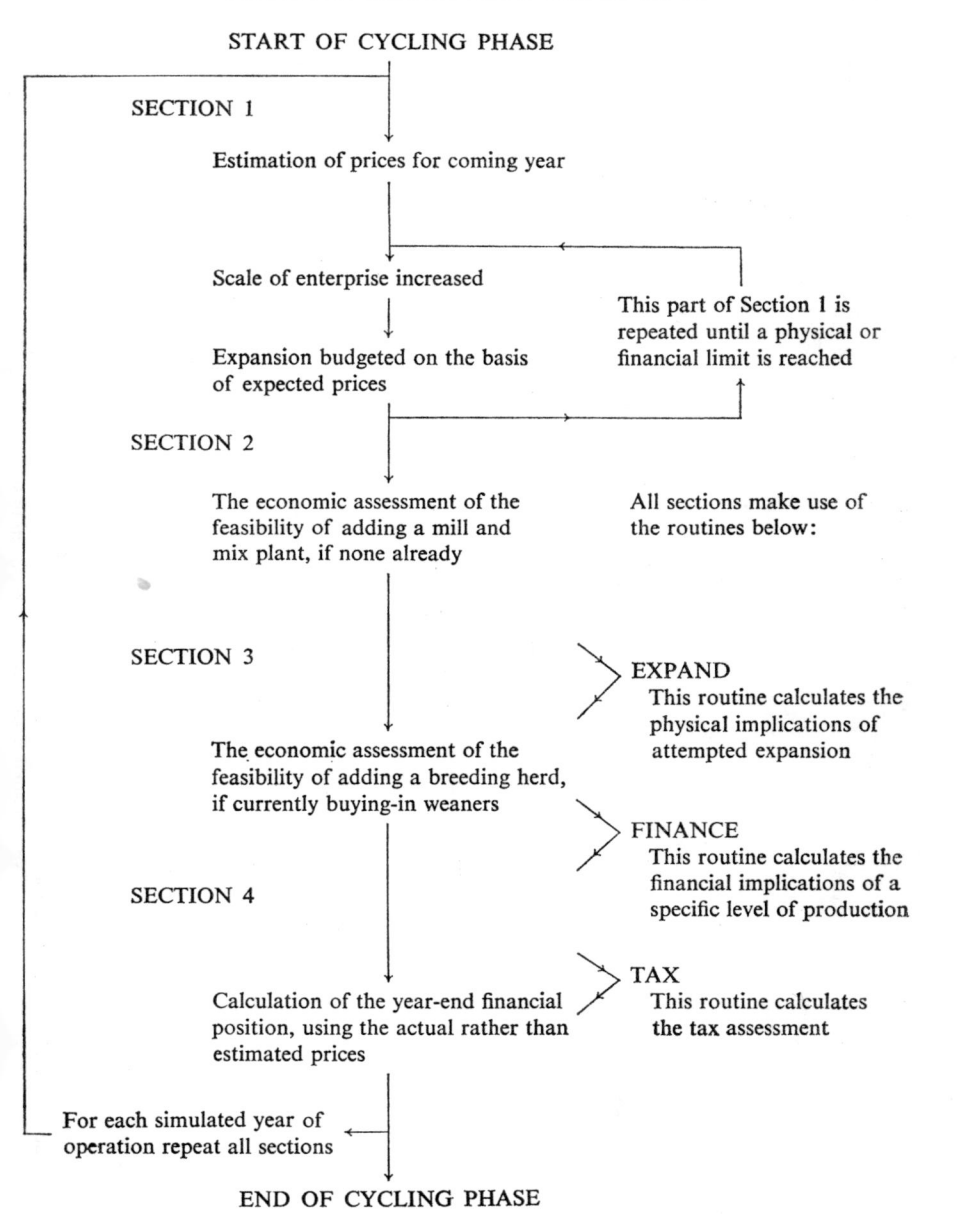

SECTION 1

Estimation of prices for coming year

Scale of enterprise increased

Expansion budgeted on the basis of expected prices

This part of Section 1 is repeated until a physical or financial limit is reached

SECTION 2

The economic assessment of the feasibility of adding a mill and mix plant, if none already

All sections make use of the routines below:

SECTION 3

The economic assessment of the feasibility of adding a breeding herd, if currently buying-in weaners

EXPAND
This routine calculates the physical implications of attempted expansion

FINANCE
This routine calculates the financial implications of a specific level of production

SECTION 4

Calculation of the year-end financial position, using the actual rather than estimated prices

TAX
This routine calculates the tax assessment

For each simulated year of operation repeat all sections

END OF CYCLING PHASE

Fig. 3. The cycling phase of the computer model of firm growth.

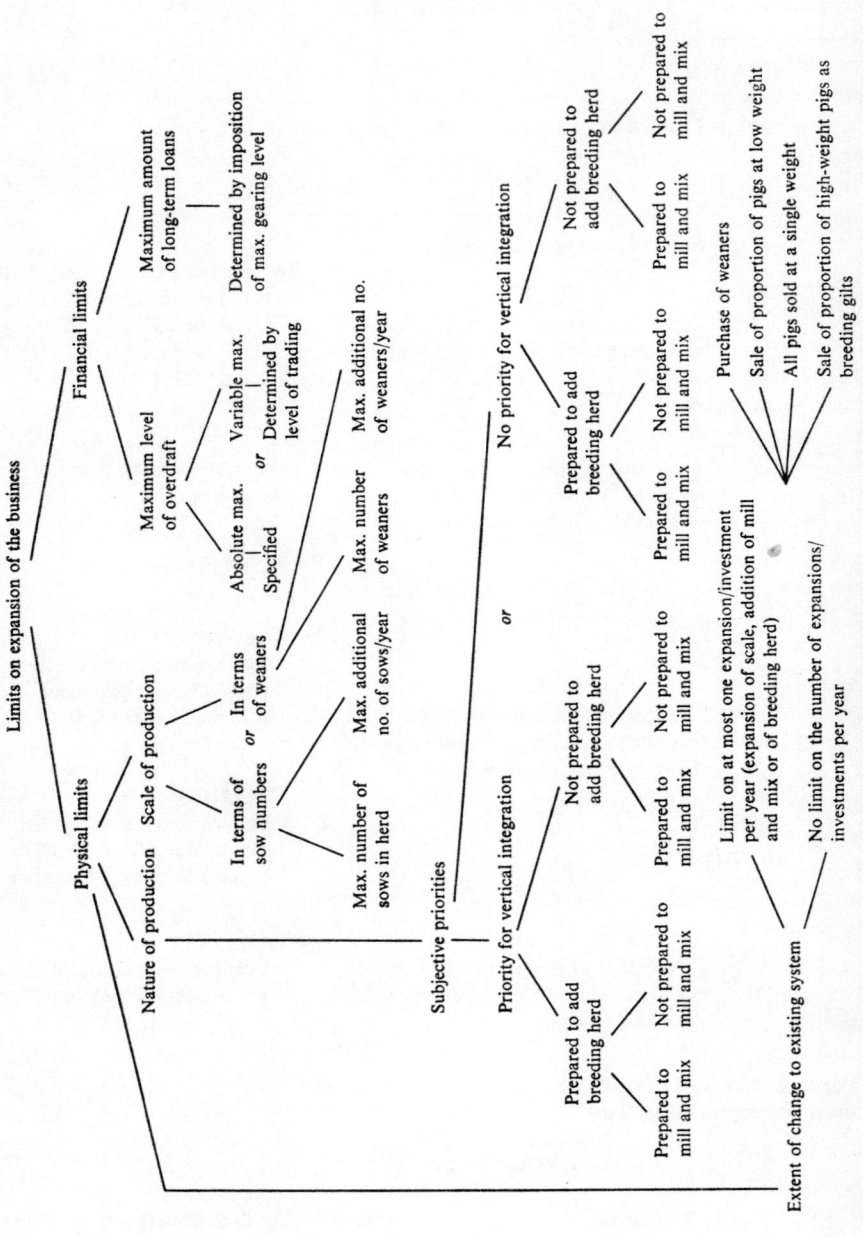

Fig. 4. The limits which can be imposed on expansion of the firm in the model of firm growth and the possibilities for specifying different forms of production system.

was essentially one of flexibility. The first version of the model contained a relatively high level of detail and a working version was produced in a matter of a few months' work. However, as soon as this model was tested in the field it became obvious that a very much higher level of flexibility and generality was required. An indication of the range of options that were found to be needed in just one sector of the model specification is shown in Fig. 4. The logical problems of incorporating this level of flexibility into any type of model are considerable.

A further aspect of the model's initial limitations was that of presentation. In order to gain acceptance by users it was found to be essential to provide output which conformed as closely as possible to standard accounting and budgeting conventions. The output from the model is therefore considerably more complex (a total of about 70 variables for each year of simulated development) than was originally thought necessary for deriving logical conclusions. In addition to development budgets of the physical expansion of the unit, the model produces complete sets of accounts, each table containing the results for the total run of the model. In this way it is easy to identify sudden changes in the balance sheet, profit and loss account, cash flow and other statements and to trace-back the reasons for the change. It also makes it easy for the users to check for themselves the intermediate calculations and transfers between accounts.

The initial state of the business is specified in considerable detail, including the age, number and type of buildings available, loans outstanding and the current cash position. Time-series of future prices, expected levels of technical performance and any physical, financial or subjective constraints on possible future development are also specified. The model then calculates the implications of trying to grow under these assumptions. The calculations involved are relatively straightforward, but an indication of the computational work involved in such a generalised model is given by the flow chart of a small part of the feeding routine that is shown in Fig. 5.

It has been found that although the accounts and tabular results are very useful in convincing the farmer of the validity of the calculation processes, certain major trends can be better appreciated by the use of graphs.[9] It is therefore possible to select key variables, such as equity, after-tax profit, loans and total assets, to be printed out in graphical form.

7.4.3. Research and Extension Use of the Model

The use of the model for research into growth processes has operated at two levels. First, it has been useful in a negative way in confirming the

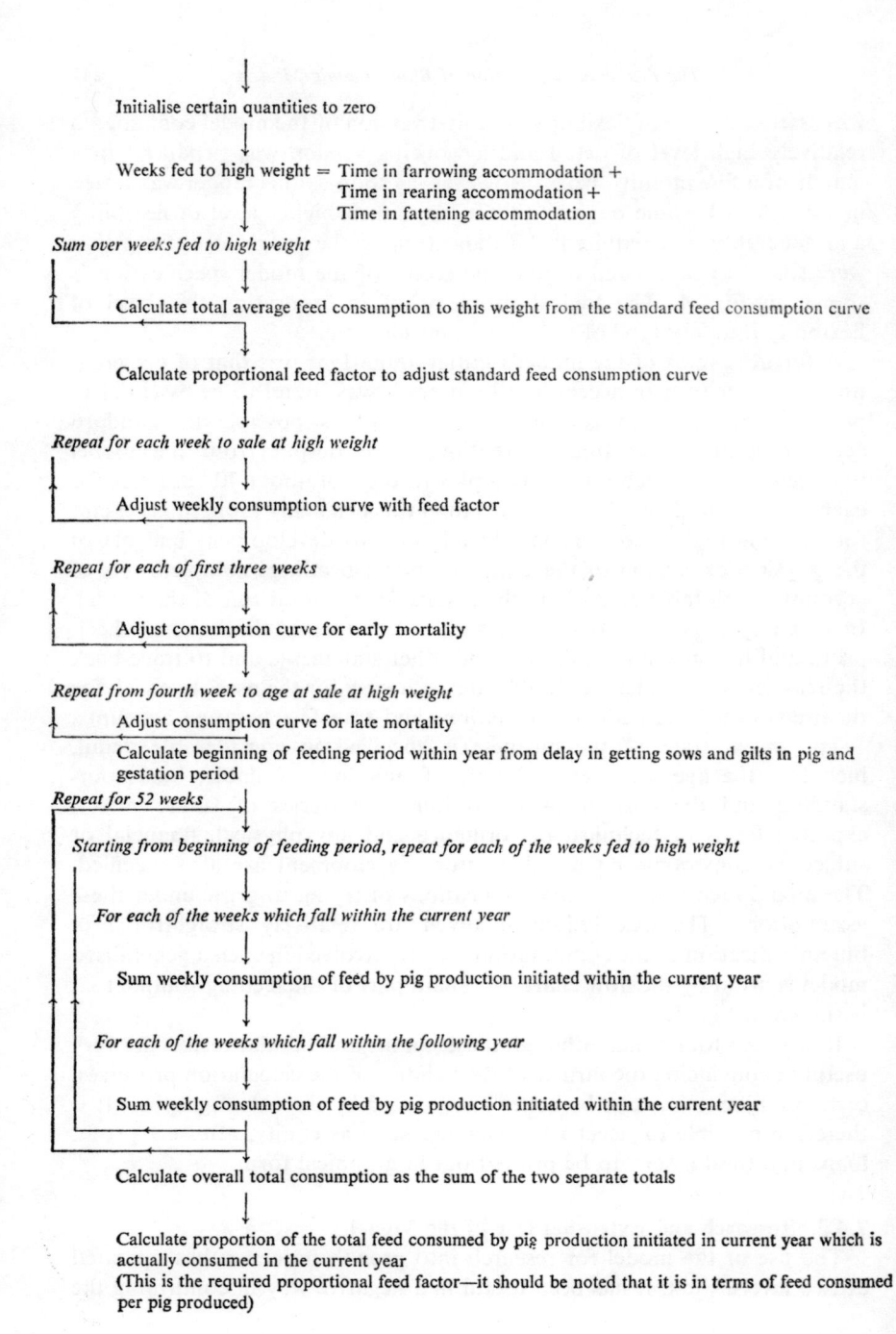

```
                    │
                    ▼
        Initialise certain quantities to zero
                    │
                    ▼
        Weeks fed to high weight = Time in farrowing accommodation +
                                   Time in rearing accommodation +
                                   Time in fattening accommodation
Sum over weeks fed to high weight
┌───────────────────┐
│                   ▼
│       Calculate total average feed consumption to this weight from the standard feed consumption curve
│                   │
└───────────────────┤
                    ▼
        Calculate proportional feed factor to adjust standard feed consumption curve
                    │
                    ▼
Repeat for each week to sale at high weight
┌───────────────────┐
│                   ▼
│       Adjust weekly consumption curve with feed factor
│                   │
└───────────────────┤
                    ▼
Repeat for each of first three weeks
┌───────────────────┐
│                   ▼
│       Adjust consumption curve for early mortality
│                   │
└───────────────────┤
                    ▼
Repeat from fourth week to age at sale at high weight
┌───────────────────┐
│                   ▼
│       Adjust consumption curve for late mortality
└───────────────────┤
                    ▼
        Calculate beginning of feeding period within year from delay in getting sows and gilts in pig and
        gestation period
Repeat for 52 weeks
┌───────────────────┐
│                   ▼
│       Starting from beginning of feeding period, repeat for each of the weeks fed to high weight
│                   │
│                   ▼
│           For each of the weeks which fall within the current year
│                   │
│                   ▼
│           Sum weekly consumption of feed by pig production initiated within the current year
│                   │
│                   ▼
│           For each of the weeks which fall within the following year
│                   │
│                   ▼
│           Sum weekly consumption of feed by pig production initiated within the current year
│                   │
└───────────────────┤
                    ▼
        Calculate overall total consumption as the sum of the two separate totals
                    │
                    ▼
        Calculate proportion of the total feed consumed by pig production initiated in current year which is
        actually consumed in the current year
        (This is the required proportional feed factor—it should be noted that it is in terms of feed consumed
        per pig produced)
```

original overall thesis that the growth process is peculiar to the particular firm and to the economic environment in which it operates.[4] The complex simulation model was not able to add anything to the 'theory' of growth over and above the general conclusions that had been obtained from the very simple mathematical model. It clearly demonstrated that essentially the same patterns of growth could arise through totally different combinations of the same highly detailed factors. No one system of production could be said to be superior, all had their advantages and limitations, according to circumstances.

Secondly, having established the importance of always considering the detailed firm situation, it was possible to use the model at a different level. Research at this second level was carried out into the relative importance of key factors affecting fundamentally different types of pig production systems, in a variety of situations. The considerable flexibility of the model and the absence of any internal data or fixed relationships provided an ideal experimental tool for answering a variety of specific problems.

The advisory or extension role of the model is to demonstrate to a farmer, through the development of a range of forward budgets under different assumptions, the *extent* of the effect of these factors on the growth of his particular business.

Such a model is essentially no more than a glorified calculator. The data which it requires is considerable, but it is all relatively simple and readily available for any specific situation. The model can be used to re-assess rapidly the relative profitability of different systems of production as feed and pig prices change, and can provide up-to-date centralised information of this kind to pig advisers and farm management extension workers. The research version of this model was completed in 1971 and a FORTRAN version handed over to the Meat and Livestock Commision at that time. A COBOL version was subsequently developed by them and has been available to farmers on a fee-paying, trial basis for the past 18 months.

The approach does not attempt to predict what will happen in the future. It does, however, through the automation of the traditional forward

---

FIG. 5. Example of the calculation of a single feed factor. The example depicted is that for determining the amount of feed fed in the current year to high-weight pigs which were produced from the unit's breeding herd and as a result of production initiated in the current year. (Similar calculations are required for those factors concerning low-weight pigs, feed consumed in the following year and for feed consumed by pigs bought in as weaners.)

budgeting procedure, offer the opportunity to explore a wide range of highly relevant problems, both in a research and in an advisory context. The original objectives of the study were specific and relatively limited. It is believed, however, that in creating this model a reasonable compromise has been achieved in creating a tool of some direct practical use to agriculture.

7.5. THE DAIRY FORECASTING MODEL

7.5.1. Introduction

The following discussion relates to the dairy model of Street[10,11] which is used as an example of some of the practical problems likely to be encountered when producing models for extension purposes. Many of the problems encountered in the structuring of bioeconomic models both with respect to abstraction of the real world and the representation of the biological response processes are common to both the pig model previously discussed and to this dairy model. This section reviews, in greater detail, the practical problems found in the derivation of a model which is usable as an extension tool. It illustrates the difference between this type of model and dairy models of the type described by Crabtree[12] which are oriented primarily towards research use.

7.5.2. Background and Objectives of the Dairy Model

The dairy model is essentially problem-oriented. As such it must be considered in terms of the problems which it was designed to solve. Historically, milk production systems have evolved in relation to changing technological and economic circumstances. In the latter case in the UK the price environment for the inputs and outputs of the milk production process has been fairly stable and predictable. In practice there exists a multitude of dairy systems[13] which can be operated. They show wide physical differences, yet can yield apparently similar economic performances. With a period of rapid change and price instability facing agriculture it is important to review existing systems and consider their economic viability under these likely new conditions. The new conditions may well dictate modifications to current production policies if satisfactory profit margins are to be achieved. There may no longer be time for slow evaluation in this respect, but a need for rapid change to exploit short-run market conditions. It is in this area that models can help the decision-maker to assess rapidly the various options available in terms of projected economic situations. The policies most likely to affect the milk producer, and over which he has considerable discretionary control, concern feeding, stocking rate, grass management and conservation, breed and age structure of the herd, culling and replacement and the date and spread of calving. In the main, the manipulation of these factors is a tactical policy, but may have repercussions on how the various elements of the whole production system should be integrated and will involve costs as well as potential gains. Figure 6 shows the way in which factors of production may be linked in a conceptual model to define the interactions within a dairy system and indicates the dynamic interdependency of the various factors which affect output and the complexity involved in changing a system even when the fixed costs are ignored.

Clearly, the best combination of policies for change in the dairy unit will depend on the particular farm size, its capital and organisational structure and its economic and climatic environment. However, for any farm where change is envisaged, a model is required which can lead the decision-maker to conclusions about problems concerning the interaction of the parameters shown in Fig. 6 and the way they impinge on profit. Without this information it is impossible to estimate return on capital for the various fixed-cost configurations which might be considered. The model was therefore designed to help answer the following questions:

(i) How do the factors illustrated in Fig. 6 interact to affect system performance?
(ii) How does the financial performance of a particular system design behave in respect to different price environments for inputs and products?
(iii) What pattern of cash flows does a particular system generate?
(iv) What influence does managerial efficiency in terms of the input/output relationship have on physical performance indices and the financial viability of given system designs?
(v) What production targets must be met if a particular strategy is to be successful?

These are essentially the problems which the researcher would also wish to ask, so a model which could answer these questions would be useful in both extension and research. Reference to Fig. 6 shows that to evaluate answers to questions of the type posed above, complex budgets are required. These budgets involve intricate phasing of the physical and financial aspects of those inputs and outputs which show a high degree of

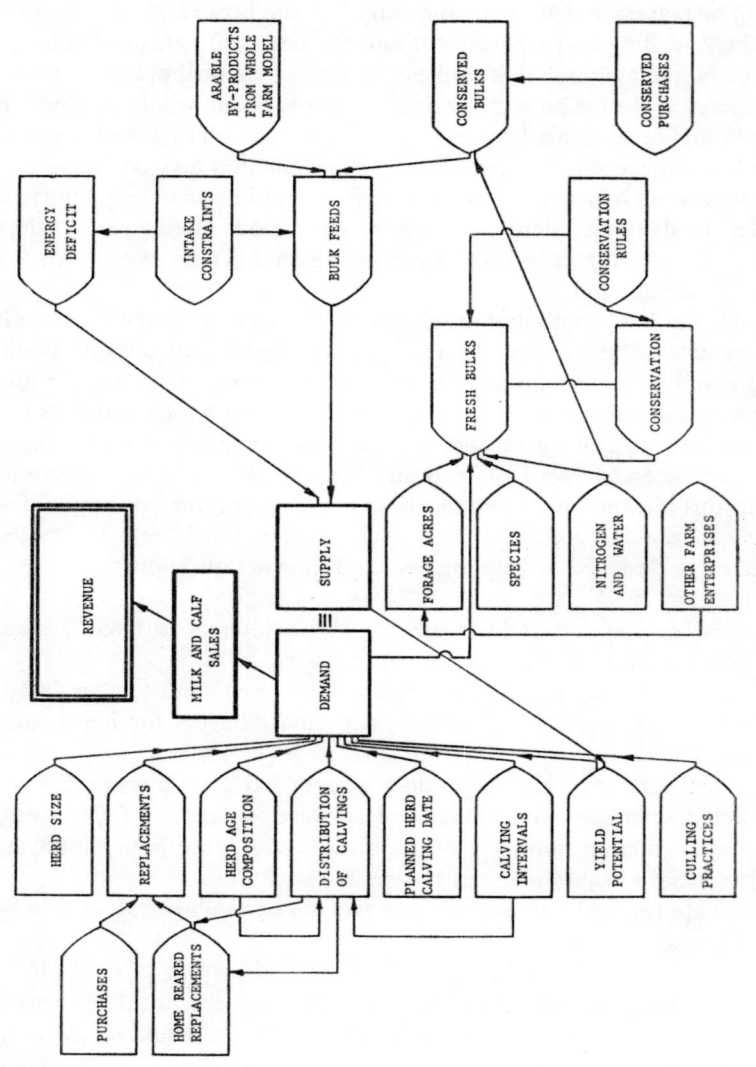

Fig. 6. A conceptual model of the factors influencing feed supplies and the demand made on them by a dairy herd.

seasonal variability, as well as the logical structuring of the many interrelationships and interdependencies of the components of the system. The dynamic nature of this problem dictates that it can only be adequately handled by the simulation approach.

7.5.3. The Structure and Function of the Model

7.5.3.1. General

The form of model shown in Fig. 6 is an example of a 'skeleton' simulation model[16] for milk production. This was designed to contain the basic parameters* of a dairy system and the basic logic of the interdependencies shown in Fig. 6. The model only becomes functional when combined with actual farm data. This enables the model to be applied to a wide range of situations, to include qualitative modification to data defining the included response processes and, importantly, to remain useful over a long period of time. In this form the model is no more than a calculator allowing the user discretionary control over the input parameters and their respective levels. These parameters include such factors as:

   (i) the date and spread of calvings;
  (ii) forage acres;
 (iii) herd size;
 (iv) forage conservation and grassland management;
  (v) feeding policy;
 (vi) breed and age structure of herd;
(vii) culling and replacement practices; and
(viii) availability of feeds from arable enterprises.

The model is therefore a mechanised budgeting method which produces financial and physical indices in response to sets of assumptions input by the user. The usefulness of the model depends on the quality of these inputs. Given a set of data assumptions this simulation can be considered essentially as a means of specifying and manipulating the relationship between feed supplies and the demand made on these supplies by the dairy herd and is basically structured in the manner of the conceptual model of Fig. 6. The herd manager can effect some control on both the supply of, and demand for, forage and other feeds. When necessary, and desirable, supplies can be augmented by the purchase of feeds or, alternatively,

---

* Including constants defining the general shape of lactation curves, percentage 365-day yield increments for delayed conception, etc.

yields can be permitted to fall to result in a balance between demand and supply. The manner in which demand is equated with supply has a major impact on the total organisation of the enterprise, influencing labour requirements, and fixed and variable costs, as well as output and revenue. The input can be formulated to investigate these various possibilities.

The model is structured to permit both the demand and supply sides of the relationships to be modified as required for each simulation experiment. By repeatedly specifying new sets of data assumptions[11] alternative situations can be compared and production methods chosen with associated forecasted performance targets.

7.5.3.2. Structure

The basic structure of the model is described in the simplified flow diagram of Fig. 7. This diagram illustrates the basic simulation module and the peripheral programme that is necessary when a model is introduced into an extension environment.

The basic simulation module (outlined in Fig. 6) is contained within the time loop of Fig. 7 and this section has four major elements:

(i) A sub-routine (HERDCD) which for each period of time updates the current state of the herd by the implementation of culling, calving, the introduction of replacements (new additions) and the adjustment of lactation number and lactation stage for the cows in the herd.

(ii) A sub-routine (BULKSM) which simulates the seasonal supply of all the various categories of feed.

(iii) A sub-routine (LACTAT) which simulates lactation and feeding for each individual cow and for the assumptions laid down in the data input.

(iv) A sub-routine (ANALYS) which calculates physical and financial performance indices and sorts them for later use by the report writer.

Essentially (i), (ii) and (iii) can be discussed in terms of the supply/demand relationship of Fig. 6. These relationships and their formation are described elsewhere.[10] The subsequent discussion provides a brief review of the included relationships.

(a) *Demand or nutrients.* The demand for feeds is simulated in sub-routine LACTAT. The basic building block for determining the demand

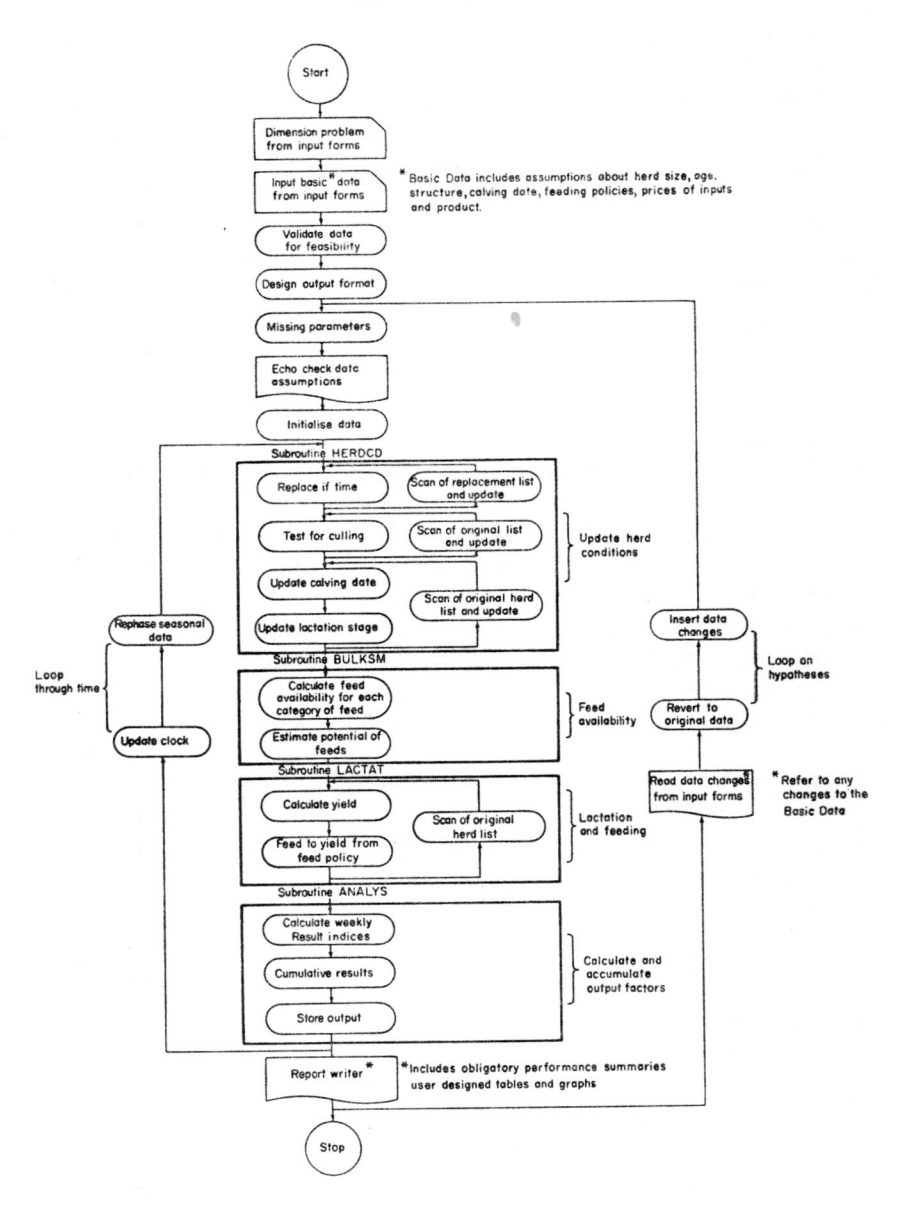

FIG. 7. A simplified flow diagram of the dairy forecasting model.

for nutrients is the lactation curve[15] adjusted for age, calving date, pregnancy, etc. The aggregate nutrient demand for the herd at any time is defined by the age and calving distribution of the herd. The resultant of the distribution of heifer replacement calvings, the subsequent calving intervals and the criteria used for culling and replacement determine the herd calving distribution. Age, calving date and the pattern of feeding in turn affect the actual shape of the individual lactation curve. Hence, the calving distribution is important and determines both the seasonal distribution of aggregate milk production and the revenue derived from the sale of the milk. To enable the simulation of lactation for each cow and for each time period, sub-routine HERDCD is used to update the stage of the herd with respect to cow age, calving date, lactation stage, etc.

(b) *Supply of Nutrients.* Nutrients can be supplied from a number of sources which can be categorised into bulk feeds and concentrates (Fig. 6), either of which may be farm-produced or purchased. When grass is the primary method of supplying bulk, it may be considered as being consumed either as the fresh product or as conserved material. For each method, nutrient losses between production and consumption occur. Also the frequency and severity with which a pasture is defoliated influences both the quantity and quality of grass production. These factors, plus the provision made for conservation, can affect the seasonal pattern of forage availability and its quality which in turn influences the level to which bulk feeds can be incorporated in the ration at any time. Assumptions about feeding include types of feed, feeding potential and quantities available and their cost per unit are input as data. Strategies can also be included which allow the group feeding of cows. Feed supplies in any time period are simulated by sub-routine BULKSM.

7.5.3.3. Operation of the Simulation Section of the Milk Production Model
For each simulation run, a complete set of data assumptions is required. This is input via data input forms[11] and is used to supplement the data built into the model. Where insufficient data are supplied, missing parameters are automatically entered by the model. The time period for calculations in the model is one week. For each week of the simulation run appropriate calculations are carried out and after completion of the calculations in each week the time is advanced and calculations repeated for the following weeks until a run of the desired length is completed (see Fig. 7). For each time period the nutrient demand for each cow is calculated based on its age, calving date and conception period length

and by reference to the appropriate lactation curve. The nutrient demand for all classes of feed is aggregated for the week and compared with the supplies of each class available. Milk sales for each week are computed and the ration cost calculated from the seasonal costs of the components of the ration. Indices are then computed on a weekly basis and accumulated.

After the calculations for each week are completed, the status of the system, as reflected by various rates and levels of the parameters of the system, is updated by adjusting livestock and feed inventories, Culling and replacement policies are also implemented and the lactation stages of existing cows incremented. Thus, dairy systems are generated with their relevant seasonal resource requirements and revenues.

7.5.3.4. Testing Alternative System Designs

By modifying the data input* to each simulation run, a variety of management strategies can be investigated rapidly and inexpensively by the input of the appropriate data which is basically of three types:

(i) 'Skeleton data'—stored as compiled data in the program. Whilst this data can in certain instances be changed using the input forms available with the program e.g. the (constants describing the shape of the lactation curves) the program always reverts to this data for each new run since it is upon this data that it relies to infill missing parameters.

(ii) 'Basic data'—a data specification which forms the 'base plan' for a series of hypotheses about different system configurations and price environments.

(iii) 'Data change information'—sets of data referring to changes to the basic data assumptions ((ii) above) and each set causing a re-run of the model.

For each of the plans a sequence of the physical and financial performance indices is generated for use as targets for monitoring purposes. The robustness of a particular system design to a changing economic environment or reduced response efficiency can also be tested.

7.5.3.5. Other Elements of the Model

Whilst the simulation aspect of the model is the most important element, there are other considerations which should be viewed with respect to

---

* Data input in this respect may define the economic environment, the system design or the efficiency of the included biological response processes.

Fig. 7. Without these, it would be impossible to use such a model for extension purposes. These other elements of the model are discussed under the following four headings.

(*a*) *Data specification.* As previously mentioned, models must be formulated so that they can be used by users with differing levels of requirements. This is especially true with respect to the specification of the data input. Some users will have comprehensive records, understand their meaning and the limitations of their accuracy. Many potential users will not be so well endowed with information or enlightened as to its value. There will also be a wide range of users of the model. For the model to operate effectively for extension purposes it must be designed as a complete information system [14] and be capable of handling this diversity of use and information availability. In this respect it is essential that a model should include:

(i) Facilities to reduce user data requirements to a minimum for the use of the model that he has in mind. For example, the skeletal data mentioned above should be in compiled form and not input by the user.
(ii) Unambiguous and specific data input requirements contained on forms which are easy to fill in and which limit actual data errors, prevent the misplacing of data within a specified field and minimise 'punching' errors.
(iii) Facilities to check this data for inconsistencies and feasibility so that user time can be saved in error checking, and computer time saved by eliminating abortive runs.
(iv) Methods by which missing or unavailable data may be infilled but with the associated writing of reports informing the user of the assumptions and their implications.

The dairy model has been structured with these points in mind. A manual [11] with associated input forms has been produced. Forms and cards have been coded to enable the co-ordination and the mechanisation of card punching and job processing at the computer installation. Within the program, feasibility checks are made and data assumptions reported and, where possible, missing parameters infilled (see Fig. 7). Thus, the model can be tailored precisely to a particular farm situation with a complete data specification, or used as a model with almost completely assumed data, or employed at any point between these two extreme cases.

(b) *Output generation.* The output of the models should be meaningful to the different categories of users. This means each user should be able to design his own output obtaining only a few key indices or a complex set of indices if he so requires. This facility is important if computer output is to become meaningful to producers.[17] Models should always provide checks on the data assumptions that have been used for a particular run. The dairy model provides three types of output:

(i) 'Echo checks' on all data assumptions, both input by the user and assumed.
(ii) An obligatory summary of financial and physical performance.
(iii) User-requested output as graphs and tables, the quantity and layout of which is at his discretion.

In the case of the user-requested output, up to 40 indices are tabulated and plotted against time. The user by the use of simple output request forms can design his output to give as many tables and graphs as he wishes. This is of particular importance in extension work where users require to use the model to answer different questions. This implies that they will need different performance indices and need to be able to compare these with other indices in a specific way.

(c) *Iterative facilities.* Since models of this type are designed to explore alternative system designs and to study the performance of these designs within different economic environments, it is essential that they have simple update facilities to enable the users to specify a series of runs with a single submission to the computer and with minimum effort.

Figure 7 illustrates, for the dairy model, how this is achieved. The model reverts to the original data assumptions for each new hypothesis. By submission of the appropriate input form, any single item of data, or any set of data items in the original hypotheses, can be changed and a re-run made.

One of the major problems of any simulation model for extension purposes is the adviser and computer time spent in searching the profit surface for better strategies. This is especially true when jobs are 'batch processed' and errors in form filling, data derivation, etc., are inevitable. These cause serious delays in conclusions being reached. One way to alleviate this problem is to build automatic search routines into the model where new financial targets are set and the model automatically tries to find a method of satisfying the new objective. The dairy model discussed in this paper is at present undergoing modification in this respect. The procedure can be seen in Fig. 8. This approach may be useful in many enterprise models where the bulk of advisory investigations are concerned with only marginal changes to existing systems.

(*d*) *Handling the processing problems.* One of the major problems confronting the model builder in extension work is defining the input and program design so that it can be handled by the normal procedures used by a commercial computer installation and within the available budget. The following criteria have been incorporated in the dairy model and must be met if this objective is to be satisfactorily achieved:

(i) Data forms must be designed so that they can be used by personnel who may be unfamiliar with computer requirements, but they must also adhere to the normal requirements laid down by system analysts for data. This requires that the following conventions must be satisfied: (a) The job and batch number must be included and used to identify individual user input and to identify and check the validity of the input form and card order. (b) The data input should conform to normal card punching practice (for example, the punching of decimal points is unusual). Documentation should be accompanied by punching instructions for the computer data preparation department.

(ii) Output should be coded so that users may be identified and output distribution automated.

(iii) Programs should be designed to optimise within the installation job mix. This optimisation will need to consider such factors as the usage of core, peripherals and time. In respect of the latter factor it is quite likely that complicated procedures will be required to schedule the workload both for data preparation and processing, with respect to the other routine work at the installation.

(iv) Programs should be designed with the user 'budget' in mind.

(v) Separate data validation programs may be required so that jobs never halt with run time errors (included as a sub-routine in the dairy model causing the program to halt normally for fatal data errors and to report on the nature of the error). It is more usual to have separate data validation programs with the batching of validated data on to a magnetic storage device for subsequent processing.[16]

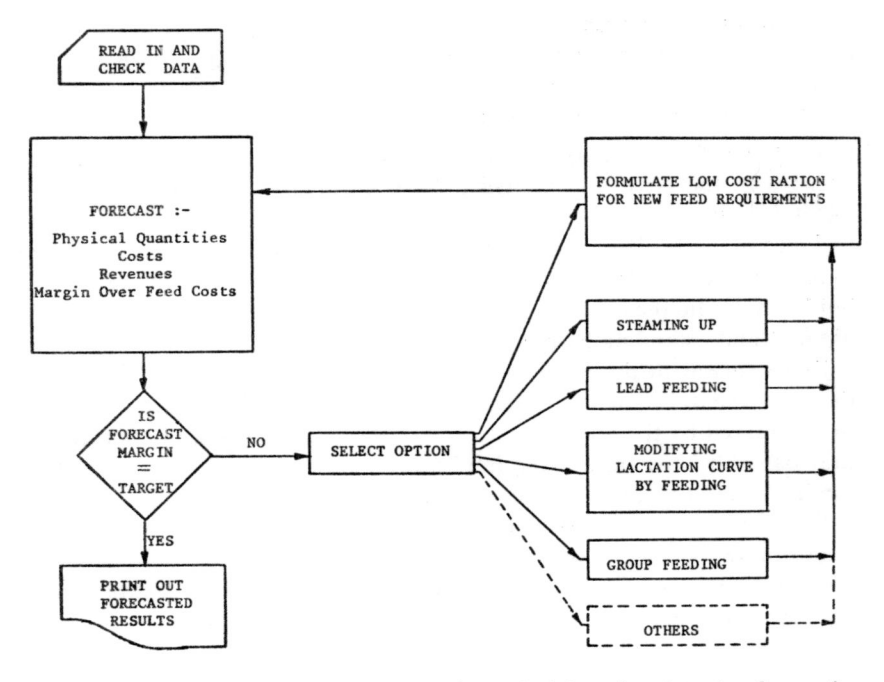

FIG. 8. Flow diagram showing the operating principles of an iterative forecasting programme.

These aspects of the modelling exercise should not be considered lightly. They form a very substantial overhead cost in terms of development effort. For many commercial models, data validation, report writing, etc., will be more costly than the actual simulatory module of the program. Furthermore, input form design and the setting-up of input and output processing systems will also be costly and organisationally complex.

7.5.3.6. Validation of the Models

It is recognised that the validation of the models is a difficult procedure. Experience has shown that with this type of model, the most satisfactory validation procedure is achieved by the use of historic time-series data. Within this historical decision framework, actual performance is compared with that predicted by the model. In this respect the model builder is able to validate the basic logic of the model for controlled conditions and accurate data.

The precise technique of validation for extension purposes is one that has not been fully developed in the field of model application. It is an especially difficult task for models of this type since many of the errors, or divergence between predicted and actual performance, are likely to be due to data specification errors, inadequate comprehension of the model, or changes in the environment which have been unpredictable. These are inevitable once one moves out of the region of controlled experimentation and into the area where management ability and environment have a major impact on the output derived from a given set of inputs. One can only hope to validate the model for controlled conditions and rely on 'sensible' use and understanding by potential users.

Experience with the practical use of models has, however, shown that they are able to isolate key decision areas, the effects of which are unlikely to be masked by the above-mentioned effects. Perhaps more importantly, they can show the farmer the implications of a given strategy since the model is able to simplify elements of the system so that the interactions within it become understandable. Thus, absolute accuracy is not one of the objectives of the modelling exercise. In this instance the objectives are the determination of sensitive decision areas, the ranking of alternative strategies, and the production of targets for the chosen plan.

7.6. SUMMARY AND CONCLUSIONS

This paper has attempted to illustrate ways in which effective models of complex bio-economic systems can be constructed. It has been argued that superficially complex, but impracticable, models of agricultural systems can be constructed relatively easily. The real difficulties lie in making them both practical and useful, and these should not be underestimated.

Two models of bio-economic systems, which have now been in use for some years by researchers and extension workers, have been used to illustrate some of the problems. Both are designed for application to specific farm situations and are relatively large and complex. As has been illustrated in the case of the firm growth model, very much simpler versions would have been satisfactory for demonstrating the general principles of the system's behaviour and for drawing some very general conclusions. Such simple models, however, would have been incapable of being applied to specific farm problems.

The complexity of the models arose not from the introduction of sophisticated relationships but from the need to provide detail and adequate flexibility. Both are essentially only high-speed calculators.

They are based on the principle of forward budgeting of feed, stock, milk or cash from the present point in time and are designed to be able to answer specific, but important, questions.

More theoretical models of the same systems would possibly have included such interrelationships as those between, say, the housing type and feed conversion ratio in pigs, or stocking rate and potential grass growth in milk production. Such relationships are difficult to determine and are not easy to modify for individual farm situations. Their values in practice are confounded by the climatic environment and by the levels of managerial efficiency found on the individual farm.

Both of the models, therefore, simply develop strategies based on input and performance information at levels assessed by the adviser to include these effects. These information levels may be changed at will, as they are provided as data to the models. This simplistic approach allows subjective information to be included by the user and avoids many of the complex internal adjustment procedures required if experimental response relationships are to be used.

Although the models require considerable quantities of data, it is of a relatively readily available type. Their well-defined, simple structure makes it possible to check the accuracy of the models easily and this helps in obtaining the confidence of the users.

The essence of the approach which we are advocating is that of simplicity and practicality. To our knowledge, the application of the more theoretical 'systems' or cybernetic concepts in agriculturally related fields has not produced any results of practical significance, in spite of considerable efforts in this area. We would argue that there should be a reduction in the amount of general systems research and in the practice of developing more and more complex systems models almost for their own sake. Simply to recognise the complexity of reality, and then to set out to mirror it, irrespective of the time and cost involved, is unlikely to be a productive exercise.

Models should be constructed to meet limited, well-defined objectives and there has to be a greater recognition of this need for relative simplicity. There is, in fact, a strong argument for producing less general programmes than the ones which have been described here. By restricting each package to a single specific enterprise or problem, such as, for example, the expansion of a pig fattening herd, many of the problems of providing generality within a single programme would be overcome. Interactive computer models of such bio-economic systems would also greatly facilitate the study of system response and the testing of hypotheses.

Future models, constructed according to the philosophies contained in this paper, could, we believe, be extremely useful to the agricultural industry in many areas. The flexibility, simplicity and independence of internal data of such bio-economic, calculator models would ensure their practical usefulness in a financial services role, within new research programmes or as the basis of extension work, over periods of many years into the future.

REFERENCES

1. Charlton, P. J. & Thompson, S. C. (1970). Simulation of agricultural systems, *Journal of Agricultural Economics*, **21**, 373–89.
2. Charlton, P. J. (1973). Effective interdisciplinary research: an elusive goal, *New Zealand Journal of Agricultural Science*, **7**, 151–8.
3. Meadows, D. L. (1970). *Dynamics of Commodity Production Cycles*, Wright-Allen Press, Cambridge, Mass.
4. Charlton P. J. (1973). Computer simulation of firm growth, unpublished Ph.D. thesis, University of Reading, England.
5. Penrose, E. T. (1959). *Growth of the Firm*, Blackwell, Oxford.
6. Walker, O. L. & Martin, J. R. (1966). Firm growth research opportunities and techniques, *Journal of Farm Economics*, **48**, 1522.
7. Olsson, N. (1971). A multi-period linear programming model for studies of the growth problems of agricultural firms. II. The model, *Swedish Journal of Agricultural Research*, **1**, 155.
8. Charlton, P. J. & Street, P. R. (1970). Some general problems involved in the modelling of economic systems on a digital computer, in: *The Use of Models in Agricultural and Biological Research* (ed. J. G. W. Jones), Grassland Research Institute, Hurley, pp. 50–62.
9. Charlton, P. J. (1972). Financing farm business growth, *Farm Management*, **2**, 60–70.
10. Street, P. R. (1971). Synthesis of grassland systems for dairy cows, with particular reference to exploitation of seasonal price schedules for liquid milk, unpublished Ph.D. thesis, University of Reading, England.
11. Street, P. R. (1973). Nottingham University dairy enterprise simulator, University of Nottingham, Dept. of Ag. Publication.
12. Crabtree, J. R. (1970). Towards a dairy enterprise model, in: *The Use of Models in Agricultural and Biological Research* (ed. J. G. W. Jones), Grassland Research Institute, Hurley.
13. Street, P. R. & Seabrook, M. J. (1973). Dairy farming in the East Midlands, University of Nottingham, Dept. of Ag. Publication.
14. Blackie, M. J. & Dent, J. B. (1973). Budgetary control: review and reconstruction, *Farm Management*, **2**, (5), 261–71.
15. Wood, P. D. P. (1967). Factors affecting the shape of the lactation curve in cattle, *Animal Production*, **11**(3), 307–16.

16. Blackie, M. J. & Dent, J. B. (1973). A planning and control system for the small farm, *Zeitschrift für Operations Research*, **17**, B173–B182, Physik-Verlag, Würzburg.
17. Pearse, R. A. & Street, P. R. (1972). Report on a conference on business planning and control techniques, *Journal of Agricultural Economics*, **23**(3), 335–7.

8

The Study of Agricultural Systems: Application to Farm Operations

G. F. DONALDSON*

International Bank for Reconstruction and Development (World Bank), Washington, DC, USA

8.1. INTRODUCTION

The evaluation of alternative ways of organising farm operations, especially those associated with mechanised cereal growing, has been much studied in a systems context. Such studies began in the endeavour to analyse the investment choices confronting individual farmers in selecting field machinery. Those reported here involve system simulation models of a Monte Carlo type. These were used to assess the adequacy of available machinery in various farming situations. They were also applied in assessing the implications of continuing mechanisation for family farms, and the sector as a whole, in terms of labour displacement and farm size adjustments. Similar models were used to assess machinery capacity requirements in various locations, in order to provide guidelines to manufacturers for machinery development and distribution.

The problem was not seen in a systems context at the outset. Rather it was seen as a vexing farm investment problem, and several conventional investment and budgetary tools were applied first (Donaldson[1]). This usefulness was restricted by the variable timeliness effects and their related costs. The search for a means of taking account of a large number of interacting variables, many of which could best be taken into account as discrete probabilities, led to the use of systems concepts and simulation models. The system simulation approach proved a convenient means of studying this subject, as these and the many similar studies will testify.[2-18]

---

* The views expressed here are those of the author and should not in any way be attributed to the IBRD.

8.2. APPROACH AND CONTEXT

The mechanisation of farm operations is a predominant feature of technological change in agriculture. Its progress can be seen in the increasing use of larger and more complex machines on farms, and in the reduction of the farm labour force. Less easily observed are the changes in farm organisation and management that are a necessary concomitant of this continuing mechanisation. Only the changing structure of agriculture, with fewer farms and fewer people but more machines and increasing output, betrays the changes that must be occurring in the farm business.

The studies discussed here are set in this practical context of dynamic farm adjustment and environmental uncertainty. The purpose at the outset was wholly normative in order to provide forecasts that might help to improve the accuracy of decisions concerning farm machinery investment, although in the course of events the models were also used to give some insight into the actual behaviour of farmers and the farm sector.

Conceptually, the purpose was to provide generalised information for use in decision-making. The models do not reflect the conditions in any particular situation. This approach recognises that, in addition to encountering uncertainties in the process of technological change, individual farm businesses are relatively small so that the cost of searching for information on individual problems very often exceeds the cost of a possible decision error. Consequently there is a need, on the part of extension services and farm consultants, for analyses that are generalised so that the results can be applied to more than one farm. Thus these models relate to different sizes of machinery, used over a range of acres, with varying soil types and other locational factors. The aim is to assess the expected performance or capacity of a set of machinery in these various circumstances.

In a farm operations context, 'capacity' is determined by the rate of work achieved and the time over which the operation is continued. For a farm machine these are determined by a number of constraints and variables, each of which is characteristic of the particular time and working situation being considered.

The *rate of work* is affected by the operating characteristics of the machine, the yield and conditions of the crop or product being processed, the weather or environmental conditions in which the operation is undertaken, and by various operating decisions that may be taken by the operator. The rate of work achieved in machine operation is thus a stochastic variable dependent on the interaction of numerous other variables. These may include any number of the physical, biological, environmental and human features, each of which may vary from place to place, year to year, and in many cases field to field and day to day.

Similarly, the *time available* for the operation is dependent upon the adequacy and the extent of the machinery system under the operating conditions, the biological tolerances related to the particular product and operation involved, the prevailing weather or ambient conditions and, again, certain operating decisions made by the farmer. When determining the time to be taken, there are 'trade-offs' involved, just as there are when considering the rate of work of a machine. The actual time available is certainly not rigidly defined by outside effects but can very often be extended—but at a cost. It may involve a loss in yield through shelling or bird damage, or a loss of produce quality. Alternatively, it may result in lower market prices, or an increased possibility of not completing the operation; or a direct cost, such as grain-drying, to rectify the effects of untimely operations.

These many factors may combine and interact in many different ways. In some instances the one effect may influence both the machine rate of work and the time available for the operation. In particular, weather effects, such as a fall of rain, can affect both. However, the many variables that are involved are largely independent and each combination tends to create a different set of effects on the operation and on costs.

8.3. SIMULATION OF FARM OPERATIONS

The objective of the two models outlined below is to assess the operating capabilities and costs of alternative machinery systems used in field operations associated with cereal production. Model I simulates the *harvesting* of cereals and Model II the *seeding* of cereals. In their construction the interacting variables were quantified and the relationships between them specified as completely as available data would allow. In doing this the first step was to identify the entities of concern in the farm operations and establish the variables that were to characterise the models.

8.3.1. Harvest Model Components

8.3.1.1. Location

Due to data limitations the model was first restricted to the south-east corner of England.[19] Subsequently, a comparative study was made over four locations in Saskatchewan, Canada (Donaldson[20]). These four sites are roughly equidistant in a 500-mile arc from south-west to north-east and cover a range of soil and climatic zones, characteristic of much of the cereal-growing region, from the relatively light soils and dry summers of the south-west to the relatively heavy soils and moister summers of the northeast. Weather data for each location covered a period in excess of 30 years. Weather data for southern England is illustrated in Table 1.

8.3.1.2. Machine Systems

A range of six harvesting systems is evaluated, each comprising one or more combines and a grain drier. It is assumed that auxiliary equipment is available to permit full extension of the capability of these main units. The combine range considered is based on composite machines representing the models produced by the major manufacturers. The driers considered were of the continuous-flow type with a rated capacity matching that of the combines.

In addition to combines and driers, another machine variable is included in the model to provide some flexibility in the combine capacity available. In practice the combine capacity on a farm can fairly readily be extended by one or more actions. The existing combine may be used at a faster rate

TABLE 1
Distribution of rain-free days

Number of dry days	Percentage dry days	Frequency	Number of dry days	Percentage dry days	Frequency
19	33·9	1	34	60·7	2
20	35·7	1	35	62·5	1
21	37·5	1	36	64·3	1
22	39·3	1	37	66·1	3
23	41·1	–	38	67·9	3
24	42·9	1	39	69·6	3
25	44·6	3	40	71·4	1
26	46·4	1	41	73·2	2
27	48·2	2	42	75·0	2
28	50·0	2	43	76·8	–
29	51·8	1	44	78·6	1
30	53·6	2	45	80·4	1
31	55·4	–	46	82·1	–
32	57·1	2	47	83·9	1
33 (Mean)	58·9	1			

Note: Shaw weeks 39 to 46 (23 July–24 Sept.) over 40 years 1932–1961 at East Malling.

of work or in unfavourable working conditions, thus providing extra capacity at the cost of higher grain losses. Alternatively, another machine may be purchased, rented or borrowed, a derelict machine may be pressed back into service, or a custom operator (perhaps a neighbour with surplus capacity) may be employed. In order to simulate this flexibility in the system, additional combine capacity is provided in the model in order to handle any crop not harvested in the available time. This is charged at a penalty rate equal to the average custom rate.

8.3.1.3. Acres Harvested
Because the acreage of cereals grown per farm varies, the analysis was made for successive 25 acre intervals, from 25 to 2500 acres (1200 for the British version).

8.3.1.4. Grain Drying
Harvest operations were assessed with and without a grain drier. Despite the fact that artificial grain-drying is widely used in many temperate farm production regions throughout the world, it is far from being an established practice in Canada. Consequently, one of the main questions to be answered by this study is whether or not artificial grain-drying is an economic proposition. To do this, it was necessary to consider the harvest operation with and without the aid of a grain drier. The inclusion of a drier in the harvest process so altered the sequence of operations, however, that it was necessary to modify the simulation model substantially. Though two versions were run in the computer, they contain common basic routines, and most of the variables are used in both.

8.3.1.5. Biological Tolerances
The variables considered in this category include (a) the time of ripening of the crop, (b) the yield, (c) the rate of shelling loss from the mature crop, and (d) the grade loss associated with weather damage to the standing crop (Canada only).

The *time of ripening* determines the date on which harvesting can begin, and hence the weather conditions encountered. The British version takes no account of sequential weather effects over the harvest period, so that a single starting date (23 July) is used for all years, whether or not a grain drier is used. In the Canadian version the starting date is regarded as stochastic (determined by weather previous to the harvest period), based on some 25 years' observations. The actual starting dates ranged from 9 August to 5 September at Location 1 (south), and from 19 August to

15 September at Location 4 (north). For each location any particular date within the range was given a probability based on its recorded frequency.

The date of ripening of successive crops or fields on any one farm also determines, *inter alia*, the length of time for which the crop remains standing once it is harvest-ripe. This in turn influences the shelling loss and the grade loss that occurs in the standing crop. This effect was introduced into the model by adjusting the proportion of the crop that was subject to shelling and grade losses as harvest progressed.

The cereal crop yield affects the harvest in two ways. First, it is assumed to affect directly the rate of work achieved by the combine. Second, it determines, in combination with the time of ripening and the prevailing weather effects, the amount of lost revenue due to grain shelling and quality loss. Consequently, crop yield is a parameter that appears in several identities within the model.

The yield obtained is regarded as location-dependent, and in order to simplify the assessment it is assumed that the crop to be harvested is all wheat, and the range of yield for each area was based on the average yields recorded over time for each location. Since yield is also stochastic (again dependent largely on weather prior to harvest), each yield level within the range at all locations can be given a probability based on the recorded observations. It should be noted that the recorded data are (a) based on yields obtained subsequent to harvest losses, and (b) average yields for each location. Accordingly, the losses calculated for each location may be underestimated, and the degree of variability allowed may be less than is actually encountered on any one farm.

The shelling losses that occur in the standing crop are obviously affected by many factors. Clearly these will include the yield, but the proportional loss is also influenced by (a) the crop variety, since some varieties hold the grain more tightly (and are consequently considered harder to thresh), (b) the prevailing weather, particularly wind, (c) whether or not the crop is swathed, since losses in the swath are expected to be less than in the standing crop, and (d) the length of time the crop is standing after harvest-ripeness is reached.

Unfortunately, the measurements necessary to quantify this variable are not readily available for all locations, though some measurements have been made at Swift Current and Milfort Research Stations: Using these, together with data obtained from a thorough evaluation of shelling losses (made in Sweden for a number of wheat varieties over a 15 year period[21]), an average grain loss over successive days was estimated. These data, shown in Table 2, were compared with more limited British data and found to be similar.

The grain quality effects, or *grade losses*, that arise in the unharvested crop due to weather damage are assumed to relate directly to rainfall. Three grade dockages, of equal value, are allowed in the Canadian model calculated at 4 cents per bushel on the unharvested crop yield after two, four, and six inches of rain, respectively.

8.3.1.6. Weather Constraints

The weather variables considered in the model include (a) the occurrence of rain-free days, (b) the level of precipitation on rainy days, (c) the relative humidity and time available for combining on rain-free days, and (d) the rate of drying of the wet grain.

It is assumed that harvesting can proceed only on *rain-free days*. In order to permit the widest possible variation in the time and duration of

TABLE 2

Projected shelling losses for wheat

Day	Percentage loss	Value loss (£)	Day	Percentage loss	Value loss (£)	Day	Percentage loss	Value loss (£)
1	0·30	0·12	22	0·30	0·12	43	0·20	0·08
2	0·25	0·10	23	0·35	0·14	44	0·15	0·06
3	0·30	0·12	24	0·30	0·12	45	0·20	0·08
4	0·30	0·12	25	0·35	0·14	46	0·15	0·06
5	0·30	0·12	26	0·30	0·12	47	0·20	0·08
6	0·25	0·10	27	0·35	0·14	48	0·15	0·06
7	0·30	0·12	28	0·30	0·12	49	0·20	0·08
8	0·25	0·10	29	0·25	0·10	50	0·55	0·22
9	0·25	0·10	30	0·25	0·10	51	0·50	0·20
10	0·25	0·10	31	0·25	0·10	52	0·55	0·22
11	0·25	0·10	32	0·25	0·10	53	0·55	0·22
12	0·25	0·10	33	0·25	0·10	54	0·55	0·22
13	0·25	0·10	34	0·25	0·10	55	0·50	0·20
14	0·25	0·10	35	0·25	0·10	56	0·55	0·22
15	0·15	0·06	36	0·35	0·14	57	0·40	0·16
16	0·15	0·06	37	0·35	0·14	58	0·40	0·16
17	0·15	0·06	38	0·35	0·14	59	0·40	0·16
18	0·10	0·04	39	0·40	0·16	60	0·35	0·14
19	0·15	0·06	40	0·35	0·14	61	0·40	0·16
20	0·05	0·02	41	0·35	0·14	62	0·40	0·16
21	0·25	0·10	42	0·35	0·14	63	0·40	0·16

harvest, rainfall data were collected for the period 1 July to 30 November. The daily rainfall for this period at the four locations was obtained from historical weather records covering 40 years from 1922 to 1961 in Britain and 35 years from 1931 to 1966 in Canada. Days when less than 0·01 in of rain were recorded are considered rain-free and regarded as potential operating days. Days when rainfall exceeding 0·01 in was recorded are regarded as rainy days on which combining is not possible.

The *level of precipitation* recorded in the form of rain was used to adjust the number of available days in order to allow for a drying-out period subsequent to rain. To do this, information was obtained from farmers' diaries as to the days on which combining took place. These were compared with the rainfall data in order to assess how long a delay occurred before combining continued. The scale used in the model was to lose one day for each 0·10 in up to 0·30 in, and subsequently one day for each additional 0·50 in. The kind of data used are shown in Table 3.

Within the rain-free operating days, the time available for combining is further restricted by the grain moisture content and surface moisture on the straw, both of which are related to the *relative humidity*. Each day a mature crop passes through a cycle of moisture content in concert with the ambient humidity. This has been described by Arnold[22] and analysed by Crampin & Dalton.[23] At certain temperatures this results in the accumulation of moisture on the straw, in the form of dew, which makes combining virtually impossible. Consequently, the amount of time available for combining in any one day varies from place to place and from one time of year to another. Using the same farm records mentioned above, together with data from weather records for each location, an estimate was made of the average *daily hours available* for combining at each site. In Britain these averaged about 10 h per day, while in Canada they ranged from 12 h per day at Location 1 to 9 h at Location 4. Again, these are approximate figures, since the actual hours available will vary from the beginning of harvest to the end, as the season progresses. To a large extent the shorter hours at Location 4 reflect the later harvest starting data as much as a difference in climate.

Finally, weather conditions affect the *grain moisture content* of the crop and therefore the amount of drying that may need to be done. To assess the need for drying, the relationship between grain moisture content and the prevailing weather effects needs to be established. Such a relationship has been explored in English conditions at the National Institute of Agricultural Engineering.[22] The data consisted of detailed hourly readings of grain and air moisture content, recorded 24 h a day over four consecutive

TABLE 3
Cumulative frequency distributions for rainfall by successive five-day periods, 1931–66

Location 1 (Swift Current)

	August						September				
	5–9	10–14	15–19	20–24	25–29	30–3	4–8	9–13	14–18	19–23	24–28	29–3
Rainy days												
0	0·42	0·42	0·36	0·33	0·36	0·39	0·39	0·36	0·53	0·25	0·42	0·50
0–1	0·61	0·75	0·67	0·58	0·64	0·61	0·69	0·61	0·67	0·61	0·72	0·83
0–2	0·78	0·92	0·89	0·83	0·78	0·80	0·94	0·92	0·83	0·78	0·94	0·97
0–3	0·92	1·0	0·97	0·92	0·92	0·97	1·0	1·0	0·89	0·92	1·0	1·0
0–4	1·0		1·0	0·97	1·0	1·0			0·97	0·97		
0–5				1·0					1·0	1·0		
Precipitation (in)												
0–0·25	0·43	0·57	0·74	0·46	0·61	0·45	0·59	0·56	0·41	0·59	0·76	0·78
0–0·50	0·71	0·76	0·87	0·71	0·65	0·64	0·77	0·87	0·53	0·81	0·95	1·0
0–0·75	0·81	0·90		0·79	0·70	0·82	0·91	0·91	0·70	0·85		
0–1·0	0·86		0·91	0·83	0·74		1·0		0·76			
0–1·25	0·90	0·95	1·0	0·88	0·78				0·82	0·92	1·0	
0–1·50		1·0		0·96	0·83			0·96	0·94			
0–1·75	0·95				0·91	0·95		1·0				
0–2·0	1·0			1·0						1·0		
0–2·25					0·96	1·0			1·0			
0–2·50												
0–3·0					1·0							
0–5·0												

harvests. By selecting periods with various proportions of rain-free days from the years recorded, it was possible to build up a pattern of grain moisture content which is assumed to represent the pattern existing in any period with a similar proportion of rain-free days. The pattern of grain moisture content thus obtained is shown in Fig. 1, in the form of cumulative time at successive moisture levels, the form in which such data are most useful in simulation procedure.

8.3.1.7. Machine Performance

This includes (a) the rate of work achieved in operation, and (b) the amount of time lost during an operating day. For Canada, both these variables were quantified on the basis of measurements and information obtained from field test records kept by the Saskatchewan Agricultural Machinery Administration (AMA). The *rates of work* expected for combines were estimated using a combination of two methods. The first involved analysis of the rates achieved by machines under field test conditions. To do this, the acreage harvested each day was divided by the time the machine was operated. This was then assessed in relation to yield using regression analysis. This implied that a basic rate of 9·5 acres per

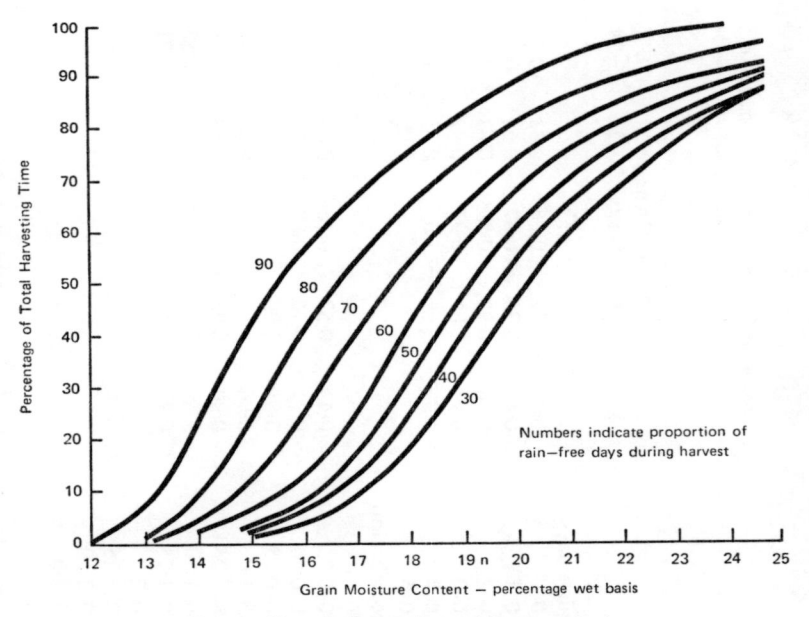

FIG. 1. Cumulative grain moisture content.

hour (with no yield) is reduced by 1·2 acres per hour for each additional 10 bushels yield.

Since sufficient data were not available for all machines, a further set of expected rates of work were determined, using the relationship:

$$R = 3((W/192) + (B \cdot L^{1\cdot 5}/38\,600) + (S/7400))$$

where R = rate of work (tons per hour), W = cylinder width in inches, B = body width in inches, L = straw walker length in inches and S = combined chaffer and sieve area in square inches.

This value was related to a range of yields to give rates of work per acre. When compared with the rates obtained by regression analysis of the test data, this formula was found to overestimate the rates expected for low yields and to underestimate those for high yields. By comparison of the rates obtained by both methods (for the 'medium-size' combine only), a correction factor was developed to adjust the rates obtained using the formula. This was applied to the expected rates calculated from the formula for the other sized combines. The adjusted rates are used in the simulation as deterministic values, related to the crop yield in each case.

For each five-day period considered, however, the acres harvested per day are also affected by the *lost time* due to stoppages for breakdowns, maintenance, adjustments and operator relief periods. Using the AMA records, the frequency and duration of lost time were estimated as a frequency distribution for use in the model (see Table 4).

For Britain, these variables were assessed as one overall variable based on farm diaries and log books. These data are illustrated in Fig. 2.

8.3.2. Cereal Harvest Model Specifications

The model is designed to simulate up to six harvesting systems over a range of acreages, locations and seasons. The variables included are:

Biological tolerances—variable yield
　　　　　　　　　　—variable ripening date
　　　　　　　　　　—grade losses
　　　　　　　　　　—shelling losses
Weather constraints—harvest time limit
　　　　　　　　　　—variable wet days
　　　　　　　　　　—variable precipitation
　　　　　　　　　　—lost days

Machine performance—variable rate of work
—operating hours per day
—variable lost machine time

For each location, each machine system is run over a specified 1000 trials or 'iterations' at each successive 25-acre level. For each iteration,

TABLE 4
Cumulative Distribution of operating time lost

Operating time worked (%)	Frequency of observation	Cumulative frequency	Operating time worked (%)	Frequency of observation	Cumulative frequency
		(Mean = 73 per cent)			
16	2	0·009	69	4	0·356
19	1	0·013	70	6	0·382
23	1	0·018	71	7	0·413
27	1	0·022	72	6	0·440
29	2	0·031	73	8	0·476
30	1	0·036	74	7	0·507
31	1	0·040	75	4	0·524
35	1	0·044	76	5	0·547
42	1	0·049	77	3	0·560
43	3	0·062	78	13	0·618
44	1	0·067	79	4	0·636
46	1	0·071	80	5	0·658
47	1	0·076	81	3	0·671
48	2	0·084	82	8	0·707
50	3	0·098	83	10	0·751
51	1	0·102	84	5	0·773
52	2	0·111	85	7	0·804
53	1	0·116	86	7	0·836
54	3	0·129	87	2	0·844
55	3	0·142	88	3	0·858
56	2	0·151	89	2	0·867
57	1	0·156	90	3	0·880
58	2	0·164	91	3	0·893
60	1	0·169	92	5	0·916
61	3	0·182	93	4	0·933
62	4	0·200	94	1	0·938
63	4	0·218	96	3	0·951
64	3	0·231	97	3	0·964
65	7	0·262	98	1	0·969
66	1	0·267	100	7	1·000
67	9	0·307		—	
68	7	0·338		225	

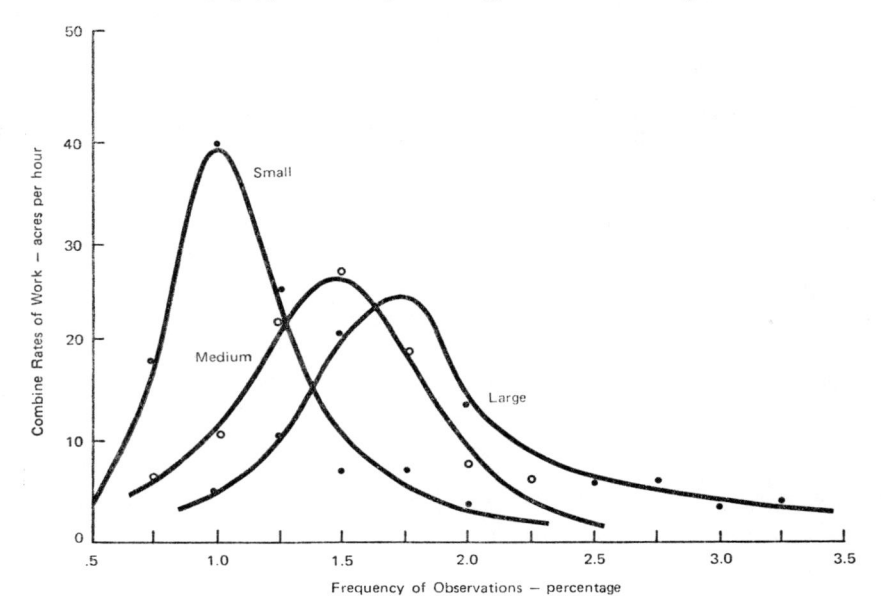

The Study of Agricultural Systems: Application to Farm Operations 279

FIG. 2. Distributions of combine rates of work.

or year, the harvest starting date and the yield per acre are selected from their respective distributions. The rate of work for the system is then found in correlation with yield.

Having established the yearly variables, harvests are simulated over successive 25-acre intervals up to a specified limit that is varied according to the size of the machine system. Each harvest is based on a series of consecutive five-day periods as independent quanta of the total time taken to complete the harvest. The days and five-day periods number from 1 July. The starting date determines the starting period, which is always regarded as a full quantum even if the starting date does not occur on the first day of that period.

For each five-day period, the proportion of available hours per day actually worked (this simulates time lost due to breakdown and machine maintenance), the number of wet days during which no work is done, and the level of precipitation, are selected at random from their respective distributions. The actual number of working days in a five-day period depends upon the number of wet days, the precipitation, and the number of non-working days carried over from the previous period.

Wet days are automatically lost days, and if the number of wet days exceeds two, an extra lost day is added on that account. Lost days are also caused by excess precipitation in the period. If the number of days lost on all accounts exceeds five days in a five-day period, no work is done in that period, and the excess is carried over to the next period as lost days.

When combining is possible, work done is calculated from the number of working days, hours available, proportion of available hours actually worked and the rate of work of the system in that year. The remaining unharvested acres are found and the cost of shelling and grade losses are calculated on that acreage. Subsequent periods are considered until the harvest is completed or until all available time has elapsed. If harvest has not been completed after a certain number of periods, custom combining is employed to augment the owned system so that in subsequent five-day periods the rate of work is effectively doubled, and a penalty charge is incurred equal to the custom rate. If, at the end of the maximum time allowable, the harvest is still not complete, the remaining acres are assumed to be harvested at a penalty rate equal to the cost of custom combining.

When a grain drier is used, wet days are automatically lost days, but no additional days are lost for a series of wet days because the drying facility allows combining to continue as soon as the straw surface is dry. Similarly, no additional days are lost due to the level of precipitation until 0·50 in is recorded, and one additional day is lost for each successive 0·50 in in a five-day period. This particular lag effect is assumed to represent the soil moisture restraint which can prevent combining even if the grain is dry.

No harvesting is considered possible if the grain moisture content exceeds 22% of the dry weight of the grain, and no drying is considered necessary until the grain moisture content exceeds 16%. Once grain has to be dried, it is assumed necessary to reduce it to 14% moisture. All other variables are calculated in the same way.

A grade loss occurs when the cumulative precipitation exceeds a given amount, and the cost incurred is equal to the remaining amount of grain multiplied by the grade dockage factor. A maximum of three grade losses is allowed for each harvest.

Shelling losses are calculated on a day-to-day basis, when the cost incurred is equal to the remaining amount of grain multiplied by the shelling loss factor for each elapsed day. It is assumed, for the purpose of calculating shelling losses, that the amount of grain decreases linearly within a five-day period.

When the harvest has been completed, the cost of the harvest is formed from the cumulative sum of the fixed and variable costs of the system, the cost of custom combining and the grain loss penalties. The total cost is then computed for successive 25-acre levels, and accumulated over the number of trials, according to the identity:[24]

$$TC_{ai} = F_c + V_c(N - N_c) + C_c(N_c) + L_s + L_g$$
or $$TC_{ai} = F_c + V_c(N - N_c) + F_d + V_d(N_d) + (H_d.N_d(M\ 0{\cdot}14)) + C_c(N_c) + L_s + L_g$$

where TC_{ai} = total cost at acreage a for machine system i, N = acreage harvested, N_c = acreage harvested by custom services, N_d = acreage of crop dried, F_c = fixed costs of combining, V_c = variable costs of combining, F_d = fixed costs of drying, V_d = variable costs of drying, H_d = cost of heat for drying, M = grain moisture content before drying, C_c = custom combining charge, L_s = cumulative value of shelling loss and L_g = cumulative value of grade loss.

From this, the average total cost per acre and the average marginal cost (averaged over the 25-acre increment) are computed, in the form:

$$ATC_a = TC_{ai}/N$$
$$AMC_a = TC_{ai} - (TC_{ai} - 25)/25$$

and for each identity the standard deviation is calculated, and the variance distribution over the total number of trials is determined. After cycling over the requisite number of five-day periods, all levels of crop acres, and the full complement of harvests (trials), machine systems, and locations, the sequence stops and the output is printed.

8.3.3. Model-Based Experiments

Additional output was obtained by running the models with different data and assumptions. Experiments conducted in this way include the use of (a) different cost values to simulate the holding of a combine for a longer period before resale, (b) different costs to simulate the use of a second-hand machine, (c) different cost and rate-of-work parameters to simulate the introduction of a prototype machine, (d) restrictions on the starting and finishing dates to simulate competing activities, (e) different weather constraints, such as shorter delays after rain, (f) different lengths of harvest to simulate competing activities, (g) substitution of additional combine capacity for drier capacity, and (h) alternative harvesting policies, in order to simulate a 'minimum-cost' versus 'minimum-time' strategy on the part of the farmer. The results of these are presented in Figs. 3–13 (see Section 8.4.1).

8.3.4. Cereal Seeding Model Components

8.3.4.1. Location

Only two locations were considered using this model, both in Canada, Swift Current (Location 1) in the south-west and Melfort (Location 4) in the north-east.

8.3.4.2. Machine Systems

Although a wide range of field equipment might constitute a tillage and seeding system, there is, at most points in time, a model system that is most popular and widely accepted. On this basis, a single combination of equipment items was chosen as representative of the whole. This included a tractor, discer, cultivator and harrow, with the assumption that these three implements would be used in a sequence of tillage and seeding operations.

The five tractor sizes used were chosen by classifying the tractors marketed by major manufacturers according to their production characteristics and then taking the mean of the relevant parameters for the five most common categories. The tractor sizes considered were 38·4, 56·0, 68·2, 96·4 and 123·3 PTO horsepower, respectively. Using estimated draft requirements for the three implements, and assuming the effective drawbar horsepower to be 65% of maximum PTO horsepower, a set of optimum sized implements was calculated for each tractor. The optimum sizes were then adjusted to coincide with the nearest size available on the market (see Table 5).

8.3.4.3. Acreage Seeded

The same acreages were considered as for the cereal harvest models, *viz.* successive 25-acre increments up to a maximum 2000 acres.

TABLE 5
Composition of tillage and seeding machinery systems

	Systems				
	I	II	III	IV	V
Tractor (PTO HP)	38·4	56·0	68·2	96·4	123·3
Discer (ft)	8	12	16	21	28
Cultivator (ft)	10	13	16	23	29
Harrow (ft)	36	52	64	92	120

8.3.4.4. Soil Types

It is assumed that the major determinant of field operating capacity for any machine system is the condition or 'tractability' of the soil. This is determined largely by soil moisture content and is thus susceptible to weather effects, but also by physical soil type. To assess the effect of this variable, two soil types were considered and identified as a 'sandy soil' and a 'medium-to-heavy soil', respectively.

8.3.4.5. Biological Tolerances

The biological variables included in the model are (a) the starting date of spring tilling, (b) the yield effects associated with seeding progressively later than the starting date, and (c) the final seeding date. It is recognised that each of these is influenced by weather effects, but these are regarded as being outside the scope of the model.

Using data recorded by the field staff of the Saskatchewan Department of Agriculture, it was estimated that tilling could commence about seven days prior to the earliest seeding date. Thus it is assumed in the model that the *spring tillage starting date* was seven days earlier than first seeding at each location.

After modifying the recorded dates to allow for this assumption, a distribution of the occurrence of the starting dates was produced to serve as a basis for selecting a starting date in the model. The adjusted dates ranged from 10 April to 30 April at Location 1, and 16 April to 12 May at Location 4.

It has long been postulated that there is a narrow optimum time range for seeding cereals, and that seeding outside that range will result in lower yields. Since the likely causes are weather-related and weather varies as each season progresses, these factors manifest themselves as a *yield penalty for untimely seeding*. Using existing data from the research stations at Swift Current and Melfort, together with experimental results from stations in North Dakota and Montana, the effect of seeding time on yield was investigated. On the basis of these data a schedule of yield loss factors was derived, showing the expected variation in yield from the expected maximum. In the model this was expressed in the form of positive and negative additions to yield, over successive days.

An additional charge is made as a *harvest timeliness penalty*. The time of seeding affects the time of harvest. The later the harvest period, the worse the weather encountered. The extra cost involved was calculated by running the harvesting model with the harvest starting date fixed at two dates—first 15 August, then 14 September. The difference in the cost per acre was then divided by 30 days and this figure was applied as a penalty charge. Since the spread in seeding dates is at least halved by the time the crop is ready to harvest, the loss factor was applied to every third day after the 'optimum' seeding date.

Since the yield penalty increases progressively once it begins, it is reasonable to suppose that there will be a *last seeding date* after which seeding will be unprofitable, due either to yield penalty or the risk of a frozen crop.

TABLE 6

Yield penalties relative to seeding date in dollars per acre

Seeding date	Swift Current	Melfort	Seeding date	Swift Current	Melfort
April 17	3·55	8·60	May 17	0·07	0·24
18	3·23	8·10	18	0·11	0·18
19	2·92	7·60	19	0·14	0·12
20	2·60	7·10	20	0·18	0·06
21	2·42	6·60	21	0·24	0·00
22	2·24	6·10	22	0·31	0·06
23	2·06	5·60	23	0·37	0·12
24	1·88	5·10	24	0·43	0·18
25	1·71	4·60	25	0·50	0·24
26	1·53	4·10	26	0·56	0·29
27	1·35	3·82	27	0·62	0·40
28	1·17	3·54	28	0·68	0·50
29	0·99	3·26	29	0·75	0·60
30	0·81	2·98	30	0·81	0·70
			31	1·21	0·80
May 1	0·75	2·70			
2	0·68	2·42	June 1	1·62	0·90
3	0·62	2·14	2	2·02	1·00
4	0·56	1·86	3	2·43	1·10
5	0·50	1·58	4	2·83	1·20
6	0·43	1·30	5	3·24	1·30
7	0·37	1·20	6	3·64	1·95
8	0·31	1·10	7	4·05	2·60
9	0·24	1·00	8	4·45	3·25
10	0·18	0·90	9	4·86	3·90
11	0·14	0·80	10	5·26	4·55
12	0·11	0·70	11	6·03	5·20
13	0·07	0·60	12	6·79	5·85
14	0·04	0·50	13	7·56	6·50
15	0·00	0·40	14	8·32	7·15
16	0·04	0·29	15	0·09	7·80

It is assumed, therefore, that all seeding stops on 15 June at both locations.

In establishing *crop yields* upon which to base the penalties, it seemed appropriate to use something higher than the average yield obtained in each area since this figure would reflect the losses we are trying to estimate, as well as the variation in husbandry from one farm to another. After comparing district and research station yields it was decided that the yield at Location 1 is 25 bushels per acre and at Location 2, 37 bushels per acre (see Table 6).

8.3.4.6. Weather Constraints

The weather effects applied in the model were determined, using a method developed by Rutledge & MacHardy.[25] This employs daily minimum and maximum temperature and precipitation, and monthly averages of wind velocity, dew point, sunshine hours and day length. Using these data, soil moisture content is computed. The Rutledge–MacHardy study related moisture content in the top three zones of a medium soil or the top two zones of a sandy soil to *effective tractability*, using conventional tractors. They were able to establish critical moisture levels in these respective soil zones, above which cultivation was expected to be impossible. By examining the weather for the period 1931–60, using this relationship, the probability of a day being unsuitable for cultivation was calculated for each day from 1 April to 15 June (the assumed last day of seeding). These values were used in the simulation with the assumption that a non-working day in terms of cultivation is a nonworking day for all similar operations (see Table 7).

8.3.4.7. Machine Performance

The operating performance of the alternative sized machinery systems is assumed to be adequately described by (a) estimated rates of work, related to the width of the respective implements and the speed at which they are pulled, (b) a time loss factor, representing turning and adjustment losses, and maintenance and repair time, and (c) a constraint on the hours operated per day, depending on the number of operations employed.

The rate of work was calculated for all five sizes of the three types of implement, using the formula:

$$C = \frac{R.W.e}{8 \cdot 5}$$

where C = capability in acres per hour, R = forward speed in miles per hour, W = width of cut in feet and e = efficiency factor to allow for

TABLE 7

Probability of a non-workday, medium to heavy soil, Melfort (Saskatchewan) 1931-60

Date	April	May	June	July	August	September	October
1	1·00	0·77	0·40	0·37	0·27	0·37	0.73
2	1·00	0·70	0·40	0·30	0·23	0·47	0·70
3	1·00	0·73	0·37	0·37	0·17	0·37	0·67
4	0·97	0·70	0·40	0·27	0·30	0·37	0·70
5	0·97	0·57	0·37	0·23	0·20	0·40	0·73
6	0·97	0·53	0·37	0·23	0·30	0·40	0·77
7	0·97	0·47	0·40	0·20	0·13	0·43	0·77
8	0·97	0·43	0·27	0·37	0·23	0·57	0·73
9	1·00	0·53	0·17	0·33	0·20	0·47	0·70
10	1·00	0·57	0·13	0·23	0·10	0·47	0·70
11	0·97	0·50	0·13	0·23	0·13	0·50	0·70
12	0·97	0·37	0·20	0·17	0·20	0·53	0·70
13	0·97	0·37	0·20	0·23	0·17	0·57	0·73
14	0·97	0·33	0·23	0·37	0·13	0·53	0·73
15	0·97	0·17	0·37	0·33	0·20	0·57	0·77
16	0·97	0·23	0·40	0·23	0·23	0·50	0·77
17	0·97	0·30	0·33	0·20	0·20	0·60	0·77
18	0·93	0·23	0·37	0·13	0·23	0·63	0·70
19	0·97	0·30	0·33	0·17	0·17	0·60	0·73
20	0·97	0·20	0·33	0·20	0·20	0·57	0·77
21	0·93	0·17	0·27	0·33	0·13	0·63	0·80
22	0·90	0·20	0·17	0·27	0·17	0·63	0·80
23	0·87	0·17	0·27	0·17	0·17	0·63	0·83
24	0·90	0·13	0·20	0·20	0·27	0·60	0·83
25	0·83	0·20	0·37	0·20	0·23	0·63	0·83
26	0·83	0·23	0·40	0·13	0·23	0·67	0·83
27	0·77	0·17	0·60	0·13	0·17	0·67	0·87
28	0·73	0·17	0·50	0·23	0·23	0·67	0·90
29	0·77	0·33	0·47	0·27	0·20	0·67	0·90
30	0·67	0·37	0·33	0·10	0·33	0·77	0·90
31		0·37		0·20	0·30		0·90

turning and other time losses, including removal of blockages and filling seed and fertiliser boxes.

For the *cultivator and discer*, the size was adjusted to suit the different tractor sizes, assuming a draft requirement of 250 lb per foot of width, with the drawbar horsepower of the tractor being 65% of the PTO horsepower at a forward speed of 4 mph. It is also assumed that an efficiency factor of 82·5% for the cultivator and 60% for the discer is applicable.

For *drag harrows*, a draft of 50 lb/ft and a forward speed of 5 mph with an efficiency factor of 82·5% was assumed.

8.3.4.8. Fixed and Variable Costs
These costs were estimated for each machine in all systems, using data from a variety of sources.

8.3.5. Seeding Model Specifications
This model is constructed to simulate five alternative tilling and seeding systems over any acreage range in any two locations. The variables considered are:

    Biological tolerances—crop yield
                        —variable starting date
                        —yield effects
                        —harvest timeliness penalty
    Weather constraints—seeding time limit
                        —variable soil tractability (related to weather)
                        —soil types
    Machine performance—rates of work

For each location every machine system is used over 1000 trials, each representing a single season. At the beginning of each trial a starting date is selected from the specified distribution. The three implements are used over the same acreage in sequence. The whole area is cultivated first at the given rate for each system, operating 10 h per day. Once cultivation is completed, the area is seeded, using the discer, seeding for four successive working days, then stopping to harrow that area. The seeding and harrowing sequence is repeated until the job is complete.

If the selected starting date is later than a specified date, then the first cultivation is omitted and a penalty charge is incurred, related to the estimated loss of yield resulting, and no variable costs are incurred for cultivation. If the seeding operation is not completed by a second specified date, then the rate of work is doubled, to simulate the working of a double shift, and an additional variable cost of $2 per hour is incurred, representing the opportunity cost of the extra labour. If seeding is not completed by a third specified date, then all operations stop and a cost is incurred equal to net revenue from the unseeded acres.

The tilling and seeding sequence is simulated over successive 25-acre intervals up to a specified limit. Each spring operating period is based on a series of individual days which together comprise the total available days. The probability of being able to work on any one day (based on soil tractability as determined by the soil moisture budget) is determined, using random numbers as for selecting rain-free days in the harvesting model.

Over consecutive days, beginning on the starting date, the cost of the yield timeliness effects is accumulated from the array of positive and negative yield effects. In addition, the cost of the harvest timeliness effects is accumulated, using the estimated daily additional cost derived from the harvesting model.

When tilling and seeding have come to an end, the cost of the operation is formed from the cumulative sum of the fixed and variable machine costs, and the cost of the yield and harvest penalties. The total cost is then computed for successive 25-acre levels, and accumulated for the number of trials, according to the identity:

$$TC_{ai} = F_s + (V_k \cdot N_k) + (V_w \cdot N) + (V_h \cdot N) + L_y + L_h$$

where TC_{ai} = total cost at acreage a for machine system i, F_s = fixed cost of system, V_k = variable cost of cultivating, V_w = variable cost of wide-levelling (discing), V_h = variable cost of harrowing, N = acreage seeded, N_k = acreage cultivated, L_y = cost of yield-time effects and L_h = cost of harvest-time effects. From this, the average and marginal costs per acre are calculated. The standard deviation of these values is computed for each 25-acre level and all values, together with the range and frequency of the average cost, are printed.

8.3.6. Model Experiments

Experiments were conducted on this model by varying the rates of work to allow for differences in efficiency between the different-sized systems. Since the rate of work is a very significant variable in these models, in that it is multiplied by very large numbers (acreages), the deterministic values employed must be considered inadequate to characterise the various systems. Some of the disadvantages associated with using these figures are overcome by experimenting with adjusted figures. Some of the results are shown in Fig. 14–16 (See Section 8.4.1).

8.4. INTERPRETATION OF MODEL OUTPUT

The output from the models is in the form of average and marginal costs for each successive 25-acre level of use. Since each model includes a range of machine sizes, each experiment yields a family of such curves. Given each experiment is based on a large number of trials (usually 1000), we can plot the distribution of the cost variation about the mean average or marginal cost for alternative systems over each acreage.

The cost computation takes account of fixed costs, direct variable costs and the opportunity costs associated with each particular system in the form of lost revenue. In some cases revenue gains are offset against the direct costs of each system. Thus the cost curves reflect all the variation in profitability for these particular farm operations. In other words, the average revenue will equal marginal revenue and the curves will be linear and horizontal. The types of cost curves obtained are shown in the accompanying Figs. 3–16.

In using these results, consideration of one curve is useful only for a short-run decision, such as choosing a level of production given the availability of one system. In considering the long-run question of which machine system best suits a particular production level, the costs of interest to the decision-maker are represented by the long-run cost curve delineated by the least-cost range of the successive short-run curves. In addition to the cost curve, however, we have the density function for the average cost at each acreage level. Thus the decision-maker can examine expectations for the cost variation that might occur from one year to another. Since this represents a situation of decision choice between probabilities it could be analysed further using a Bernoullian framework, but this has not been pursued here.

8.4.1. Results of Experiments

Figure 3 shows the typical average and marginal cost curves obtained. These are for a harvest with minimal time constraints due to competing operations. The scale advantages of being able to use larger machine systems are apparent, but these are offset somewhat when contract services and older or secondhand machines are included.

The effect in cost terms, of modifying the time available for harvest as may be due to competing activities, is shown in Fig. 4. The idiosyncratic crossing over of the average cost curves reflects inadequate specifying of the contract costs at the higher acreages. This was intentional on the grounds that contracting for large acreages is not considered a feasible alternative.

Altering the cost bases for the fixed (or overhead) costs has a predictable effect illustrated in Fig. 5. This provides some evidence in support of the desirability of examining alternative expenditures of a medium-term capital nature.

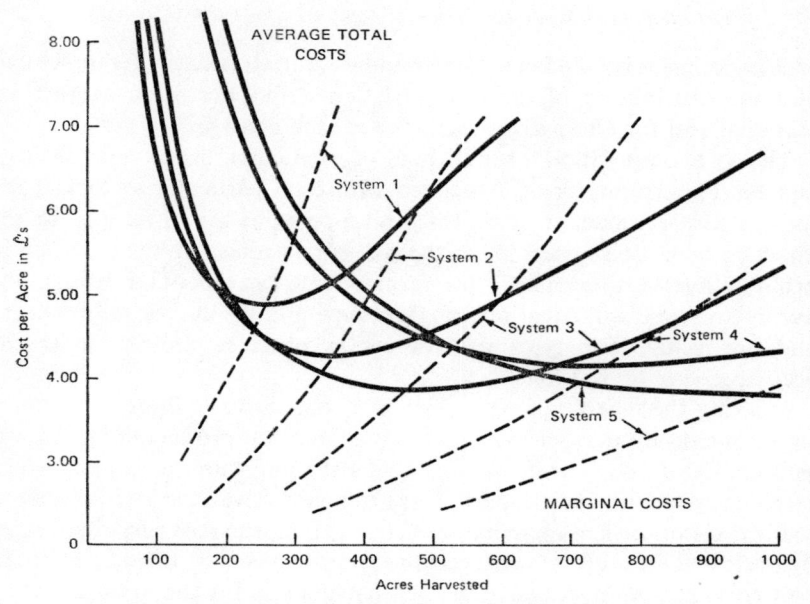

FIG. 3. Cereal harvest costs—Britain, nine-week harvest.

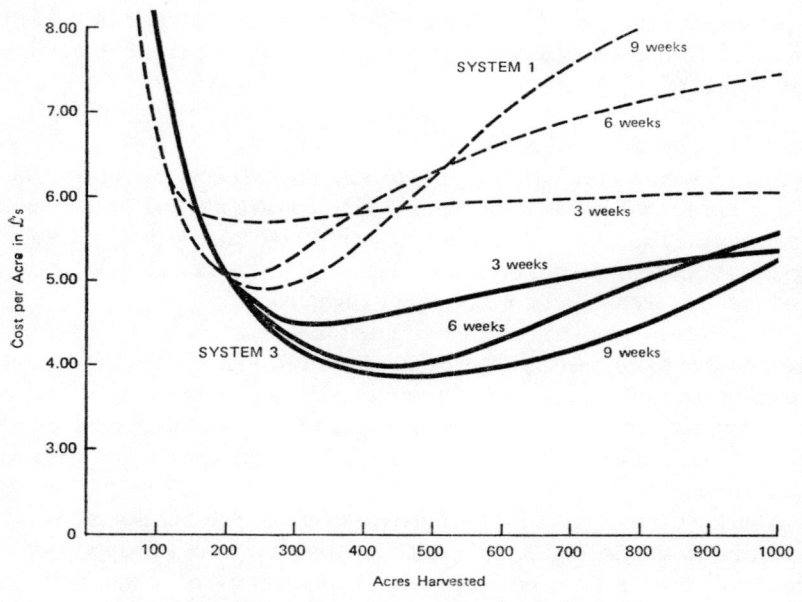

FIG. 4. Cereal harvest costs—Britain, modified harvest time.

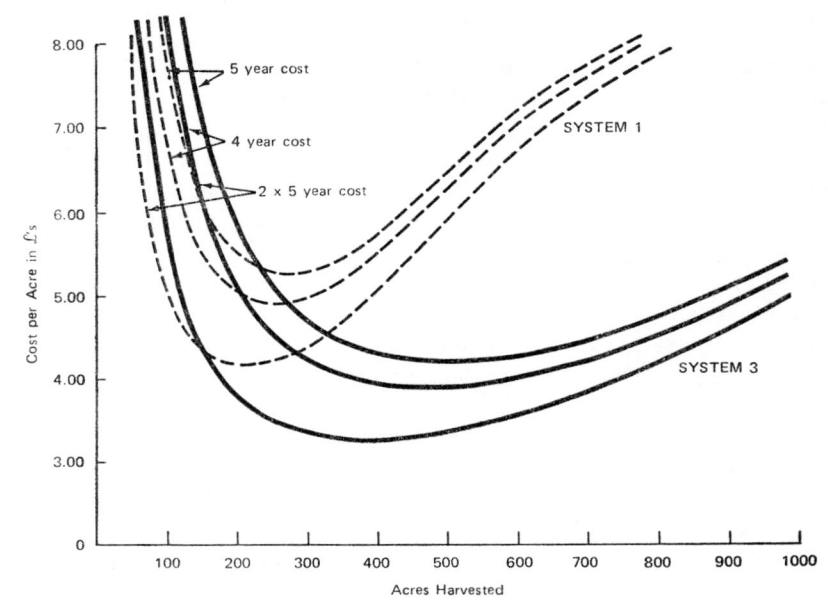

FIG. 5. Cereal harvest cost—Britain, alternative cost bases.

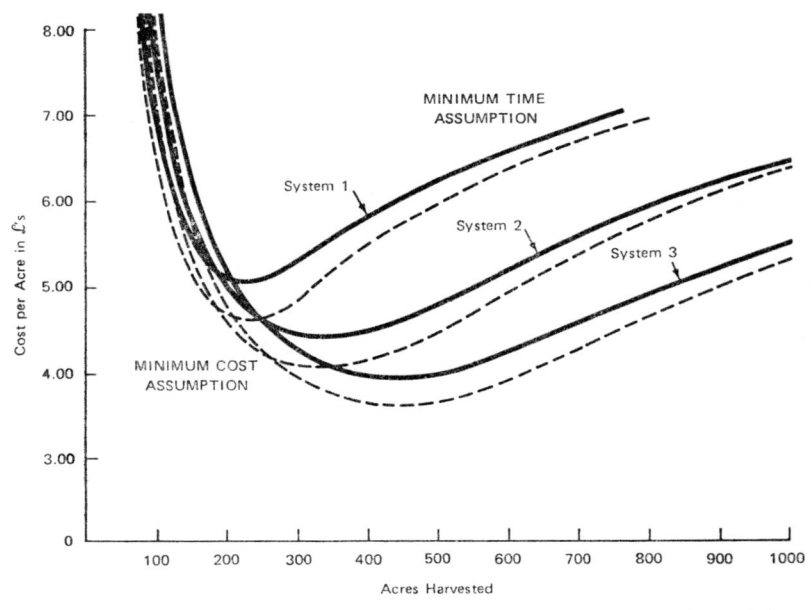

FIG. 6. Cereal harvest costs—Britain, minimum time vs. minimum costs, six-week harvest.

The effects of pursuing the 'minimum time' or 'minimum cost' management strategies are suggested by the results shown in Fig. 6. The difference between the costs for the two strategies is greater for the longer harvest, since the 'minimum cost' strategist encounters even lower drying costs. The costs that might be encountered if the drier were abandoned altogether are shown in Fig. 7. The variability of the 'without drying' costs is also much greater than for the alternative.

The kind of cost variation that might be encountered is shown by the distribution curves, printed sideways on the cost curves, presented in Fig. 8. The increased variability encountered as the capacity of the machine system is extended helps to explain farmers' tendencies to select larger machines than static analysis would suggest was appropriate.

The somewhat different shaped cost curves obtained for the Canadian situation are shown in Fig. 9. The lower slope of the marginal cost curves indicates the much less severe operating constraints on harvest in the Canadian as opposed to the British situation. The fact that some of the machine systems never provide a least-cost alternative raises the possibility that the manufacturers' range is not well suited to this location. However,

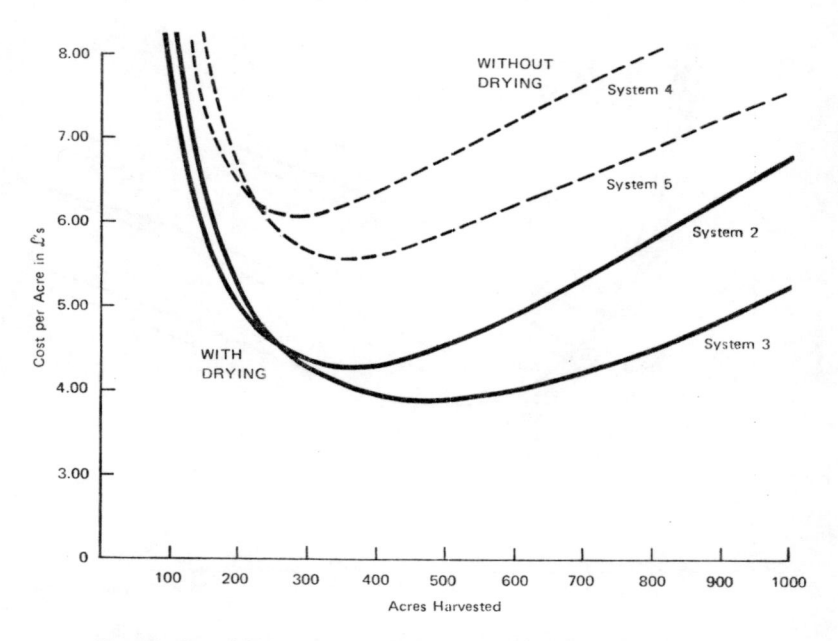

FIG. 7. Cereal harvest costs—Britain, combine–drier substitution.

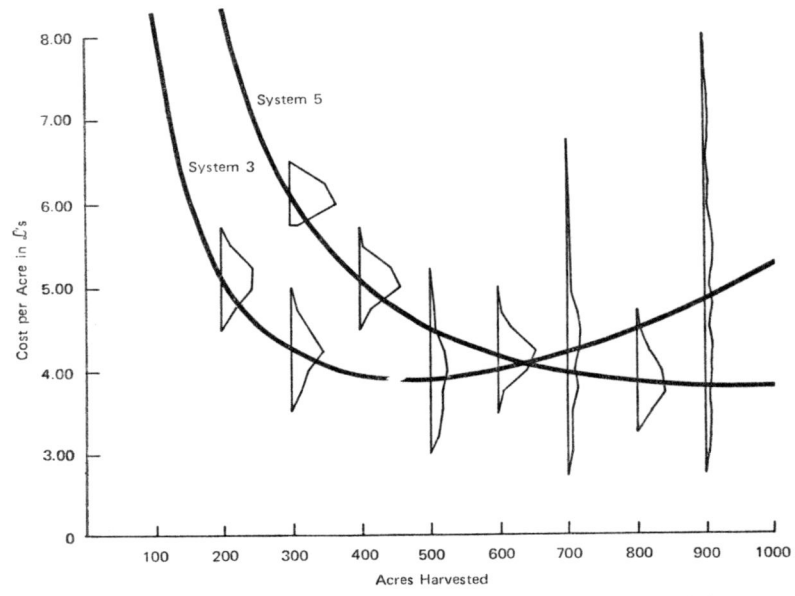

FIG. 8. Cereal harvest costs—Britain, variance distribution, nine-week harvest.

a review of the cost density curves indicates that all four systems may provide minimum risk alternatives at various levels of use.

The differences in costs between locations is shown in Fig. 10. Clearly the more northerly location has the more serious constraints. The difference in order for the other three locations is largely a response of the weather pattern—Location 1 gets more summer storms than those further north. The differences in risk confronting the farmer at different locations are implied by Figs. 11 and 12. Not only does the cost variation increase with extension of acres harvested, but it increases more rapidly in the more difficult location. The overall higher average cost at Location 4 reflects the higher crop yields.

The benefits from including drying in the harvest operations at Location 4 are suggested in Fig. 13. The benefits are much lower at the other locations. The opportunity benefits to be obtained by 'delayed shelling' are also shown here. Such a delay can be obtained by swathing the crop before combining. Alternatively, it might be obtained by plant breeding for a reduced shelling propensity.

The nature of the cost curves obtained from the seeding model are shown in Figs. 14 to 16. The more even sequential juxtoposition of these average

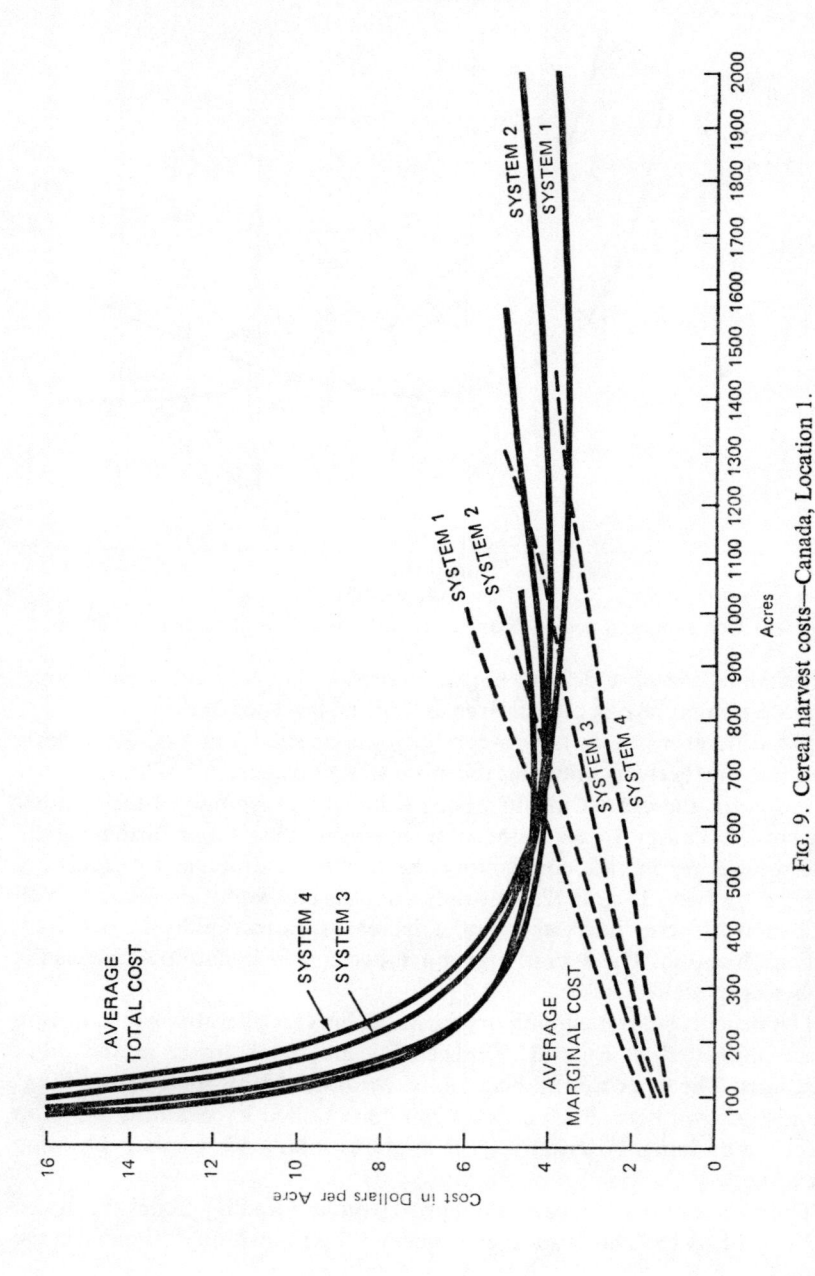

Fig. 9. Cereal harvest costs—Canada, Location 1.

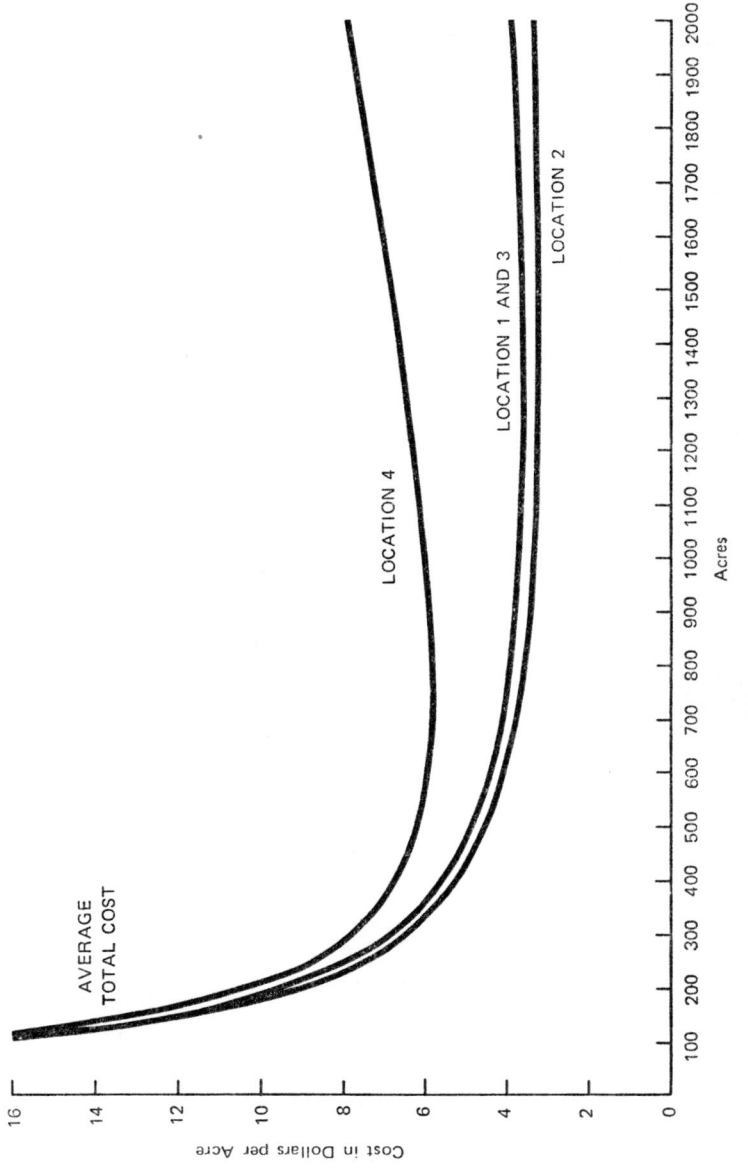

The Study of Agricultural Systems: Application to Farm Operations 295

Fig. 10. Cereal harvest costs—Canada, System 4.

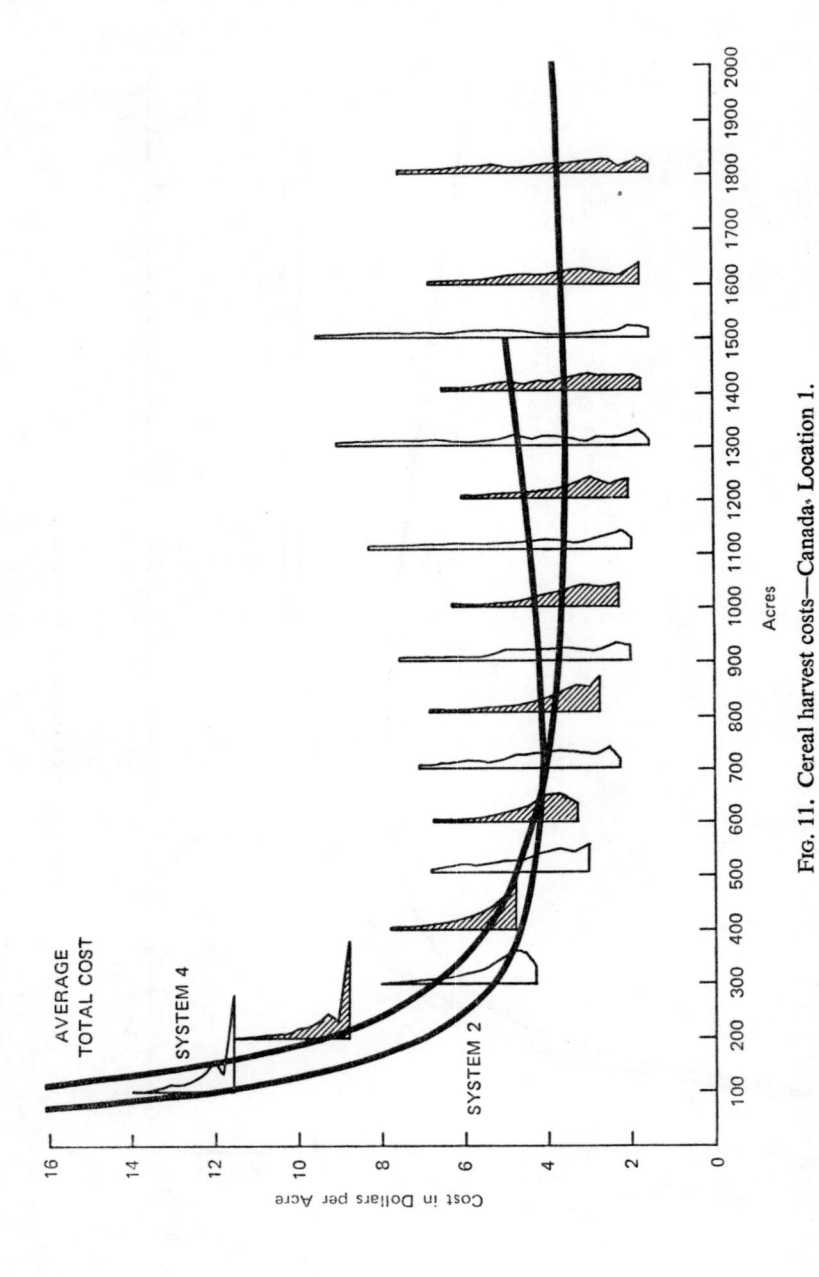

Fig. 11. Cereal harvest costs—Canada, Location 1.

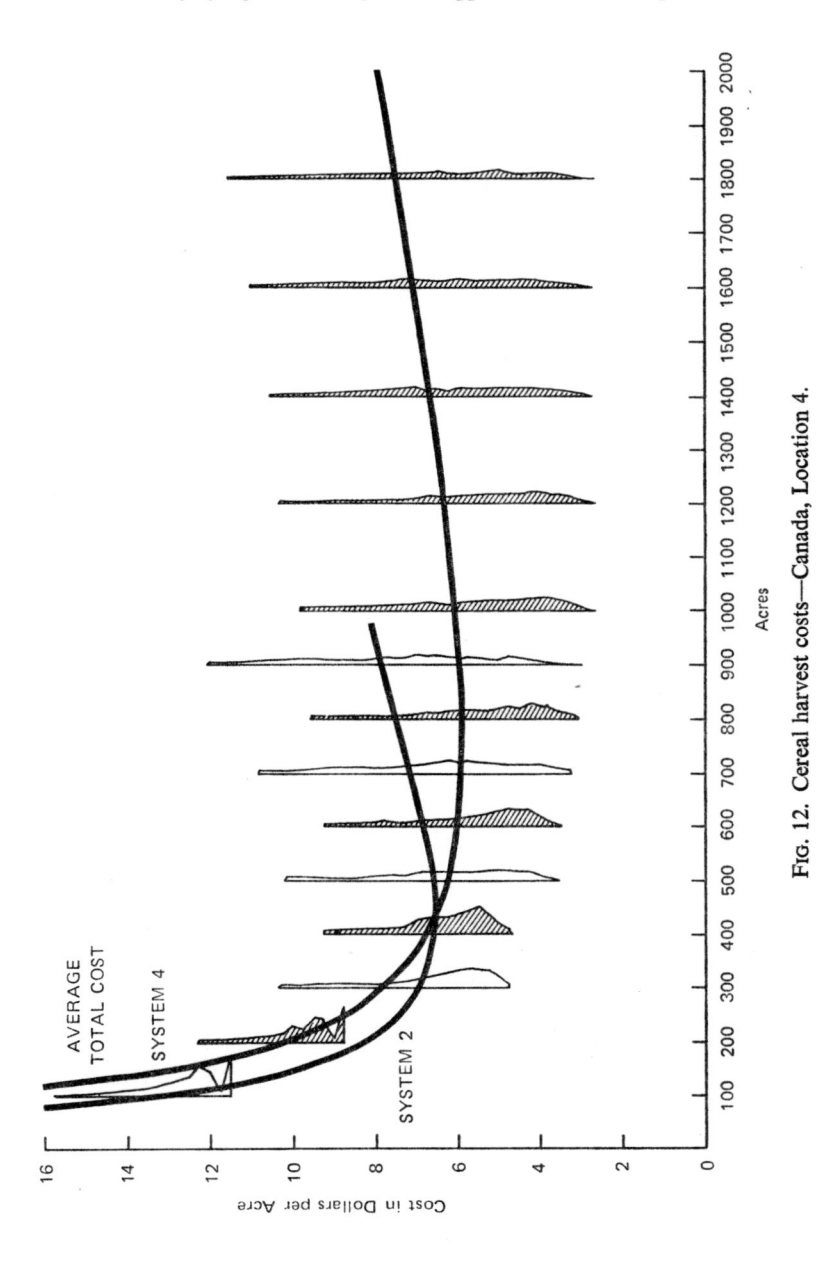

The Study of Agricultural Systems: Application to Farm Operations 297

FIG. 12. Cereal harvest costs—Canada, Location 4.

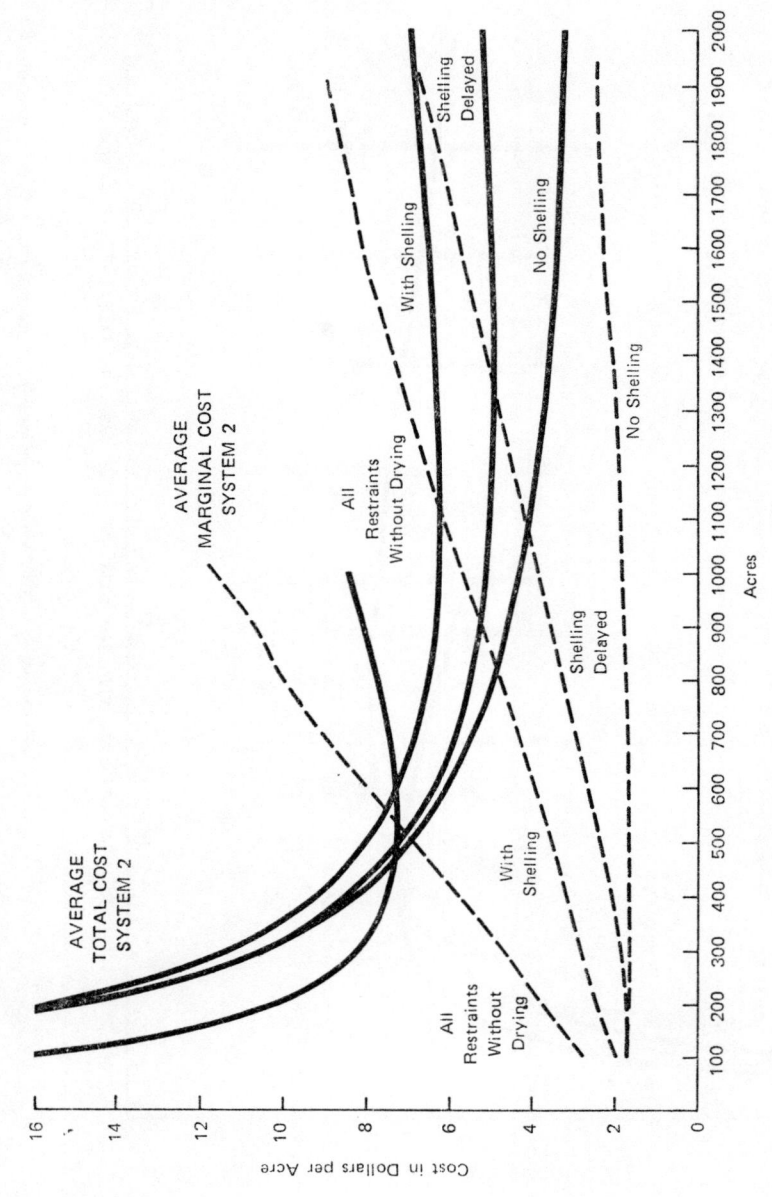

FIG. 13. Cereal harvest costs—Canada, including grain drying, Location 4.

cost curves, relative to those for harvesting, is partly a reflection of the relatively fixed rates of work used in the model, but also suggests the adequacy of the alternative systems to the particular location. The much greater slope of the marginal cost curves further suggests the acreage specific character of these alternative systems. Comparing the cost curves from the harvest and tillage models respectively, it seems that the more critical decision in terms of cost incurred may relate to the tractor and tillage system. However, the alternatives seem somewhat more clear-cut.

The relatively lower cost variability at comparable acreages is suggested by Fig. 14, though the risk still increases as the machine system is extended. However, there is also a slightly higher variability at low acreages which reduces as the system is extended. This effect is created by the assumption that seedbed preparation will start on the same date regardless of the acreage to be cultivated and is unlikely to occur in practice.

The cost differences on different soil types in the same location is illustrated in Fig. 15. The relative importance of this variable to farm management advisers is evident.

The cost variability is indicated by Fig. 16. The reduced risk associated with having 'excess capacity', or a machinery system larger than the minimum cost alternative, can be seen from the distributions about these cost curves.

8.4.2. Model Validation

Validation began, as is appropriate, at the outset of the modelling procedure and probably never ends! To some extent everyone who makes use of the results goes through their own assessment of the validity of the model. The results were also verified against intuitive expectations, and an *ex post* comparison was made between the 'optimal' or minimum cost alternatives and the machine systems actually found on farms.

The variables to be included in the models were chosen after many discussions with research station scientists, farm economists, extension personnel, and farmers. Just as the scientist searches for *a priori* postulates that are acceptable, so the model-builder seeks to incorporate those aspects of reality that are intuitively recognised to be significant in the system.

In the subsequent stage of quantifying the variables, every effort was made to check the accuracy of observations contained in the data obtained. Where possible, an attempt was made to identify the relationships involved, by applying statistical tests and by reference to other studies where such tests have been used. By empiricist scientific standards, the extent of

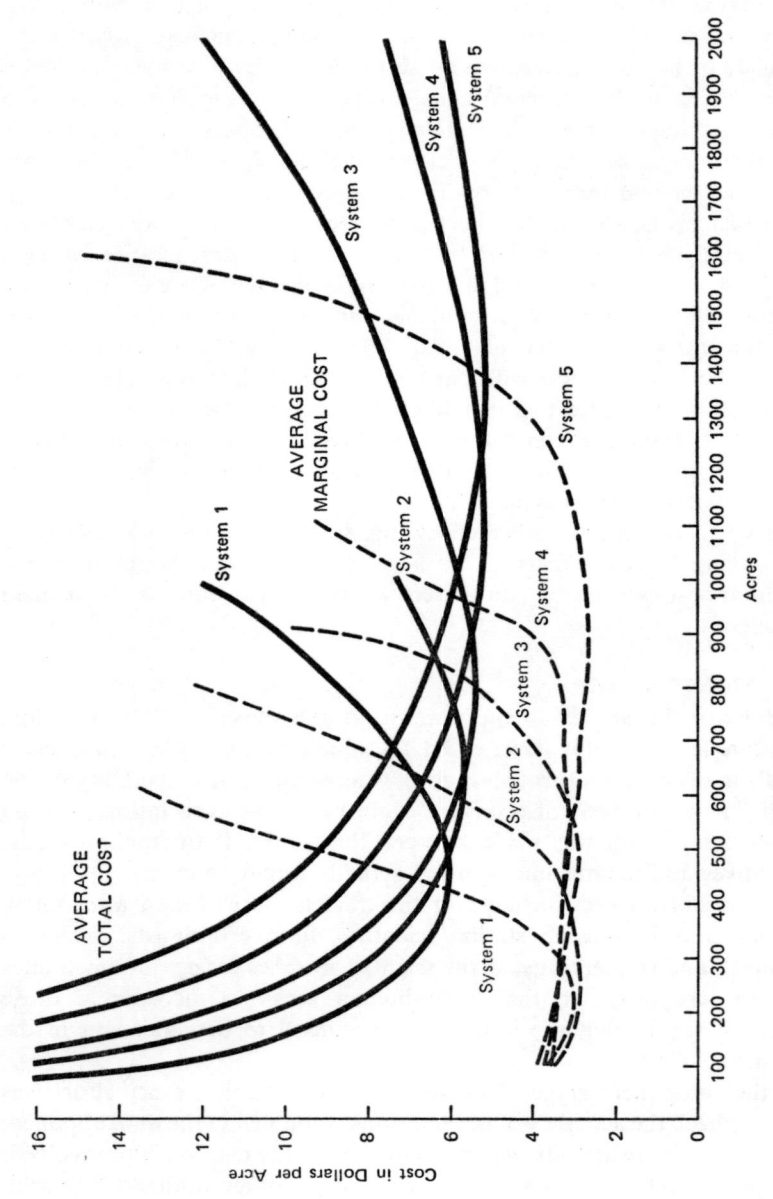

FIG. 14. Cereal, tillage and seeding costs—Canada, Location 1.

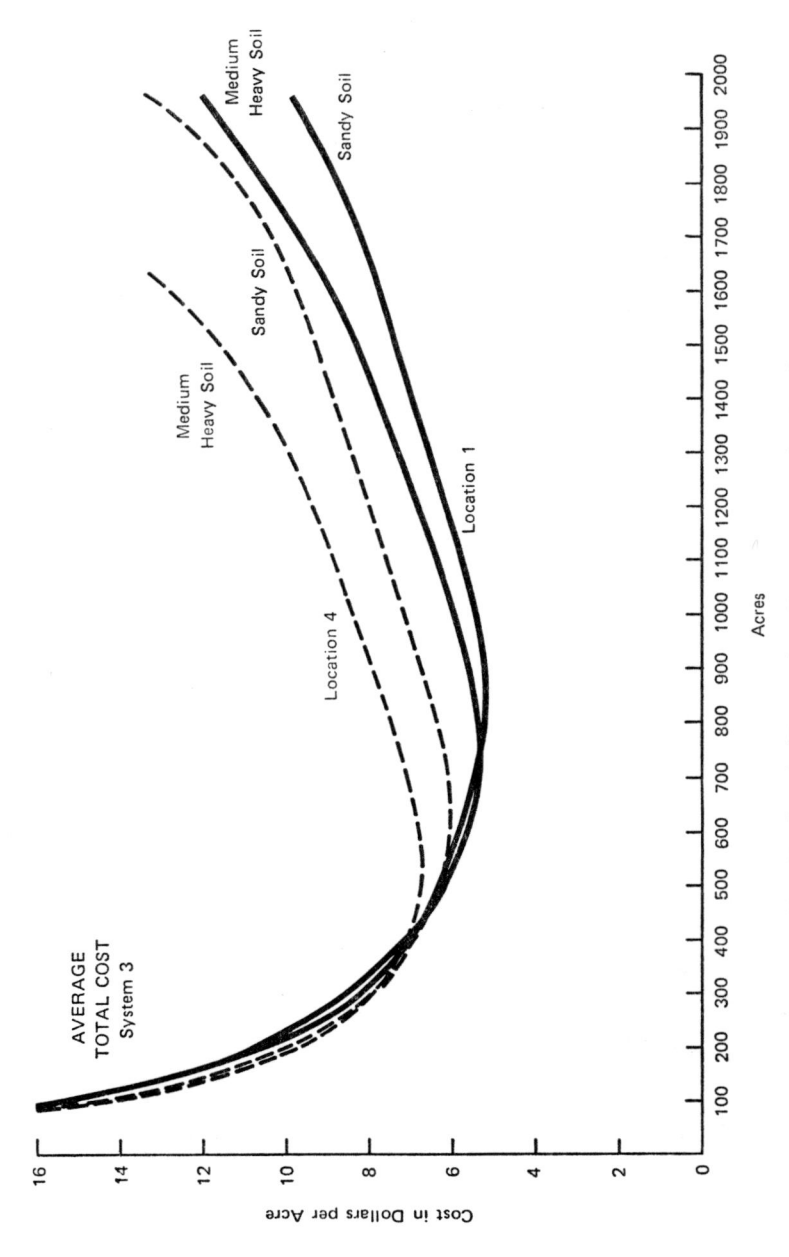

FIG. 15. Cereal, tillage and seeding costs, Location 1 and 4.

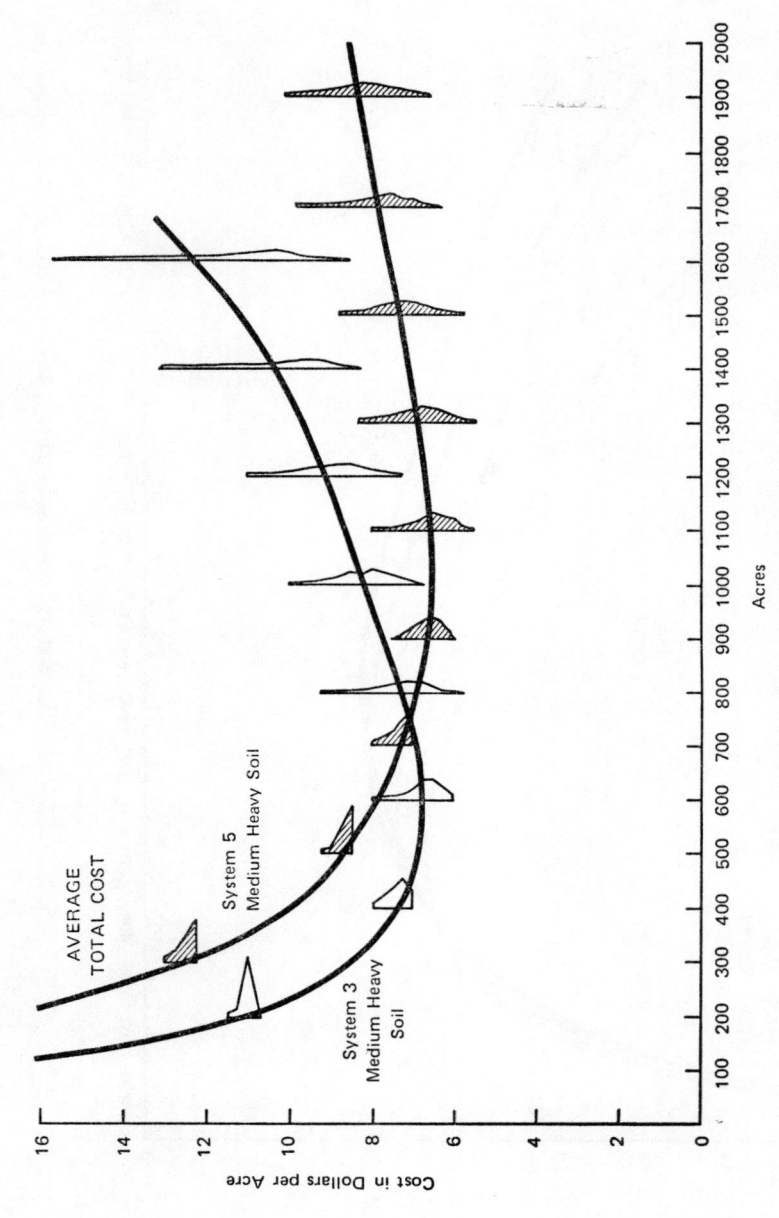

FIG. 16. Cereal, tillage and seeding costs, Location 4.

statistical testing is inadequate, but the form in which the variables considered are introduced into the model helps to overcome this inadequacy. The distributions used are regarded as discrete and are based on unmodified historical data and not on fitted mathematical expressions. Where operating relationships had been little explored, an experimental approach was used in reaching values that fitted reality (for example, the weather-lag effects in the harvesting model).

Once the models had been built and the first results obtained, the *ex post* validation was made. To do this, for the harvesting model in the British case, information on the size of combine owned and the acreage of cereals harvested on farms was obtained from farm survey records. Using the minimum cost ranges from the study, a comparison was made between the optimal acreage range (on a minimum cost basis) and the acreage harvested. The results of this comparison are shown in Fig. 17.

Certain assumptions made in the course of this comparison may, to some extent, explain some of the observable differences. For instance, the combines in the model were all self-propelled, whereas many in the sample, particularly in the small category, were tractor-drawn, and so their operating capacity might be different. Similarly, it is assumed that the acreage harvested in the years of the survey (which covered different farms in five different years) is the acreage for which the machine was purchased. This may not be so, since some may be planning expansion and others cutting back on their cereal acreage. Again, it is assumed that the combines in the survey can be accurately represented by cost estimates relevant to new machines in 1968. Finally, the use of the minimum cost alternative as the 'mark' for comparison takes no account of the risk aversion of the farmer. The wide variation in the degree of capacity utilisation suggests, too, that farmers are individualistic in their preferences or that they have difficulty in obtaining the right size of machine to do the job. The degree of coincidence of practice and projections as seen in Fig. 17 nevertheless gives some support to the validity of the model.

The tillage model is the most inadequate. Although the soil water budget and tractability relationships seems well identified and reliable, the enforced use of deterministic rates of work greatly reduces the value of the model, especially when these rates are multiplied by such large acreages.

The results presented above, like the output of all models, should be regarded with circumspection. On the other hand, some additional information of a general kind can be obtained from the models, and some comparative evidence is valid.

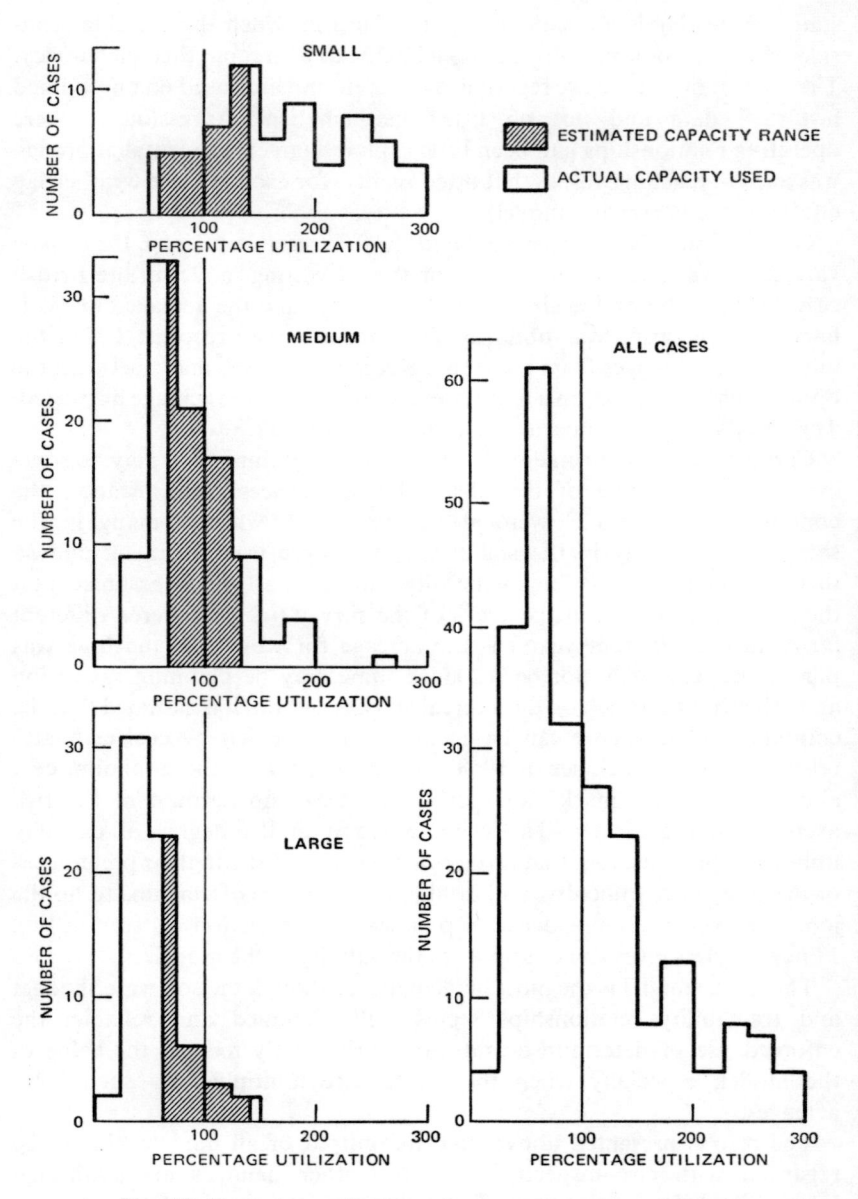

FIG. 17. Percentage utilisation of estimated combine capacity.

8.4.3. Data Problems

In all cases the adequacy of the model is determined by data constraints. Where historical data is required, we can make better provision for data collection in future and better analyses of existing records. Where experimental data can be used, as for all the technological variables, there is scope for the collection of specific data for such a model. As in research generally, the value of models is that they clearly indicate the particular data that might usefully be recorded. The models described above suggest the usefulness of farm machinery testing.[26] The use of this type of approach would therefore seem to be particularly useful in developing countries in two respects. First, because of the paucity of data and the need to restrict data collection to a minimum in order to keep costs down. Secondly, because the implications of new technology (whether transferred or local, modern, old or intermediate) for yields, costs, reliability of output, input needs, employment or other side effects, are very frequently not clear until the interactions of the components of the real world agricultural system have been observed. This can be done most effectively using systems analysis.

REFERENCES

1. Donaldson, G. F. (1966). A guide to decisions on optimum combine capacity, *NAAS Quarterly Rev.*, no. 74.
2. Candler, W. (1960). The Purdue corn harvesting, drying and storage simulator, Seminar on Ration Formulation and EDP Decisions, Oklahoma State University, Stillwater.
3. Cloud, C. C., Frick, G. E. & Andrews, R. A. (1968). An economic analysis of hay harvesting and utilisation using a simulation model, University of New Hampshire Agricultural Experiment Station Bulletin, No. 495, Durham, N. H.
4. Dalton, G. E. (1971). Simulation models for the specification of farm investment plans, *Journal of Agricultural Economics*, 22(2), 131–41.
5. Dalton, G. E. (1974). The effect of weather on the choice and operation of harvesting machinery in the United Kingdom, *Weather*, 29(7), 252–60.
6. Donaldson, G. F. & McInerney, J. P. (1967). Combine capacity and harvest uncertainty, *Farm Economist*, 11(4).
7. Donaldson, G. F. (1968). Allowing for weather risk in assessing harvest machinery capacity, *American Journal of Agricultural Economics*, 50(1).
8. Frisby, J. C. & Bockhop, C. W. (1966). Influence of weather and economics on corn harvesting machinery systems, American Society of Agricultural Engineers, Paper No. 66–306.
9. Gemmil, G. T. (1969). Approaches to the problem of machinery selection, unpublished M.Sc. thesis, Department of Agriculture, University of Reading.

10. Hunt, D. R. (1969). A systems approach to farm machinery selection, *Journal of Institution of Agricultural Engineers*, **24**(1).
11. Jose, H. D., Christensen, R. L. & Fuller, E. (1971). Consideration of weather risk in forage machinery selection, *Canadian Journal of Agricultural Economics*, **19**(1), 98–211.
12. Link, D. A. & Bockhop, C. W. (1964). Mathematical approach to farm machinery scheduling, *Transactions of the ASAE*, **7**(1).
13. MacHardy, F. V. (1966). Programming for minimum-cost machinery combinations, *Canadian Agricultural Engineering*, **8**(1).
14. Ryan, T. J. (1973). An empirical investigation of the harvest operation using systems simulation, *Australian Journal of Agricultural Economics*, **17**(2), 114–26.
15. Sorensen, E. E. & Gilheany, J. F. (1970). A simulation model for harvest operations under stochastic conditions, *Management Science*, **16**(8), B459–B565.
16. Tanago, A. G. (1973). Haymaking machinery selection under risk: a simulation approach, University of New England Farm Management Bulletin No. 17, Armidale, New South Wales.
17. Van Kampen, J. H. (1971). Farm machinery selection and weather uncertainty, in: *Systems Analysis in Agricultural Management* (ed. J. B. Dent and J. R. Anderson), Wiley, Sydney; pp. 295–329.
18. Rutherford, I. (1973). The use of a deterministic cost model for advisory work, *Proceedings of the Symposium on Systems Applications in Agricultural Engineering*, National Institute of Agricultural Engineering, Silsoe, England.
19. Donaldson, G. F. (1970). Optimum harvesting systems for cereals: an assessment for South-East England, Economics Department, Wye College, University of London.
20. Donaldson, G. F. (1970). Farm machinery capacity: an economic assessment of farm machinery capacity in field operations, Royal Commission on Farm Machinery, Study No. 10, Queen's Printer, Ottawa.
21. Fajersson, F. & Krantz, A. M. (1965). Studies on resistance to shattering in varieties of winter and spring wheat at Weibullsholm during the period 1952–1965, *Agri. Hortique Genetica, Landskrona*, **23**(3–4), 101–71.
22. Arnold, R. E. (1955). Cereal moisture contents in the field, *Farm Mechanisation*, **12**(69).
23. Crampin, D. J. & Dalton, G. E. (1967). The determination of the moisture content of standing grain from weather records, *Journal of Agricultural Engineering Research*, **16**(1), 88–91.
24. Duckham, A. N. (1966). The role of agricultural meteorology in capital investment decisions, Dept. of Agric., Report No. 2, University of Reading.
25. Rutledge, P. & MacHardy, F. V. (1968). Influence of weather on field tractability in Alberta, *Canadian Agricultural Engineering*, **10**(2).
26. Donaldson, G. F. (1970). Farm machinery testing: scope and purpose in the measurement and evaluation of farm machinery, Royal Commission on Farm Machinery, Study No. 8, Queen's Printer, Ottawa.

9

Economic Analysis of Farms

S. C. THOMPSON

CANFARM Data System, Canada Department of Agriculture, Guelph, Ontario, Canada

9.1. INTRODUCTION

The largest single problem facing farm operators since the early 1960s has been the flow of cash. Short- and long-term borrowing needs have taken over from land and labour as the most limiting constraints on the farm business. Correct phasing and recycling of funds at the firm level therefore assumes a new importance in the analysis and planning of enterprises. The time scale involved in cycling funds can vary from the monthly flows of a livestock farmer or horticulturalist to the investment problems of new buildings for grain or livestock. Technical husbandry problems, while still forming a major portion of the work carried out by farm advisers, can be sufficiently well identified for effective remedial measures to be taken. Technical relationships are also sufficiently well known for planning decisions based upon them to be made with reasonable confidence—at any rate, given that a problem exists, no better information is available for solving it!

9.2. A FARMER'S VIEW OF SYSTEMS

Many farmers are aware to their cost that 'whole farm' advice purveyed to them in the 1960s—especially computerised advice—fell short of what they really needed in order to decide upon a viable course of action. Many more farmers are not aware of the degree to which decision aids can successfully be used. More importantly, they wish to reserve to themselves that final decision which integrates all the parts into a whole. Consequently the type of help most often requested by farmers is for market information on machinery, feed, and seed costs. Their decisions appear frequently to be of a partial nature—they replace individual parts of a system rather than the system as whole.

Partial decisions do allow the farmer flexibility to follow the farming environment, where improvements to machines and husbandry techniques occur more by stealth than by revolution; and they do tend to smooth out cash outflows and inflows. The danger exists, however, that tactical problems may obscure long-term targets, and a farmer operating an unprofitable system may descend beyond help before he manages to evolve out of that system.

Is there then a place for a systems approach in planning aids developed at the farm level? Market research would suggest not. Farmers and advisers are asking extension services for advisory aids such as: machinery replacement, crop insurance, least-cost rations, least cost fertiliser mix, cropping timeliness, silo capacity, tractor size (to match present system) and livestock break-even analysis. Very few requests have come from the farmer himself, or his advisers, for services which combine livestock or crop enterprises together in a systematic way or on an aggregate basis[1] as shown in Fig. 1.

CANFARM
FARM PLANNING
WORK IN PROGRESS

Crops:
  Budgeting
  Quota grains
  Machinery replacement
Stock:
  Budgeting
  Least-cost rations
  Dairy planning
  Beef planning
Whole farm:
  Maxiplan (LP)
  ADC cash flow
  Management game
  Loan calculator

MAJOR REQUESTS FOR
FARM PLANNING, 1974

Crops:
  Insurance
  Share/cash rent
  Machinery planning
  Timeliness
  Fertiliser mix
  Silo capacity
Stock:
  Dairy, beef, hog rations
  Break-even investment
  Dairy one-year plan
  Dairy investment plan
Whole farm:
  Cash forecast
  Management games
  Retirement/estate planning

FIG. 1.

9.3. POLICY IN AGRICULTURE

Farmers and advisers, however, are not the sole customers for farm management aids. A rising trend is shown in the number of agricultural institutions which require farmers to supply them with information in a standard and comparable way before their services will be made available to the farmer. Banks and agricultural credit agencies are using computerised cash flow forecasting packages to assess loan repayment ability. Ministries of Agriculture are beginning to use computer packages in special programmes such as agricultural adjustment schemes and retirement planning, while marketing boards require the farmer to make detailed (and computerisable) calculations regarding the allocation of market quotas.

It is ironic that those management aids which approach the system as a whole should be initiated for forensic purposes, such as the French IGER recording scheme, rather than by the farmer as management aids in their own right.[2] Government organisations in countries with developed agricultural advisory networks have tended to sustain the systems approach, even in the face of contrary market research, and most are developing interlocking services along the following lines.[3]

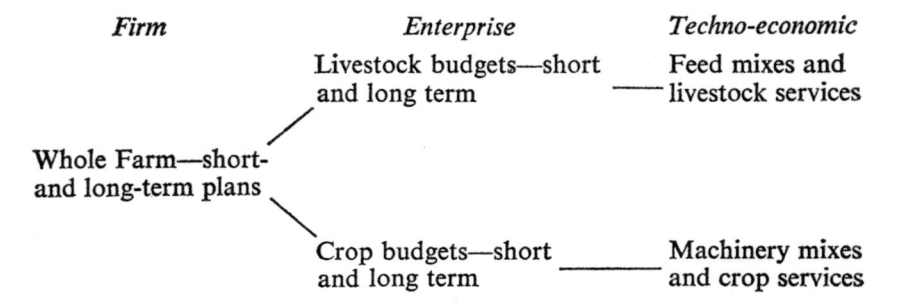

Firm	*Enterprise*	*Techno-economic*
	Livestock budgets—short and long term	Feed mixes and livestock services
Whole Farm—short- and long-term plans		
	Crop budgets—short and long term	Machinery mixes and crop services

A similar format of interlocking programmes is also beginning to emerge in Eastern Europe, where governments are not so concerned with market research or extension advice. State farms and cooperatives are learning to use computers in planning the most efficient way to meet production targets set at regional level, and not surprisingly are using techniques similar to those used in the West—feed mix programmes, least-cost machinery complements, critical path methods for harvesting, and the like.[4]

Governments, whether they be in the East, West or Far East, are already heavily involved in the planning of agriculture, the manipulation of production, and the formation of stable farming environments within trading boundaries. What, then, can an individual farmer hope to achieve by taking the initiative in planning his farming system? Is he not increasingly placed in a position where his responsibility is largely to respond in a predictable manner to agricultural policy in the form of grants, subsidies, insurance schemes and production quotas? Responding in a predictable manner would of course require him to understand fully some of the more complex policies, so that it would seem entirely natural (to give a hypothetical example) for the unveiling of a new policy on milk production in the EEC to be accompanied by a computer program, written by the policy-making agency, aimed at helping individuals to make the best decision in line with that policy.

9.4. THE ADVISORY SERVICES IN AGRICULTURE

Farmers do not, however, readily submit to increasing regulation by government, neither have societies in general achieved an agricultural environment sufficiently stable to discourage individual initiatives; though it should be noted in passing that such stability has existed for example in Denmark in the 1950s and early 1960s, when a system of bacon production and marketing was perfected and remained successful for a decade before succumbing to a changed environment. Planning, both at the partial and whole-farm levels, therefore remains important for the farmer who wishes to exercise his freedom of choice and control his own destiny. Extension services in North America are trying to respond to market research findings by building the partial programmes demanded by farmers: and Canadian forecasts of clientele numbers are encouraging, roughly 2000 qualifications in 1974 and 4000 in 1975, from a farming population of 300 000, roughly the same size as Britain's. In addition to these farmer-oriented services, Canadian credit agencies expect to reach another 3000–4000 farmers per year by way of mandatory cash flow forecasting programmes[5] (see Fig. 2).

Far less farmer use, however, is foreseen for the whole-farm services being offered by Western countries. Linear programmes for whole-farm extension have been developed in almost all European countries and in North America, and in many cases were the first computerised aids to be released to advisory services. Forecasts of low farmer response to whole-

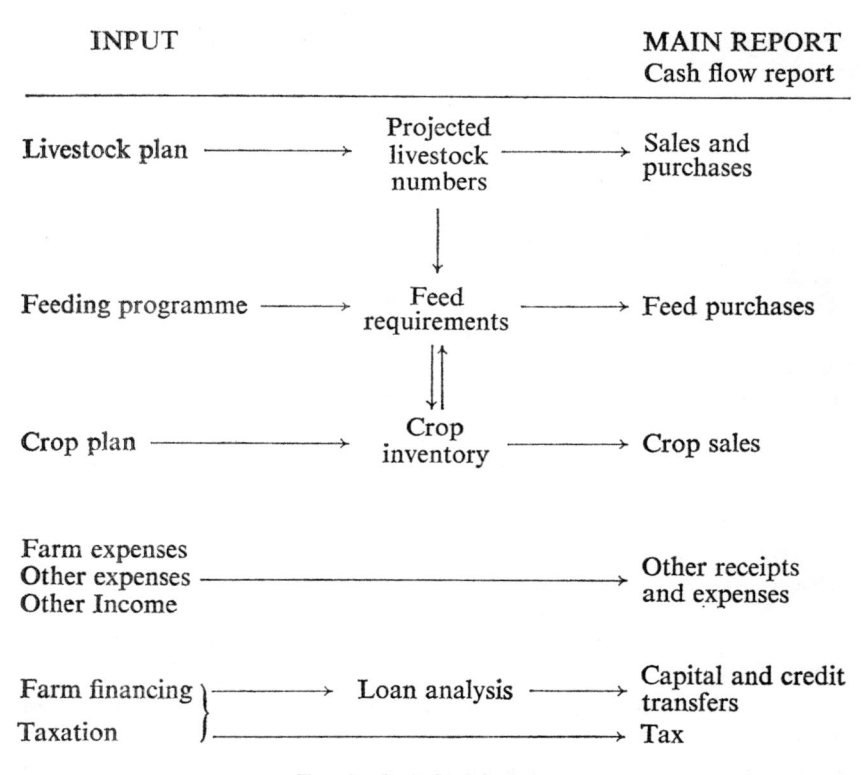

```
INPUT                                          MAIN REPORT
                                               Cash flow report

Livestock plan    ──►   Projected
                        livestock      ──►    Sales and
                        numbers                purchases
                            │
                            ▼
Feeding programme ──►   Feed           ──►    Feed purchases
                        requirements
                            ↕
Crop plan         ──►   Crop           ──►    Crop sales
                        inventory


Farm expenses
Other expenses  ─────────────────────────►    Other receipts
Other Income                                  and expenses


Farm financing ⎫ ──►  Loan analysis  ──►   Capital and credit
                                            transfers
Taxation       ⎭ ─────────────────────►    Tax
```

FIG. 2. Cash flow forecaster.

farm linear programming have not, however, deterred these organisations from developing LP packages, along with programmes to calculate crop and livestock enterprise budgets from disaggregate data.

A working definition of what extension services in general mean by a computerised farm planning service may help to clarify this apparent anomaly. Most of their services are 'capable of aiding farmers by providing quantitative (rather than qualitative) solutions to individual management problems'. This definition in no way rules out the use of such services in group extension, or for research or policy purposes—and it is in the forms of group extension and the production of advisory pamphlets that an initial use is seen for the various whole-farm and enterprise services currently being built and tested. Increased individual use of these services is foreseen for 1976 and beyond, but an important prerequisite will be the provision of standard data for the different enterprises and regions of each country in order to relieve the farmer of the quite large burden of providing all input data himself.

9.5. INFORMATION GATHERING

The provision of data for decision-making brings the consideration of the systems approach at farm level almost full circle—or at least serves to point out a missing link in that circle. Many extension services in North America offer computerised farm and enterprise accounting systems to their clients, while a number of European countries have chosen to promote manual methods of book-keeping by accountancy firms. In the present state of development, both computerised and manual accounting services concentrate on the aggregate aspects of the farm business, such as records for tax purposes.[2] Information at this level of aggregation is not useable at all for the many partial decisions facing a farmer, neither is it detailed enough for use by itself in enterprise and whole-farm planning services. Few farmers are prepared to keep routine farm records at the level of detail required for planning, though most are willing to keep records at the level necessary for monitoring and controlling a plan.

This distinction between the type of information going into a plan and the information coming out of it for monitoring purposes has been recognised by economists for many years, and extension services are now beginning to build data banks for use in advisory work. West Germany for example has begun work on a bank of agricultural production functions designed to take more than 30 man-years.[6] Canada, which operates a farm accounting system for 10 000 farmers, is assembling a bank of data based on the records of those farmers who agree to the release of their data for public use.

Even a data bank on the West German scale, however, is unlikely to fulfil all the requirements of farm planning packages and farmers will continue to rely heavily upon daily market information, farm commodity futures, and the farming press.

9.6. COSTS AND BENEFITS

The essential administrative problem facing agencies involved in building advisory services is that of trying to forecast whether or not a service will be a success before any major resources are committed to it. Budget controllers are becoming increasingly concerned with potential benefits from public service programmes, and the agricultural sector is being called upon to justify its programmes using criteria similar to those used by industries such as transport and communications. Three levels of concern can be identified:

(i) Costs and benefits for the farmer.
(ii) Costs and benefits for the adviser.
(iii) Overall costs and benefits of the programme.

The enormous difficulties of forecasting benefits at each of the three levels—especially if those benefits are to be expressed in money terms—cannot be overstated. Yet, with experience, forecasts can be made which are at least comparable one with another. If the concept of costs and benefits is accepted, it becomes important to measure (or forecast) such items as:

(i) Charge per problem and per farmer per year.
(ii) Benefits in money terms to the farmer.
(iii) Development cost.
(iv) Farmer hours to complete and understand the first and subsequent sets of inputs and outputs.
(v) Adviser hours to become fully trained, and to convey the service to the farmer.
(vi) Cost of testing, and gathering of standard data.
(vii) Number of users and frequency of use of the service.
(viii) Turnround time for mail in and remote terminal services.

Costs at all three levels include all manpower costs: money amounts for the farmer's time as well as manpower costs for testing, training, advisory, and processing time. Benefits will include any savings to the extension agency or farmer (or researcher) over any current method of achieving the same end. Any standard data (collected from, or donated by, farmers) should be assessed for its benefits in group extension research or policy matters, and any research or policy use of the programme itself should also be assessed.

On a technical note for those concerned with accountancy, costs and benefits cannot be forecast in absolute terms, but in marginal terms—along the lines of a partial budget: 'new costs' less 'costs saved', and 'new benefits' less 'benefits forgone'. Costs and benefits accrue as flows over time, rather than lump sums at a specific point in time; any analysis of cost and benefits would therefore take account of time lags by converting these money sums to their present values.

Few advisory services appear willing to publicise the results of any such analysis, though there is general agreement that the benefit:cost ratio to the average farmer is likely to be high by industrial standards, certainly above 3:1. For example, some preliminary estimates are shown in Table 1. Would such agencies also be prepared to test their extension programmes against such commonly accepted industrial standards as an internal rate of return of 15% or more, and a payback period of four years or less?

TABLE 1

Predicted benefit:cost ratios and rates of return for some CANFARM management advisory services

Service	Benefit:cost ratio	Rate of return (%)
1. Feed mix	2·7:1	38
2. Maxiplan	2·5:1	40
3. Crop budget	3·4:1	103
4. Livestock budget	3·0:1	50
5. Loan calculator	2·2:1	60
6. Prairie grain	1·5:1	22
7. Machinery planning	6·7:1	82
8. Cash flow	7·6:1	98

9.7. CONCLUSION

Farmers remain unwilling to use 'whole-farm' planning aids offered by extension agencies, even in the face of estimates by advisors that the average farmer will reap a benefit at least three times the cost—and in some instances up to 10 or 15 times the cost, even when the farmer's own labour is costed. This does not imply, however, that farmers reject the whole farm approach *per se*; rather that they see shortcomings in the computer services now available and reject the services for not being truly wholefarm oriented. The most commonly made criticisms concern the ability of the computer programme to deal with multiple objectives and, in particular, with risk.

Managers of businesses may have up to 200 identifiable objectives, and many of these are in conflict with one another. A selection of the more common criteria in an agricultural context would include: (a) short- and long-term profit maximisation; (b) fixed investment minimisation; (c) working capital minimisation; (d) return on capital maximisation; (e) income stability, and risk avoidance; (f) growth of the firm; (g) engagement in preferred enterprises; (h) influence in the community; (i) technological excellence; (j) tax avoidance; (k) high staff wages; and (l) pleasant buildings.

Attempts have occasionally been made to rank or aggregate these objectives to produce a single measure of utility which could conveniently be incorporated in the more commonly used computer programmes. But since each farmer is liable to have his own mix of preferences, it has not yet been found possible to concoct a widely acceptable utility function. So, extension officers have been provided with single objective tools, the internal mechanisms of which have not been clearly explained to them. They have in turn presented these computer printouts as being the plan to follow, instead of discussing them as only one of several possibilities open to the farmer. Viewed in this light, the benefits to the farm cannot be measured in terms of the number of whole farm plans which were implemented to the letter—very few of them are—but rather in terms of the extent to which farmers' own decision-making processes were aided by consideration of computer-produced plans. A farmer's-eye view of the systems approach to his own farm is therefore one of integrating many pieces of information, whether they be of a recording or planning nature, into a decision for which he alone is responsible. To the farmer the systems approach is not new, he has been using it for many years. What is perhaps surprising is the occasional arrogance of extension services in feeling it their duty to calculate complete solutions and to see them implemented.

REFERENCES

1. Brown, T. G. (1973). Extension farm management programming for small farms, *Proceedings North Central Farm Management Extension Workshop*, Michigan State University, USA.
2. Rowe, A. H. (1971). Computerised systems of accounting and control for farm business management, *Proceedings of the 1st International Farm Management Congress*, Farm Management Association, Kenilworth, UK.
3. US Dept. of Agriculture (1973). *Inventory of EDP Programs used in Agricultural Extension*, US Dept. of Agriculture, Washington.

4. Thompson, S. C. (1973). Agricultural planning in Bulgaria, *Journal of Farm Management*, 2(6).
5. CANFARM, (1974). Internal information.
6. KTBL, (1971), *Landwirtschaft der Zukunft*, Kuratorium für Technik und Bauwesen in der Landwirtschaft, Frankfurt, West Germany.

10

Regional Agricultural Planning

A. C. EGBERT

International Bank for Reconstruction and Development (World Bank), Washington, DC, USA

and F. ESTÁCIO

Fundação Calouste Gulbenkian, Oeiras, Lisbon, Portugal[*]

10.1. INTRODUCTION

Fortunately or unfortunately, the term 'system' can be taken to encompass almost any phenomenon we choose. When we undertook the study to be discussed, we did not conceive of it as a system analysis but merely as a method of addressing a real problem, the problem of agricultural development. The objective was to represent the development environment realistically, a necessity if one desires realistic results. But in being realistic, one is limited. No matter how complex the analysis system, a comparison with reality will show that it is a subsystem of a larger one. Because we are mere mortals with limited resources, we must be discriminating in picking the subsystem to be used for analysis. But this should be selected in such a way that it can be related to others, an objective that seems to be an underlying philosophy of this symposium.

In order to introduce our model or method of analysis, we refer to Fig. 1, which is a conceptualisation of an economic system in four dimensions, namely, production, level of processing, location and time. This figure is a poor representation in that it is a simplification of the economic system to be analysed and planned. In reality, there are a very large number of products, many unique regions, numerous levels of processing, and an infinite number of time periods. Behind each product there are many types of firms and productive factors, all interrelated. Although we have put together a relatively complex model to represent the agricultural sector, it represents only a small part of the total economic

---

[*] The views and conclusions expressed in this paper are the authors' only and should not in any way be attributed to either the IBRD or FCG.

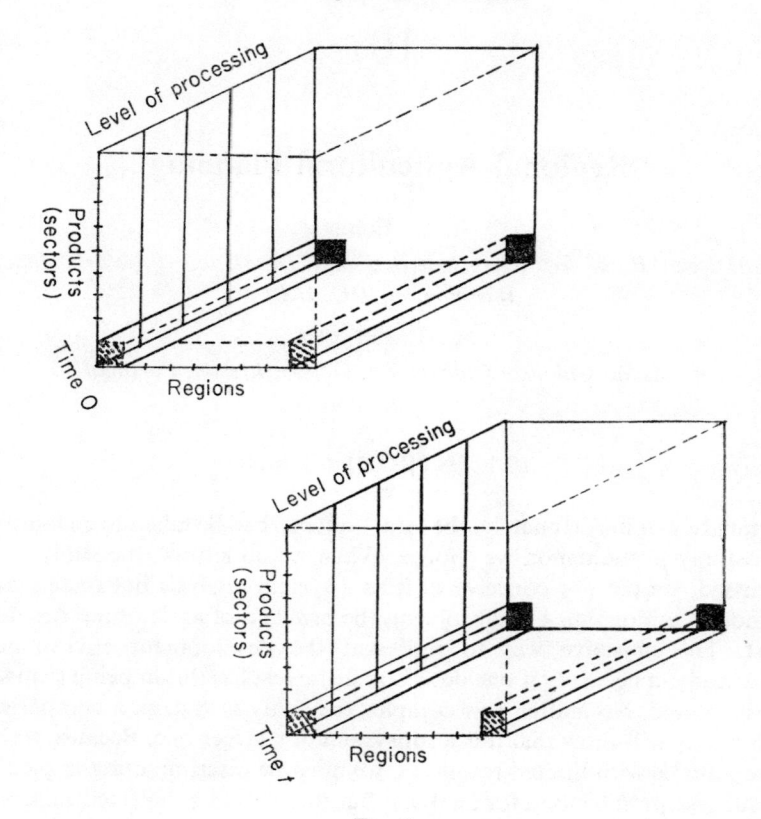

Fig. 1.

system. Moreover, only tacitly related are other interactive systems—physical, biological and social.

All these subsystems are important in economic planning, but given the present state of the arts we can only deal with one or a few at a time. For purposes of our planning analysis (referring again to Fig. 1) the agricultural and food sector is conceived as being distributed over space, extending from the basic farm production activities such as wheat and potatoes, through the processing, transportation, and marketing systems to the final consumer. Thus, we are concerned with the entire system required to get food and fibre products into the hands of households. However, as shall be seen later, some related sectors, e.g. fertilisers and processing, are analysed too but in less detail than the agricultural production sector.

Also individual farm or firm subsystems are not analysed in detail. The basic unit of investigation is land area or region.

In order to make good development plans, we believe that it is necessary to include as many as possible of the strongly linked subsystems in the analysis. What we have done in the Portugal case study, from a data assembly, resource and analytical viewpoint, is rather complex and costly.

Given the analytical framework depicted in Fig. 1, the most suitable method for empirical analysis seems to be linear programming. There are several forms of programming that could have been rationally employed, including static, comparative static, long-term dynamic, short-term dynamic (recursive) and stochastic. Because, however, of (a) time and resource availability and (b) the nature and possibilities for planning and plan implementation in Portugal, it was decided to use a comparative static approach. Essentially, this method is normative in the sense that it provides a prescription for 'what ought to be done' and not 'what will be done', in a behavioural sense. There can, however, be some overlap between the two objectives.

Our procedure in the study was first to design the model, test or validate it using a historic base year and then to make analytic projections for a year in the future. The base year, 1968, was selected for the reason that the most recent census had been taken in that year. The year 1980 was selected for the projection to give a target far enough in the future for plans to be implemented and which could include relevant economic trends. Before presenting the results of the analysis, the model is described in formal terms.

10.2. THE MODEL

The objective of the model is to determine the regional production and consumption mix of products originating in the agricultural sector that will maximise total producer and consumer surplus given:

(i) the level of consumer's income;
(ii) the cost of production, both direct and opportunity costs;
(iii) cost of transportation between markets; and
(iv) net export and import prices.

The solution to such a model is equivalent to specifying spatial equilibrium in a competitive economy as outlined by Samuelson[1] in which the selling price of each final product is equal to its supply costs of production, processing, handling, transport and selling in each region. For a single market and product the objective of the model is as defined in the graph of typical supply and demand equilibrium (Fig. 2). The triangle *PCD* represents the consumer's surplus which is the difference between the marginal value of each succeeding unit, as expressed by the demand curve DD_1, and what the consumers must pay, price *P*, for a commodity. Similarly, the producer's surplus, represented by triangle *PCS*, is the difference between the price received for the product and the marginal production cost of each succeeding unit which is expressed by supply curve SS_1. It is possible, of course, that marginal cost and average unit costs are constant with the result that the producer's surplus is zero. 'Supply curves' in the model are to be presented as step functions and can only be graphed *a posteriori* because the nature of the supply curve is only determined in the final equilibrium solution.

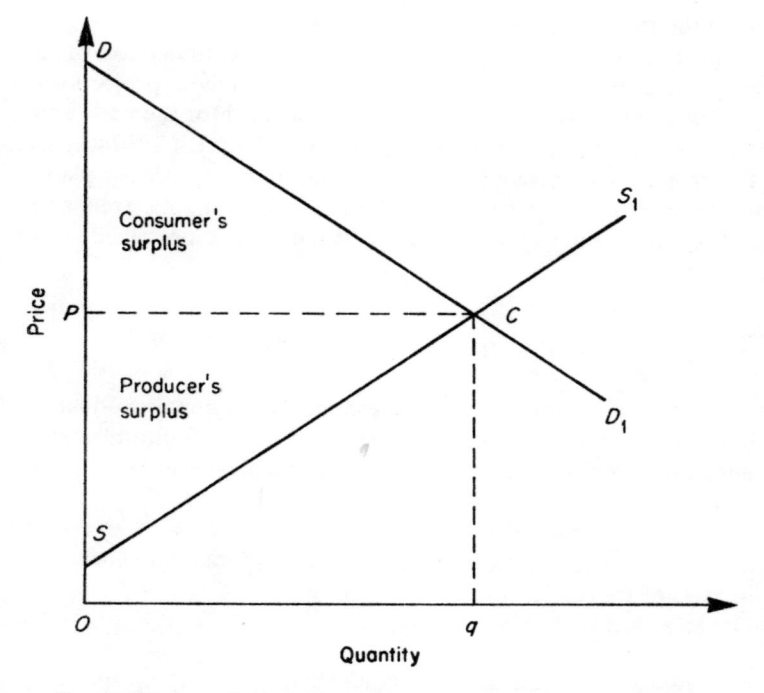

FIG. 2. Consumer's and producer's surplus in market equilibrium.

In Fig. 2 this means that the supply curve SS_1 which is determined by marginal unit cost may take different shapes and levels from solution to solution. The demand curves, however, remain the same because they are determined only by the levels of income and population which are constant or exogenous in the model. The supply curves on the other hand are functions not only of production, transport and processing costs, which are constants, but also functions of the prices of other competing products which are endogenous and variable in the analysis.

The mathematical programming analogue of this model is as follows (all exogenous or predetermined variables and matrix coefficients are labelled with an asterisk).

The objective function of the model is:

$$\text{Max} f(w) = \sum_i \sum_j S_{ij} \lambda_{ij}(S_{ij}, I_j^*)$$
$$+ \sum_i \sum_j (EP)_{ij}^*(EX)_{ij} - \sum_m \sum_j (PC)_{mj}^* M_{mj}$$
$$- \sum_p \sum_j C_{pj}^* P_{pj} - \sum_i \sum_j (IP)_{ij}^*(IM)_{ij}$$
$$- \sum_m \sum_{\substack{j \\ j \neq j'}} \sum_{j'} T_{mjj'} t_{mjj'}^* - \sum_r \sum_{\substack{j \\ j \neq j'}} \sum_{j'} T_{rjj'} t_{rjj'}^* \qquad (1)$$
$$- \sum_i \sum_{\substack{j \\ j \neq j'}} \sum_{j'} T_{ijj'} t_{ijj'}^*$$

where S_{ij} is the amount of product i sold in the jth region ($i = 1, \ldots, I$ $j = 1, \ldots, J$), $\lambda_{ij}(S_{ij}, I_j)$ is average utility measured in dollars derived from the ith product in the jth region which is a function of the level of income, I_j, and the amount consumed, S_{ij}. $(EP)_{ij}$ is the net price of product i exported† from the jth region and $(EX)_{ij}$ is the amount of product i exported from the jth region. $(PC)_{mj}$ is the unit cost of converting primary agricultural products, m, into consumer or final products ($m = 1, \ldots, M$) and M_{mj} is the amount of product, m, processed in the jth region. P_{pj} is the level of primary production activity, p, in the jth region. Each primary production activity may produce more than one primary product, m ($p = 1, \ldots, P$). C_{pj} is the unit cost of the pth primary production activity in the jth region. $(IP)_{ij}$ is the total import price of product i in the jth region. $(IM)_{ij}$ is the level of import of product i in the jth region. $T_{mjj'}$ is the amount of primary product, m, transported from region j to region j'. Since intraregional transport is not considered $j \neq j'$. $t_{mjj'}$ is the cost of

† The terms 'imports' and 'exports' refer to trade with other countries; the term 'shipment' refers to interregional trade.

transporting one unit of primary product m from region j to region j' ($i' = 1, 2, \ldots, J$).

In the other similar transport terms in eqn (1), the subscript, r, refers to basic resources, $r = 1, \ldots, B$, and as noted above, i refers to final products or consumer goods.

In summary, eqn (1) states that the objective of the programming model is to maximise the total net social payoff (producer and consumer welfare) of domestic sales plus the value of exports of goods originating in agriculture, given the level of consumer income in each region.

The maximisation of eqn (1) is constrained by available resources, market balance, and other conditions as follows:

$$R_{jr}^* \geq \sum_p a_{pjr}^* P_{pj} + \sum_m a_{mjr}^* M_{mj} + \sum_{\substack{j' \\ j' \neq j}} T_{rjj'} - \sum_{\substack{j \\ j \neq j'}} T_{rjj'} \qquad (2)$$

Equation (2) states that the amount of basic resource, r, used by the primary production and processing activities plus the amount shipped into region j, minus the amount shipped out, cannot exceed the fixed supply, R_{jr}.

$$0 \geq \sum_p \pm b_{pjm}^* P_{pj} + M_{mj} + \sum_{\substack{j' \\ j' \neq j}} T_{mjj'} - \sum_{\substack{j \\ j \neq j'}} T_{rjj'} \qquad (3)\dagger$$

Equation (3) states that the amount of primary product m, processed in region j cannot exceed the net amount produced in the region by production activities, P_{pj}, plus the amount shipped in, minus the amount shipped out.

$$0 \geq -b_{ijm}^* M_j + S_{ij} + (EX)_{ij} - (IM)_{ij} + \sum_{\substack{j' \\ j' \neq j}} T_{ijj'} - \sum_{\substack{j \\ j \neq j'}} T_{ijj'} \qquad (4)$$

Equation (4) states that the amount of product i sold in region j, S_{ij}, cannot exceed the amount processed in the region, plus the amount shipped in, minus the amount shipped out, minus the amount exported plus the amount imported.

$$E_i^* \geq \sum_j (EX)_{ij} \qquad (5)$$

Equation (5) states that the amount of product i exported from all regions, j, must not exceed the upper bound or limit, E_i.

$$I_i^* \geq \sum_j (IM)_{ij} \qquad (6)$$

† Production activities can use primary products produced by other production activities, as well as produce primary products, e.g. livestock activities use feed produced by crop activities, hence, b_{pjm} may be either positive or negative.

Equation (6) states that the amount of product i imported by all regions, j, must not exceed the upper bound or limit, I_i.

These import and export bounds are set to make the programming solution trade levels consistent with government trade policy and/or world trade price conditions.

$$S_{ij}, M_{mj}, P_{pj}, (IM)_{ij}, (EX)_{ij} T_{ijj'} \geq 0 \qquad (7)$$

Equation (7) is the standard linear programming constraint which states that all variables cannot be at negative levels.

In addition to these real constraints there are $(I) \cdot (J)$ pseudo-constraints that are necessary to convert this non-linear programming problem to a linear problem. The objective function is non-linear because utility functions, λ_{ij}, are included in place of fixed prices for selling or consumption activities.

$$K_{ij}^* \geq \sum_n k_{ijn}^* S_{ijn} \qquad (8)$$

This equation states that the weighted sum of product i sold in region j through all demand segments, n, cannot exceed K_{ij}, where K_{ij}/k_{ijn} is the maximum amount that can be sold through demand segment n. The upper bound, K_{ij}, would never be set above the amount for which total utility is a maximum. The range of n varies with the amount of precision desired in approximating the utility function (see Appendix I for fuller explanation).

A system diagram for a typical region of the model is presented in Fig. 3.

The specific products and resource constraints included in the model were:

Final products:

1. Wheat products
2. Maize products
3. Milled rice
4. Rye products
5. Barley products
6. Oats products
7. Dry beans
8. Broad beans
9. Chick peas
10. Potatoes
11. Tomato products
12. Wine
13. Olive oil
14. Oranges
15. Beef and veal
16. Pork and lard
17. Lamb and mutton
18. Milk
19. Butter
20. Wool
21. Non-fat dry milk
22. Cheese

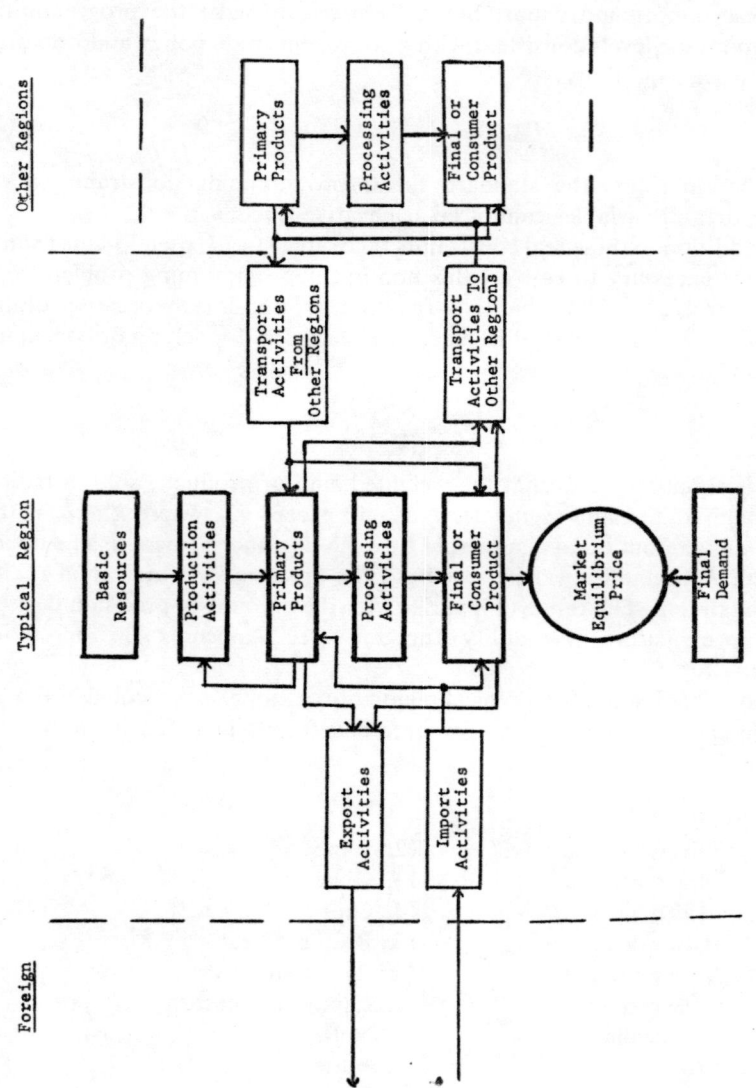

Fig. 3. Structure of linear programming model.

Intermediate Products:
1. Low-protein feed concentrates—maize, oats and barley
2. High-protein feed concentrates—oilseed meals, skim milk
3. Animal power
4. Ewe and cow milk
5. Forage

Basic products:
1. Wheat
2. Maize
3. Rye
4. Rice
5. Oats
6. Barley
7. Broad beans
8. Chick peas
9. Potatoes
10. Grapes
11. Dry beans
12. Oranges
13. Beef cattle and calves
14. Swine
15. Sheep and lambs
16. Wool
17. Ewe milk
18. Cow milk
19. Forage

Imports:
1. Wheat and products
2. Maize and products
3. Milled rice
4. Potash fertiliser
5. Beef
6. Pork
7. High-protein feeds
8. Butter

Exports:
1. Wine
2. Olive oil
3. Lamb
4. Tomatoes

Fixed resources:

Land resources:
1. Irrigated
2. Class A
3. Vineyards
4. Olive trees
5. Orange trees
6. Other land

Other resources:
7. Labour
8. Tractor power
9. Slaughterhouse capacity
10. Milk processing capacity
11. Butter and cheese processing capacity
12. Meat cold storage capacity
13. Nitrogen and phosphorus plant production capacity

Because there were no data for regional consumption, only for prices, a rather involved procedure was used to estimate regional demand functions. Consequently, we briefly describe this procedure. (A description of procedures used to obtain or derive other data as well as a set of the basic data used for programming are available from the authors on request.)

10.2.1. Demand Functions

The first step was to estimate demand functions of the form:

$$P_{ij} = f(S_{ij}, I_j)$$

where S_{ij} = per capita consumption of commodity i in the jth region, P_{ij} = price of commodity i in the jth region and, I_j = per capita income in the jth region.

It would have been desirable to include prices of other substitute commodities in this function. However, food prices tend to be highly collinear which does not permit statistical estimation of individual elasticities. Lack of regional data on consumption presented a very serious problem in estimating regional demands. A first attempt was made by estimating demand functions for each commodity at the national level. These functions were then used to estimate quantities consumed region by region on the basis of per capita incomes and prices. The resulting consumption patterns were then compared with *a priori* expectations. These did not turn out very well. Consequently, a second round allocation of regional consumption patterns was made on the basis of regional production of crops and livestock under the rationale that foods are more likely to be consumed where they are produced. Thus a typical per capita food ration per annum was derived for each region. These rations were converted to kilograms, calories and protein consumed per person. When the value of any of these items seemed improbable, national totals were reallocated so as to move regional consumption rates into a plausible range.

After these analyses we still needed some formal expression of regional demands. Therefore, we analysed (fitted demand functions to) regional consumption patterns on the assumption that, given the present patterns of consumption, the price and income elasticities were the same in all regions. This is not very satisfactory, of course, but seemed to be the most practical course under the circumstances.

For programming purposes, and because a linear algorithm was used, the equilibrium demand (supply price equal to demand price) was approximated by using segmented demand functions (see eqn (8)). Moreover, because of the nature of the objective function the marginal area under the demand curve was used (which is assumed to be equal to marginal utility) and is equal to price as a mathematical limit (see Appendix I). The supply–demand market equilibrium can be approximated as near as desired by segmenting the utility function into smaller and smaller intervals. In general, in our programming routines we used ten segments with prices ranging from 20% higher to 20% lower than the base price.

For obvious reasons of space we skip over the validation phase of the model's development. Suffice it to say that this phase included numerous programming solutions each followed by additions of certain data and constraints. The final result was a model that performed sufficiently well that we felt it could be used for regional development planning. We now present results of that phase for the regions depicted in Fig. 4.

10.3. DEVELOPMENT PROGRAMS

For this analysis 1980 was chosen as the reference year; mainly because it is a point in the future which would permit longer-term development plans to be implemented and analysis of the relationships between long-term needs and production possibilities to be made. The development alternatives considered are the standard ones requiring both short- and long-term implementation periods. These include irrigation, improvements in livestock herds, mechanisation and the use of high-yielding seeds, fertilisers and insecticides.

Development analysis in a programming framework requires projections of the demands for agricultural products, domestic and foreign, and of certain resources. In the framework used, the domestic demands for farm products are functions of population, per capita income and prices. Because prices are endogenous in the model, projections are made of demand schedules (not particular demand levels), based on population and income growth.

Population growth, more precisely decay, was projected using the over-all country trend and regional shifts in population which were reported in the official Censuses of Population for 1950, 1960 and 1970. Even though the total country population is projected to decline slightly, populations of regions which include the large urban areas of Porto and Lisbon are projected to increase reflecting the continued migration from rural areas.

Regional incomes differ greatly throughout the country with the highly

FIG. 4. Programming regions.

rural areas of the north at the low end of the scale and the Lisbon region at the upper. Because of meagre information on shifts in relative regional income, regional incomes were assumed to grow at the same rate as the 1960 decade.

Income elasticities required to project demand schedules to 1980 were based on our studies of domestic demand for Portugal with some modifications, based on other country studies, to reflect projected changes in income levels.

It was not possible because of lack of historical data to obtain regional price and income demand functions for all products included in the analysis. For the 1968 validation phase only powdered skim milk, wool, and tomato products were *not* represented by regional demand functions. However, a national demand function for wool was derived and this demand was effective in Lisbon and all wool was assumed to be transported to this city for sale. Only the export demand for tomato products was considered and these were priced for export at 7·75 escudo per kg with an upper limit, or bound, of 138 000 tons. Powdered skim milk was assumed to be sold at 20 escudo per kg with no limit on consumption at this price. These same demands were also used for the 1980 analysis.

The number of products analysed in the model was expanded in the 1980 analysis for two reasons: (a) some products, forestry for example, were added to take into account real interdependencies in the rural areas and thus make the analysis more realistic; (b) new products, in a commercial sense, were added to determine if and how these could be fitted into the development picture. Both categories are listed below:

Additional products
1. Melons
2. Cork
3. Resin
4. Pine logs for lumber
5. Pine logs for pulp
6. Pine logs for fibre board
7. Eucalyptus logs for pulp

New products
1. Forage crops for milk cows
2. Sunflower seed
3. Safflower seed
4. Oil from sunflower and safflower
5. High-protein meals from sunflower and safflower.

For these products, perfectly elastic demands were assumed at certain prices and in some cases an upper bound was placed on the amount that could be sold at the fixed price. These details are given in Appendix III. We note that demands for feed products are not required because they are implicit (derived) rather than explicit in the programming matrix.

For productive resources, only projections of the agricultural labour force were made. These too were based on the regional trends in agricultural population in the two decades 1950–60 and 1960–70. The accelerated rate of decline in the rural labour force of the 1960s was projected to continue in the 1970 decade.

For inventories of traditional livestock, farm machinery, irrigation, agricultural product processing, and fertiliser plants, it was assumed that these would be maintained at the 1968 base levels. Thus, from any programmed increase in investments required by 1980 there must be subtracted any new investments that were put in place since 1968 to obtain net investments.

All projections and development analyses were made assuming no change in the price level. Thus all the derived prices are in terms of 1968 escudos. Relative prices for trade goods were projected to change somewhat from the 1968 base.

10.3.1. Structure of Investment Activities for Programming

Each investment activity in the programming matrix represented a package of investment goods. The farm machinery package includes tractors, ploughs, other tillage equipment and harvesting equipment necessary to carry out specific types and rotations of crop production. Improved production activities for crops and livestock were linked directly to the investment packages. Livestock investment packages include basic breeding stock, buildings and equipment.

Irrigation investment activities include the cost of building dams and the distribution system. The Department for Water Resource Development (Servicios Hydraulicos) in Portugal has had an irrigation plan on the books since the mid-1950s. The original plan consisted of about 170 000 ha. Of this total, about 100 000 has been developed or is under construction. In the analysis for 1980 the remainder of this programme provided the basis for analysing whether further irrigation development should be undertaken on the basis of expected cost and benefits in the long run.

These possible irrigation schemes and their parameters are as shown in the table below.

Forestry investment activities include cost of tree stock, land clearing

Possibilities for irrigation development

Region	Name of the irrigation scheme	Total land to be irrigated (ha)	Estimated construction period
1	Rio Lima	5 855	5
2	Villariça	3 300	4
	Macedo de Cavaleiros	2 200	4
3	Mondego	14 930	8
4	Cova da Beira	17 000	8
9	Crato	6 500	5
	Vigia	1 220	3
	Minutos	1 800	3
10	Odivelas (2 parts)	3 500	4
11	Odelouca	6 750	5

and labour for planting. Because some investment packages included items of different life expectancies and for irrigation all development could not be made in one year, all costs were expressed in present value terms, with the life of the major item in a package taken as the life of the investment.

Moreover, because investments had to be related to annual returns for programming, the present value of each investment package was adjusted by a factor equal to the present value of a one-dollar annuity. This procedure is explained in Appendix II.

10.3.2. Improved and New Production Activities

Many farms in Portugal today follow age-old traditional production practices using large amounts of hand labour and animal power and employing, for the most part, simple tools. With mass movement of agricultural labour to the cities and industrial countries of Europe, it is important to know what agricultural products will have the greatest economic advantage and what techniques should be used to produce them in the future. For this reason, new production activities in conjunction with investment activities are introduced into the programming milieu for 1980. The input–output coefficients for these activities were based on the experience of the relatively few farmers employing them and some experimental results. Although it is not certain how many farmers could and would change to new methods within the stated time span, results from the analysis produces economic goals that can be used for planning. Of course, programmes would need to be set up to encourage and assist in their adoption, certainly more concentrated and effective than those used in Portugal today.

Three variants of the 1980 model were used for analysis. Their unique features are as follows:

Variant 1. Similar to 1968 basic model (Variant A1) but incorporating modifications presented in preceding pages.

Variant 2. Same as Variant 1, except there are no upper bounds on meat cold storage capacity.

Variant 3. Same as Variant 2, except there are no upper bounds on slaughter capacity for cattle, hogs and sheep and no upper bounds on processing capacity for milk, cheese and butter.

Variants 2 and 3 are investigated because the solution to Variant 1 included very large shadow prices for those resources. In each case it would have been better to include investment activities in the matrix and let profitability determine if an investment should be made. However, data on the investment cost of these resources were not available. Hence, we used the second-best procedure of Variants 2 and 3. With the solutions to Variants 2 and 3 (as will be demonstrated later) it is possible to determine the amount of resource required to yield a certain payoff (the incremental value of the objective function). This payoff can then be compared with rough estimates of investment cost or investment cost determined at a later time, to determine if the investment would be likely to yield an acceptable rate of return.

10.4. PROGRAMMING RESULTS

All the regional details as to production, prices, consumption, trade flows and resources used, cannot be presented due to space limitations. We first summarise the investment and development types of prescriptions coming out of the model. These prescriptions are of two types: (a) those dealing with long-term investment expenditures, and (b) those dealing with improved methods of production. As noted before, these two facets are interdependent. For example, certain crop rotations cannot be used unless there is an investment in machinery needed to carry it out. A summary of the relative amount of production which would take place under improved production methods is presented in the final section.

10.4.1. Investment Programmes

10.4.1.1. Variant 1

Farm mechanisation accounts for the largest part (40%) of the investment expenditures prescribed by the Variant 1 solution (Table 1). Much of

TABLE 1
Summary of investment activities, 1980 model, Variant 1

Category	Units	Region 1	2	3	4	5	6	7	8	9	10	11	Total
Resources supplied by investment activities													
Farm mechanisation[a]	Tractor	1281·0					1342·0	1421·0		783·0	1141·0	63·0	6031·0
Milk cows[b]		44·6					59·0			2·5		2·4	108·5
Beef cows						14·5		33·9		47·8			96·2
Hogs	1000 units			3·7					0·4		11·9		16·0
Sheep			93·4					179·4	160·4	253·0		130·5	816·7
Pine	1000 ha				286·6								286·6
Eucalyptus		89·6											89·6
Value of investments (million escudos)													
Farm mechanisation[c]		99·0					96·8	90·9		53·7	76·2	2·5	419·1
Milk cows		116·0					194·7			8·2		7·9	326·8
Beef cows						12·5		28·6		41·2			82·3
Hogs				4·9					0·5		15·5		20·9
Sheep			15·4					29·6	18·8	29·6		15·3	108·7
Pine					47·0								47·0
Eucalyptus		31·9											31·9
Total cost		246·9	15·4	4·9	47·0	12·5	291·5	149·1	19·3	132·7	91·7	25·7	1036·7

[a] 50 hp tractor equivalent.
[b] Units of female stock; investment package includes other associate livestock, buildings and equipment.
[c] Includes cost of tractor and complementary equipment.

this investment would be concentrated in Regions 1, 6, 7 and 10. The reasons for this investment pattern are quite complex. For example, in Regions 1 and 10, investment in machinery is prescribed even though conventional tractors and other machinery are surplus. This occurs because it permits the use of production techniques that are sufficiently superior to traditional methods to offset the investment cost incurred. On the other hand, in Regions 6 and 7, it is more reasonable to say that investment in machinery is prescribed because labour and other sources of power are used up and it is profitable to invest in these items. These points are illuminated by the shadow prices for exhausted resources. Note that additional investment will not be programmed unless the unit cost is less than the shadow price (or marginal return). For example, in Regions 3, 5, 6 and 7, although they have positive shadow prices, the cost per investment unit is greater than the shadow price.

Milk cow herd improvement together with associated buildings and equipment also account for a large part of total prescribed investment, 32%, for Variant 1. 95% of this total is prescribed for Regions 1 and 6, regions with large concentrations of population which therefore have large demands for milk.

Some type of investment is designated for each region. Additional investments in some of these other types, beef cows, hogs, sheep and especially pine and eucalyptus forest, are precluded by the lack of some type of processing capacity. This result just emphasises the interdependency of investment decisions in a sectoral context. Many investment decisions from a sector point of view must be considered in relation to other investments and their levels. Many times they must dovetail to be productive from a country welfare viewpoint. Moreover, investments must be evaluated in relation to the expected demands for the products produced. In sectoral analysis, prices cannot be taken as fixed. Thus, the amount of profitable investment is constrained by the commodity demand.

In this vein, it is now worthwhile to look more closely at the shadow prices of the fixed resources (Table 2). The largest of these is for meat storage capacity, which in Region 4 amounts to almost 500 000 escudos per ton. This figure may be a bit misleading because it relates to capacity in the most critical season. In other words, it refers to the maximum point in time capacity. If the usual seasonal consumption pattern is maintained and capacity is expanded, higher consumption could occur throughout the year. For this reason, the shadow price or social benefit is at a very high level. Because of this high shadow price for meat cold storage it was not used as a constraint to the Variant analysis, as mentioned earlier.

Meat processing facilities, too, show high shadow prices. They also relate to critical months for slaughtering. Thus, the return indicated results from the increase in throughput for the entire year that would result from breaking the bottleneck in the critical month, again assuming no shift in seasonal production and consumption patterns.

In contrast with the results obtained for 1968 for which surplus labour was the rule for all regions except Region 1, for 1980 these surpluses disappear. In Region 2, the supply of labour is exhausted in three of the four quarters; in Regions 5 through 11, labour is constraining in two quarters. Despite the fact that labour is constraining, the marginal earnings as indicated by the shadow prices are low, the highest being about 18 escudos per hour during the fourth quarter in Region 2. Farm wages of 18 escudos per hour are relatively high by current standards, but it is not high by projected 1980 regional income per capita.

Note that further irrigation development is not prescribed by the model results. Even though the shadow price (Table 2) is quite high in Region 10, for example, it is much lower than the estimated total annual cost per hectare, namely irrigation development cost plus annual operating cost of delivering the water to the farmer. For Region 10 this total cost amounted to 4800 escudos vis-à-vis a shadow price of 2820 escudos. It must be remembered too that this shadow price per hectare must apply not only to the marginal irrigated unit but to all units of commodities which would be produced by the irrigation development. In Region 10 the irrigation development scheme totalled 3500 ha. The point is that as output from a large irrigation scheme expands, the prices of the products produced will fall. The fall depends on the total amount and package of products produced and their price elasticities of demand. (The reader may be wondering at this stage why no mention has been made about the integer or lumpy nature of investments and the need for integer programming solutions. Except for irrigation, the integer values are so small with respect to the total level of investment that the changes resulting from 'exact' integer solutions would be imperceptible. Because no irrigation development was specified in the continuous solution, it is an implicit integer solution.)

Production of forestry products—lumber, fibre board and pulp—are limited or constrained by processing capacity for every region. (Each region does not have processing capacity for each of these products but can ship to other regions which do have these facilities. However, in the solution total country capacity is used up.) As a result, the shadow prices for wood processing facilities are quite high and in many cases nearly equal to the price to the producer (Table 2). Thus, the opportunity costs are low

TABLE 2
Shadow price of resources, 1980 model, Variant I

Resources	Unit	Region										
		1	2	3	4	5	6	7	8	9	10	11
Labour 1Q	esc/hr	1·85	1·31		6·99	6·13	5·38	16·50	0·77	9·85	1·35	0·01
2Q				1·70		2·07	9·30	6·87	5·77	6·55	2·66	1·27
3Q			0·72									
4Q			17·77									
Tractor 1Q	esc/hr			60·28								
2Q												
3Q						17·61	2·73	53·22				
4Q												
Animal power 1Q	esc/hr	2·13		3·68	5·25	2·29			0·69		2·01	1·65
2Q		1·12	1·86	2·70			3·80		0·88	1·80	0·04	
3Q											0·02	
4Q												
Livestock inventory	esc/head											
Mules				723	469	696	456	725	23	247		756
Bulls			419	791	814	119	61		1102	584	656	539
Beef cows			854			1193					14	1979
Milk cows									69			
Swine						1						
Sheep												136

Regional Agricultural Planning

		Col1	Col2	Col3	Col4	Col5	Col6	Col7	Col8	Col9	Col10
Land											
Olive	esc/ha	643							725	1 330	1 139
Grape		29 836									7 433
Orange		9 125		10 466						74	
Pine		29									91
Eucalyptus											114
Cork	esc/ha	356		258							936
Irrigated					582				981	839	2 110
Class A						1 725				2 821	
Total			273		10			123		440	
Processing capacity[a]											
Beef		67 311	4 858					9 866			
Pork		99 472			96 294	55 883			77 739	70 816	90 944
Lamb and mutton	esc/ton	196 394	77 980	61 368	149 148	2 500	78 346	82 504	160 002	157 485	
Cow milk		2 560	2 116				2 641	2 648	1 076		
Sheep milk					15 790	537				15 784	
Cow milk for butter and cheese											
Meat storage[a]	esc/ton	372 217	406 708	475 412	292 731	292 731	292 731	292 731			1 089 186
Fertilizer											
Nitrogen	esc/kg		0·42	0·80		0·74					
Phosphorus											
Forestry product processing											
Lumber, pine	esc/m³	350[b]									
Pulp, pine		218	218	218	218		218	218			
Fibre board, pine		218	218								
Eucalyptus		227	227		227			227			

[a] Capacity in critical months.
[b] National constraint.

and it appears quite likely that investments in wood processing capacity would yield a high rate of return.

As noted earlier in this report, only nitrogenous and phosphate fertilisers are produced in Portugal. Traditionally, the country has exported nitrogenous fertilisers. Even if there were no additions by 1980, capacity for both nitrogen and phosphorus is sufficient for domestic needs according to the Variant 1 solution. In the analysis no consideration was given to indigenous production of potash fertilisers. This is, perhaps, something to be considered in further work.

Another image of the consequences of investments in agriculture is obtained by looking at the distribution of production among so-called conventional and new techniques (Table 3). The model prescriptions vary from commodity to commodity. Rice, oats, barley and tomatoes are all produced totally by conventional techniques. On the other hand, wheat is produced almost entirely (93%) by new techniques. 84% of total wheat production would be grown in the southern part of the country—the Alentejo—where it is possible, because of the terrain, to use large-scale equipment and other advanced technology.

TABLE 3

Percentages of products produced by conventional and new techniques, Variant 1

Product	Conventional techniques	New techniques
Wheat	7	93
Rye	34	66
Maize	67	33
Rice	100	0
Oats	100	0
Barley	100	0
Chick peas	26	74
Potatoes	79	21
String beans	100	0
Tomatoes	100	0
Beef	23	77
Pork	28	72
Mutton	29	71
Milk (cow)	39	71
Milk (sheep)	55	45
Wool	63	37
Forages	55	45

TABLE 4
Summary of investment activities, 1980 model, Variant 2

Category	Units	\multicolumn{12}{c}{Region}
		1	2	3	4	5	6	7	8	9	10	11	Total

Resources supplied by investment activities

Category	Units	1	2	3	4	5	6	7	8	9	10	11	Total
Farm mechanisation[a]	Tractor	1537·9			365·0		1149·0	1421·0		783·0	1780·0	79·0	7114·0
Milk cows[b]		57·0	39·3		7·1		44·9			2·5	19·1	2·4	133·0
Beef cows			1·4	3·9		13·9		41·5		44·9	8·6		148·2
Hogs	1000 units	8·0	17·0		0·7	5·8	4·5		0·4	6·1	40·3		71·1
Sheep								257·3	155·9	224·2		130·5	784·9
Pine					291·1								291·1
Eucalyptus	1000 ha	84·4											84·4

Investment costs (million escudos)

Category	Units	1	2	3	4	5	6	7	8	9	10	11	Total
Farm mechanisation[c]		118·0			29·0		82·0	90·9		53·7	100·2	3·5	477·3
Milk cows		148·4	33·1		18·6		147·9			8·2	63·8	7·9	394·8
Beef cows				5·1		11·7		34·9		38·7	7·3		125·7
Hogs		10·4	1·8		0·9	7·6	5·8		0·5	8·0	52·5		92·6
Sheep			2·0					36·9	18·2	26·2		15·3	98·6
Pine					47·7								47·7
Eucalyptus		30·1											30·1
Total		306·9	36·9	5·1	96·2	19·3	235·7	162·7	18·7	134·8	223·8	26·7	1266·8

[a] 50 hp tractor equivalent.
[b] Units of female stock; investment package includes other associate livestock, buildings and equipment.
[c] Includes cost of tractor and complementary equipment.

TABLE 5
Shadow price of resources, 1980 model, Variant 2

Resources	Unit	1	2	3	4	5	6	7	8	9	10	11
Labour	esc/hr											
1Q		2·65	1·41	2·16	6·38	3·65	9·16	26·17	1·11	12·60	0·96	1·61
2Q						1·85						
3Q			2·66					5·84	4·36	6·17	7·41	2·69
4Q			16·05									
Tractor	esc/hr											
1Q		59·58		54·39	5·65	20·00	34·64	136·02				
2Q						20·37						
3Q						13·63						
4Q												
Animal Power	esc/hr											
1Q		2·71	3·74	3·40	5·88		4·97		1·40	5·03	1·87	3·25
2Q				6·81	2·95							
3Q						4·15			1·03			
4Q												
Livestock inventory	esc/head											
Mules												
Bulls			862	1 524	1 185	1 135	951			1 206		1 437
Beef cows				2 560	2 182	135						885
Milk cows			142			704			90	453	38	1 422
Swine									1 259		1 765	
Sheep							39		54			134

Regional Agricultural Planning   341

Land										
Olive		602								1 091
Grape		29 894	339	689	270	44		508	1 221	7 071
Orange		9 596	5 705	4 842	6 140	11 263	1 714	7 894		
Pine	esc/hr	25	5 897	7 780	4 724	2 047		467	764	
Eucalyptus		356		164				135		47
Cork			256	89						75
Irrigated					593	51	869	1 442	952	909
Class A						4 768	242	3 539		2 451
Total			13		130			236	106	
Processing capacity[a]										
Beef		57 937	100 414		81 174	141 739	164 310	112 873	61 995	42 107
Pork		83 836	69 742		141 073	2 439	2 513	168 360	151 818	59 913
Lamb and mutton	esc/hr	189 477	157 348	140 452				2 631	952	171 256
Cow milk		2 542	2 077							
Sheep milk					15 963					
Cow milk for butter and cheese									54 859	
									149 302	
									15 818	
Meat storage[a]	esc/ton									
Fertilizer										
Nitrogen	esc/kg		0.42	0.80		4.83		4.09		
Phosphorus										
Forestry product processing										
Lumber, pine		350.00[b]	218.00		217.82		217.76	217.74		
Pulp, pine	esc/m³	218.00	218.00	218.00						
Fibre board, pine		218.00	225.23		225.08			225.05		
Eucalyptus		225.41								

[a] Capacity in critical months.
[b] National constraint.

Chick peas are produced mostly by advanced techniques in this same area because they are part of the wheat rotation. Rye and corn are mostly produced in the northern part of the country according to the model solution. However, the amount produced by new techniques is about two-thirds of total production for rye but only one-third for corn. Given the complexity of the model, these results are not easy to explain, but the explanation probably lies in the labour supply. Traditional techniques use large amounts of labour and require small current cash expenditures and capital inputs. As long as labour is abundant (surplus), traditional methods will be more economical from the model's viewpoint. This is true except in some cases where total *unit* variable (cash) and capital costs of certain new techniques are less than those for traditional techniques. Thus the new techniques replace the traditional techniques even with a zero shadow price for labour.

10.4.1.2. Variant 2

The release of constraints on meat storage capacity permits a significant increase in production, especially of livestock products, and in particular pork which was imported under Variant 1. As a result, significant additional investments are economic (Table 4). Only two of the investment categories—sheep and eucalyptus—are prescribed at a lower level than for Variant 1. The total level of investment would be increased by 22%. Investment in improved swine herds would be increased most, over 300% above Variant 1. Large meat cold storage capacity permits more of the pork consumed to be produced domestically rather than imported. Not only that, domestic pork production when unconstrained by the limits of cold storage capacity is cheaper than imported pork.

Although many of the shadow prices change in Variant 2, most of them increase, as shown in Tables 2 and 5. Those for meat and dairy product processing are relatively large compared with the cost of providing these types of facilities. (We mention again that the processing constraints are for critical months and the increase in maximum capacity has the effect of increasing throughput during all months. Thus the benefits are multiplied. In the Portugal model, two months out of the year, when production is greatest with respect to consumption, are used as a processing constraint. Hence, the benefits of removing processing constraints are multiplied by a factor of six.)

With additional investments, a larger percentage of the production of each crop is due to new or advanced production techniques (Table 6), as is expected. Investments are profitable because output by related production

TABLE 6

Percentage of agricultural products programmed for conventional and new techniques of production, Variant 2

Products	Conventional techniques	New techniques
Wheat	15	85
Rye	100	0
Maize	69	31
Rice	0	100
Oats	100	0
Barley	6	94
Chick peas	0	100
Potatoes	66	34
String beans	100	0
Tomatoes	80	20
Beef	15	85
Pork	15	85
Mutton	27	73
Milk (cow)	25	75
Milk (sheep)	66	34
Wool	61	39
Forages	51	49

techniques is profitable, not vice versa. This is another way of saying that the demand for investment goods is a derived demand. The needed increases in cold storage capacity required to implement the programmed output of Variant 2 is 11 000 tons or 40% above the base capacity of 1968, which amounted to 25 000 tons.

The social payoff to this amount of investment in cold storage facilities is a huge 2·6 billion escudos in terms of producer and consumer surplus (see Fig. 2). This represents an increase of about 8% in the social payoff compared with Variant 1. Because most of the livestock products have price elasticities greater than 1, the net value of agricultural production (i.e. total production valued at wholesale farm prices less production costs) increases by more than 5%. Because the shadow price of cold storage facilities is zero at the 36 000 ton level, this amount of investment would not be economic. In reality, one would want to stop short of this level and at a point where the social payoff would be just equal to the social cost.

10.4.1.3. Variant 3

Variant 3 also generates a programme which allows investment to proceed beyond the point where marginal social payoff is just equal to marginal cost or to a point where the shadow prices of slaughter capacity and milk processing and cold storage capacities are zero. To reach this point, the following changes in capacities would need to take place vis-à-vis Variant 2:

Processing capacity	Percentage increase
Beef and veal	9
Pork	6
Lamb and mutton	60
Cow milk	65
Sheep	54
Butter and cheese	0·3
Meat cold storage	4

Thus, slaughter capacity for lamb and mutton and processing capacity for milk should be expanded most. Some additional investment in beef and pork processing also would probably be profitable. Moreover, when processing capacity is expanded, the shadow price of investment in cold storage is again positive and would be expanded by about 4% about the Variant 2 solution.

Increases in investments resulting from Variant 3 compared with Variant 2 follow the increased processing facilities allowed. Investments in milk cows and sheep flocks increase by 22% and 46% respectively (see Tables 5 and 7). Additional livestock production requires more feed; hence, a higher level of investment in tractors and other machinery is profitable. This amounts to 16% above Variant 2.

Shadow prices of some other fixed resources increase as the livestock production restraints are removed, as assumed in Variant 3 (Table 8). Irrigation development in Region 6 has a high shadow price, 5000 esc/ha. However, Portuguese specialists believe that all potential irrigation schemes have been developed in this region. Hence, no investment activity was included in the model.

Some very significant changes occur in the method of production, i.e. conventional to improved techniques (Table 9). More of the livestock enterprises use new techniques. But rice production is shifted back to conventional techniques, probably because of the discontinuous nature of linear programming.

TABLE 7

Summary of investment activities, 1980 model, Variant 3

Category	Units	1	2	3	4	5	6	7	8	9	10	11	Total
Resources supplied by investment activities													
Farm mechanisation[a]	Tractor	1537·0	309·0		912·0		1394·0	1487·0		783·0	1767·0	63·0	8252·0
Milk cows[b]		57·1	5·1		17·8		57·9	1·6		2·1	18·7	2·4	162·0
Beef cows			48·7					80·6		25·3			154·6
Hogs	1000 units	13·4	6·0	4·3		4·4	1·7		0·4	7·3	40·3		77·8
Sheep			16·4			188·0		179·4	151·0	468·1	12·2	130·5	1145·6
Pine					277·0				1·2				278·2
Eucalyptus	1000 ha	84·4											84·4
Value of investment (million escudos)													
Farm mechanisation[c]		118·0	17·1		72·7		99·9	95·7		53·7	99·7	2·5	559·3
Milk cows		148·4	13·4		46·3		191·0	5·5		6·9	62·5	7·9	481·9
Beef cows			41·0					67·8		21·8			130·6
Hogs		17·4	7·9	5·6		5·7	2·2		0·5	9·5	52·6		101·4
Sheep			1·9			22·0		21·0	17·7	54·8	1·4	15·3	134·1
Pine					45·4				0·2				45·6
Eucalyptus		30·1											30·1
Total		313·9	81·3	5·6	164·4	27·7	293·1	190·0	18·4	146·7	216·2	25·7	1483·0

[a] 50 hp tractor equivalent.
[b] Units of female stock; investment package includes other associate livestock, buildings and equipment.
[c] Includes cost of tractor and complementary equipment.

TABLE 8
Shadow price of resources, 1980 model, Variant 3

Resources	Unit	1	2	3	4	5	Region 6	7	8	9	10	11
Labour 1Q	esc/hr	2·89	0·22	2·78	7·45	7·09	11·95	25·17	1·98	0·49	3·46	2·62
2Q						2·23				13·87	8·11	1·99
3Q			3·65					5·15	3·30	5·60		
4Q			16·03									
Tractor 1Q	esc/hr	94·19	11·75	62·74	27·71	13·35	62·56	155·66	1·31			2·45
2Q						43·39						
3Q												
4Q												
Animal power 1Q	esc/hr	2·72	4·64	3·02	5·97				1·72		2·62	
2Q				4·43	2·47							
3Q						5·32	5·34			5·86		3·82
4Q									1·19			
Livestock inventory												
Mules												
Bulls			1 074	435	851	1 276	1 262		20	1 406		1 688
Beef cows	esc/head			1 522	2 047							1 015
Milk cows									74			1 367
Swine						209		62	1 448	633	1 460	
Sheep		4				37	45		47			139

Land									
Olive	esc/ha	684		434	775	108			529	415		1 234
Grape		29 877	281	5 777	4 470	4 700	10 201		7 537		1 269	7 050
Orange		9 830	12 128	6 040	8 015	4 658	1 855	2 270	399		542	
Pine		27			164				164			66
Eucalyptus		356			73		38					73
Cork			255			577		880	1 448	940	782	904
Irrigated							5 001	359	3 938		2 734	2 515
Class A											715	
Total			26	221		194			286	144		
Processing capacity[a]												
Beef												
Pork												
Lamb and mutton	esc/ton											
Sheep milk												
Cow milk for butter and cheese												
Meat storage[a]	esc/ton											
Fertiliser												
Nitrogen	esc/kg			0.42	0.80	7.96			7.22			
Phosphorus												
Forestry product processing												
Lumber, pine		350·00[b]		218·00	218·00	217·82		217·76	217·74			
Pulp, pine	esc/m³	218·00		218·00								
Fibre board, pine		218·00		218·00								
Eucalyptus		224·78		224·60		224·45			224·42			

[a] Capacity in critical months.
[b] National constraint.

TABLE 9
Percentage of agricultural products programmed for conventional and new techniques of production, Variant 3

Products	Conventional techniques	New techniques
Wheat	15	85
Rye	100	0
Maize	55	45
Rice	100	0
Oats	100	0
Barley	14	86
Chick peas	26	74
Potatoes	47	53
String beans	100	0
Tomatoes	100	0
Beef	13	87
Pork	14	86
Mutton	24	76
Milk (cow)	21	79
Milk (sheep)	66	34
Wool	59	41
Forages	47	53

TABLE 10
Total value and indexes of production, 1968 and 1980, selected products[a]

Price base	1968	Variant 1	Variant 2	Variant 3
1968	26 992	31 683	36 928	38 829
1968	100	117	137	144
1980[b]	22 635	25 177	—	—
1980	100	111	—	—
1980[c]	22 984	—	28 938	—
1980	100	—	126	—
1980[d]	21 760	—	—	27 526
1980	100	—	—	127

[a] Wheat, rye, maize, rice, barley, chick peas, potatoes, beans, oranges, olive oil, wine, beef, pork, mutton, milk, cheese and butter.
[b] 1980 Variant 1 prices.
[c] 1980 Variant 2 prices.
[d] 1980 Variant 3 prices.

10.5. CHANGES IN THE LEVEL OF PRODUCTION, CONSUMPTION AND PRICE

Changes in the levels of investments result in changes in the levels of production. While the production of most commodities increases as investments are increased from one variant to the other, all do not. When product specific constraining resources are increased, there may be a

TABLE 11
Total commodity production, 1968 and 1980, for three variants

		1980		
Product	1968	Variant 1	Variant 2	Variant 3
	(1000 tons or litres)			
Wheat	262·3	792·7	787·6	788·4
Rye	164·3	94·9	102·9	98·3
Maize	669·9	421·4	558·6	665·7
Rice	229·0	215·4	172·5	110·0
Barley	109·8	54·4	98·5	77·3
Chick peas	8·9	10·8	10·7	10·7
Potatoes	906·3	618·0	618·0	618·0
Beans	52·7	20·8	24·6	21·0
Oranges	106·9	164·3	164·3	164·3
Olive oil	59·2	55·4	55·4	54·2
Wine	1162·9	1134·4	1123·6	1113·5
Tomatoes	750·0	1087·0	887·3	885·7
Melons	—	30·0	30·0	30·0
Beef	50·5	139·8	159·3	174·9
Pork	104·4	46·3	174·3	188·3
Mutton	25·1	35·9	35·9	54·1
Milk	354·9	658·4	640·5	742·0
Cheese	21·2	31·3	31·3	30·3
Butter	1·9	2·5	2·5	2·9
Wool	11·3	6·4	7·0	8·9
Vegetable oil	—	120·0	120·0	120·0
Cork	—	35·9	35·9	35·9
Resin	—	100·0	100·0	99·3
Pine logs ($10^6 m^3$)	—	10·2	10·2	10·2
Eucalyptus ($10^6 m^3$)	—	2·2	2·2	2·2

TABLE 12

Average prices of agricultural products, 1968 and 1980, for three variants

Products	1968	1980 Variant 1	Variant 2	Variant 3
		(escudos)		
Wheat	7.06	3·06	3·19	3·13
Rye	6·03	3·74	3·30	3·49
Maize	3·99	1·96	3·97	3·97
Rice	5·43	4·98	5·14	5·25
Barley	5·19	4·26	4·77	4·93
Chick peas	10·12	5·09	5·09	5·09
Potatoes	1·47	1·40	1·40	1·40
Beans	10·87	13·61	13·60	13·60
Oranges	7·41	3·31	3·31	3·31
Olive oil	22·10	17·92	17·92	19·16
Wine	5·83	5·89	5·87	5·89
Beef	32·60	31·28	23·02	17·08
Pork	30·89	30·59	24·73	22·39
Mutton	40·54	46·60	46·60	30·94
Milk	4·26	3·81	3·81	2·45
Cheese	61·43	31·21	31·22	37·17
Butter	44·70	23·23	23·21	23·17

reduction in the output of some product using a common resource, say labour, which constrains the total level of production.

Because of price changes among variants, it is difficult to give some overall measure of changes in the level of production. However, an attempt was made to do this (Table 10). (Note that the percentage increase in production from the 1968 base period changes markedly, depending on which price weights are used. Only products common to all analyses are used to compute total changes.)

Production in terms of 1968 prices increases by 17% for Variant 1. And since total population is expected to decline slightly, per capita production would be higher than in 1968 by a little more than 19%. On the other hand, if the prices generated by Variant 1 are used to make the calculation, total production increases by only 11%. This result is due to the fact that the equilibrium prices of a number of products decline. These are products for which production is up most. Prices of products which remain the same or go up are those with no change in production or declining production. These results flow from a number of economic factors inherent in the

Table 13
Average per capita consumption of agricultural products, 1968 and 1980, for three variants

		1980		
Products	1968	Variant 1	Variant 2	Variant 3
		(kg or litres)		
Wheat[a]	63·6	79·3	78·8	78·9
Rye[a]	11·4	9·0	9·8	9·4
Maize	27·9	27·8	24·6	24·6
Rice[a]	17·0	20·0	19·8	19·6
Barley[a]	0·2	0·3	0·2	0·2
Chick peas	1·2	1·3	1·3	1·3
Potatoes	105·7	77·3	77·3	77·3
Beans	5·3	2·6	3·1	2·6
Oranges	12·5	20·6	20·6	20·6
Olive oil[a]	3·9	3·8	3·8	3·7
Wine[a]	91·9	79·4	78·5	78·5
Beef[b]	8·3	17·6	20·3	21·9
Pork[b]	12·6	20·6	22·7	23·5
Mutton[b]	2·7	4·3	4·3	6·6
Milk	22·6	42·9	42·9	60·6
Cheese[a]	2·5	3·9	3·9	3·8
Butter[a]	0·6	0·3	0·3	0·4

[a] Processed weight.
[b] Carcass weight.

model, including changing technology of production and consumption patterns, the latter due to income elasticities of demand.

Individual commodity details of production and prices are presented in Tables 11 and 12. Production of wheat and animal products increase most. The production of feed grains; corn and barley is programmed below the 1968 level because a large percentage of livestock products are produced by the higher-grade livestock which are provided by the investment activities. This means that the grain or production per unit of feed is higher than for traditional herds. Pork production under Variant 1 is less than for 1968 because the cold storage capacity is used for beef, which means that beef has a higher net social payoff than pork.

Details of per capita consumption of agricultural products are shown in Table 13.

Interregional trade, in a relative sense, would be at a higher level in 1980 according to the model solutions (Table 14). The commodities affected

TABLE 14

Percentage of domestic production traded interregionally, 1968 and 1980 models

		1980		
Commodity	1968	Variant 1	Variant 2	Variant 3
Wheat	37	71	76	75
Rye	49	40	60	62
Maize	25	49	34	38
Rice	74	56	45	21
Oats	0	12	7	8
Barley	24	27	56	21
Chick peas	70	86	86	86
Potatoes	47	37	12	25
Beans	30	79	65	73
Oranges	69	41	41	44
Olive oil	28	29	24	31
Wine	16	13	14	14
Beef	24	47	55	52
Pork	33	86	72	64
Mutton	63	79	69	69
Milk	18	13	15	14
Cheese	30	36	47	29
Butter	5	0	0	24

most are wheat and animal products. Part of this result is due to a reduction in imports, which substitute for interregional trade, and the other part is due to regional specialisation which results from new production techniques. Note, too, that for some products interregional trade decreases —for example, rice. This is because the imports are higher than in the base year analysis.

Changes in interregional trade patterns, though somewhat academic, are important for planning transportation. We have assumed that transportation facilities would not be constraining because agriculture is a small part of the total Portuguese economy. If the model encompassed the entire economy, it would be relatively easy to include transportation constraints in the analysis.

10.6. FOREIGN TRADE

The pattern of imports changes markedly in the 1980 model results *vis-à-vis* 1968, as shown in Tables 15 and 16. There are no longer any wheat and

TABLE 15
Programme imports and exports, 1968

	Quantity (1000 tons)	Value (million esc)
Imports		
Wheat	174[a]	
Maize	119[a]	
Beef	20[a]	
Pork	3[a]	
Butter	4[a]	
High protein feeds	128	
Potash fertiliser	32	
Total value		1746
Exports		
Wine[b]	375[c]	
Olive oil[b]	22[c]	
Tomato products	138[c]	
Lamb	1[c]	
Total value		3737

[a] Imports constrained at upper bound.
[b] Million litres.
[c] Exports constrained at upper bound.

corn imports and only a small amount of beef. But imports of rice are now specified and pork imports are increased dramatically. As mentioned above, pork imports increase because there is a shortage of cold storage capacity. The patterns of exports does not change for 1980, Variant 1. The levels are higher because the constraints are larger than for 1968, in the expectation that export demand for the products would grow. For all three variants the agricultural trade balance is positive and increases from 1·8 billion escudos for Variant 1 to 4·2 under Variant 2. This latter result is due to a sharp drop in imports as exports drop only slightly under this variant. With no constraints on cold storage capacity, livestock production for domestic consumption competes with tomato exports; consequently tomato exports are lower under Variant 2.

The total value of imports specified by Variant 3 is slightly higher than for Variant 2 (Tables 17 and 18). Imports of rice, high protein feed and potash are all higher. Exports on the other hand are lower, specifically, tomato products and wine. On the average, the balance of trade is about 2% lower for the products considered. These data serve to point out that changes in critical resources can have large impacts on optimum trade patterns from a country point of view.

Table 16
Agricultural balance of payments, 1980 model, Variant 1

	Quantity (tons or 1 000 litres)	Value (1 000 esc)
Imports		
Rice	9 288	31 208
Beef	691	20 730
Pork	117 999	2 949 975
Potash	50 390	151 170
Total	178 368	3 153 083
Exports		
Tomato products	200 000[a]	1 400 000
Olive oil	25 000[a]	537 500
Wine	500 000[a]	3 000 000
Lamb	1 500[a]	78 750
Total	726 500	5 016 250
Net balance		1 863 167

[a] Constrained at export limit.

Table 17
Agricultural balance of payments, 1980 model, Variant 2

	Quantity (tons or 1 000 litres)	Value (1 000 esc)
Imports		
Rice	37 230	125 093
Beef	691	20 730
Pork	6 747	168 675
High protein feed	30 341	98 608
Potash	51 786	155 358
Total		568 464
Exports		
Tomato products	163 262	1 142 834
Olive oil	25 000[a]	537 500
Wine	495 815	2 974 890
Lamb	1 500[a]	78 750
Total		4 733 974
Trade balance		4 165 510

[a] Constrained at export limit.

TABLE 18

Agricultural balance of payments, 1980 model, Variant 3

	Quantity (tons or 1 000 litres)	Value (1 000 esc)
Imports		
Rice	79 842	268 269
High protein feed	46 370	150 703
Potash	53 784	161 352
Total		580 324
Exports		
Tomato products	162 970	1 140 790
Olive oil	25 000[a]	537 500
Wine	485 717	2 914 302
Lamb	1 500[a]	78 750
Total		4 671 342
Trade balance		4 091 018

[a] Constrained at export limit.

10.7. IMPORTANT MODEL LIMITATIONS

In addition to those pointed out in the introduction, there are the following limitations:

(i) In some regions, a few commodities are not produced at market equilibrium levels. This results because the set of prices chosen is in the wrong segment of the demand function.
(ii) The results would probably be improved if labour supply functions were used in the place of fixed availability of labour. It is doubtful that labour would be available at the level of employment specified in each region given the shadow prices derived.
(iii) There should have been investment activities for all replaceable resources.
(iv) The model is static; thus, it is not possible to give realistic time dimension characterisation of the investment programme.

Conceptually, we can deal with each of these defects; only time, resources and computer capability prevent us from doing so. Items (i) and (ii) can, however, be corrected easily and economically. Unfortunately, our funds ran out, after our original budget was exceeded a number of times.

10.8. SUMMARY AND CONCLUSIONS

A regional spatial equilibrium model was developed to determine what investments should be made and where they should be located, in the agricultural sector of Portugal through 1980. Three alternative analyses were made for that year in order to determine the returns to investments in irrigation, mechanisation, livestock herd improvement and better crop production techniques. In addition to these investments, investments in (a) cold storage plants for meats and (b) processing plants for livestock and dairy products were analysed using sensitivity analysis. Two new crop alternatives, sunflower and safflower, were considered as possibilities for import substitution, mainly for high-protein feeds.

All the investment alternatives, except irrigation development, appear to have good economic potential. The total level of investments prescribed for the three alternative analyses ranged from 1 to 1·4 billion escudos ($56 million) or about $4 million per year through 1980.

Further irrigation development does not appear economic because (a) the cost is high and (b) there seems to be large potential for increasing agricultural output from existing irrigated lands and through rain-fed agriculture using available improved production practices. These practices relate mainly to higher-yielding seeds and larger fertilisation rates. If this unexploited potential were utilised the country could produce its own needs and increase its agricultural balance of trade surplus and at the same time divert marginal lands to forestry, recreation and other use. Total cropland use declines by 48% from the 1968 level. Moreover, under each alternative projection there would be an improvement in the average diet with more protein and less starchy foods.

The programming model used for this analysis was not developed to its fullest potential because of time and resource limitation. Some regional demand functions need to be adjusted so that they define market equilibrium for all commodities. In a few regions, for a few commodities, no consumption was specified; the amount supplied would sell for a price higher than the implicit price or a greater amount could have been sold at a lower price. These defects could be corrected by a few minor adjustments in demand specifications.

More fundamental defects relate to supply functions for labour, a lack of a full range of investment activities for all replaceable fixed resources and the static nature of the model. Each of these defects is correctable to a large degree with more time and resources; even so, it would not be a perfect tool. It was found through working with this model and a similar one for Brazil that computer cost rises astronomically when the matrix size reaches a certain limit, given present computers and software. For example, increasing the matrix size from 1000 times 4500 to 1400 times 5500 resulted in a six-fold increase in real computer time.

Because of the amount of time required to assemble data for models of this type and total cost involved, we recommend that such studies be undertaken only by countries that have a well-staffed planning group, serious intentions and a good data base. Even so, simpler alternatives should be considered. Also, we have done some analysis showing that more aggregative programming models produce quite good results. But once a large model is developed it can be revised and improved at a fraction of the original cost. Moreover, they provide a tremendous amount of detailed information which cannot be obtained by any other methods known to the authors.

REFERENCES

1. Samuelson, P. A. (1952). Spatial price equilibrium and linear programming, *American Economic Review*, **42**, 283–303.

APPENDIX I

DERIVATION OF UTILITY FUNCTION FROM DEMAND FUNCTION AND CONSTRUCTING MATRIX FOR LINEAR APPROXIMATION FOR PROGRAMMING

Given demand functions of the form:

$$P_{ij} = a_{ij} - b_{ij}S_{ij} \tag{1}$$

where P_{ij} is the price of product i in region j, S_{ij} is the amount of product i consumed in region i, a_{ij} and b_{ij} are constants.

Assuming that

$$P_{ij} = \frac{dU_{ij}}{dS_{ij}}$$

where dU_{ij}/dS_{ij} is the marginal utility of consuming product i in region j.*

* This is theoretically true if marginal utility of income is constant and equal to $1.

Hence, total utility U_{ij} is obtained by integrating the demand function

$$U_{ij} = \int_S (a_{ij} - b_{ij}S_{ij})\mathrm{d}S_{ij} \qquad (2)$$

$$U_{ij} = a_{ij}S_{ij} - \frac{b_{ij}}{2}S_{ij}^2 \qquad (3)$$

For programming, eqn (3) was approximated by segmenting the function for 10 ($n = 1, \ldots, 10$) discrete levels of S_{ij} and a convex constraint.

To illustrate, assume that:

$$U_{ij} = 100S_{ij} - 0.25S_{ij}^2, \qquad (4)$$

then evaluating (4) for:

$$S_{ij}(1) = 20$$
$$S_{ij}(2) = 30$$
$$S_{ij}(3) = 40$$
$$U_{ij}(1) = 1900$$
$$U_{ij}(2) = 2775$$
$$U_{ij}(3) = 3600$$
$$\lambda_{ij} = U_{ij}/S_{ij}$$

hence,

$$\lambda_{ij}(1) = 95$$
$$\lambda_{ij}(2) = 92.5$$
$$\lambda_{ij}(3) = 90$$

Partial matrix

Objective function	=	90	92.5	95
Supply row for S_{ij}	0 ≥	1	1	1
Convex constraint	40 ≥	1	4/3	2

The coefficients of the convex constraint row are the ratios:

$$\frac{K_{ij}}{S_{ijn}} = k_{ijn}$$

or

$$\frac{K_{ij}}{S_{ij(1)}} = \frac{40}{20} = 2$$

$$\frac{K_{ij}}{S_{ij(2)}} = \frac{40}{30} = \frac{4}{3}$$

$$\frac{K_{ij}}{S_{ij(3)}} = \frac{40}{40} = 1$$

Then by proper row and column divisions, i.e. dividing the convex constraint row by 40 and then each column by the resulting coefficient in the convex constraint row, the matrix becomes:

Objective function	=	3600	2775	1900
Supply row for S_{ij}	0 ≥	40	30	20
Convex constraint	1 ≥	1	1	1

This form was used for programming because it is easily generated by the computer.

Finally, because the U_{ij} have a common constraint:

$$\Delta U_{ij}(1) = U_{ij}(1) - 0, \quad \text{for } S_{ij}(1) - 0$$
$$\Delta U_{ij}(2) = U_{ij}(2) - U_{ij}(1), \quad \text{for } S_{ij}(2) - S_{ij}(1)$$
$$U_{ij}(3) = U_{ij}(3) - U_{ij}(2), \quad \text{for } S_{ij}(3) - S_{ij}(2)$$

Hence, as $\Delta Q_{ij} \to d(S_{ij})$, $\Delta U_{ij} \to P_{ij}$.

APPENDIX II

ANNUAL RETURNS AND THE BENEFIT–COST RATIO FOR LONG-TERM INVESTMENTS

Given that the benefit–cost ratio (BCR) is

$$\text{BCR} = \frac{\sum_{i=1}^{n} R_i/(1+r)^i}{C_0 + \sum_{i=1}^{n} C_i/(1+r)^i} \quad (1)$$

where R_i is the annual return in year i to a project which has investment cost C_0 in period 0, and C_i is the annual operating cost, the following applies. If $R_1 = R_2 = \ldots = R_n$, $C_1 = C_2 = \cdots = C_n$, then (1) simplifies to:

$$\text{BCR} = \frac{R \cdot S}{C_0 + CS}$$

where

$$S = \sum_{i=1}^{n} \frac{1}{(1+r)^i}$$

or the present worth of $1 payable for n periods. Hence:

$$\text{BCR} = \frac{R}{C_0/S + C}$$

Therefore, the discounted annual benefit–cost ratio can be computed by dividing annual returns by the annual cost *plus* the investment cost divided by the present worth coefficient, S.

APPENDIX III

PARAMETERS OF PRODUCTS LACKING DOMESTIC DEMAND FUNCTION

Product	Region	Price at	Price (esc/kg)	Demand Constraint level (tons)
Melons	All	Retail	4·0	30 000
Powdered milk	All	Retail	20·0	3 000
Vegetable oils	All	Retail	21·0	120 000
Cork	1	Farm	3·07	none
Cork	2	Farm	3·93	none
Cork	3	Farm	3·07	none
Cork	4	Farm	3·07	none
Cork	5	Farm	3·60	none
Cork	6	Farm	4·07	none
Cork	7	Farm	4·67	none
Cork	8	Farm	5·80	none
Cork	9	Farm	5·20	none
Cork	10	Farm	5·27	none
Cork	11	Farm	5·20	none
Resin	All	Farm	3·25	100 000
Pine lumber	All	Farm	350·00[a]	2 720[b]
Pine for paper	All	Farm	218·00[a]	3 119[b]
Pine for fibre board	All	Farm	218 00[a]	925[b]
Eucalyptus for paper	All	Farm	252·00[a]	2 195[b]

[a] Price per cubic metre.
[b] 1000 m³ of processing capacity.

Discussion Report

J. CLARK

Economics Division, West of Scotland Agricultural College, Ayr, Scotland

and J. P. MCINERNEY

Department of Agricultural Economics, University of Manchester, England

The discussion centred primarily on the use and value of systems models in a practical context. This was seen to lie in their contribution as educational, advisory, or decision-making devices, and in the field of purpose-oriented research work. Much of the activity in agricultural systems has been directed to this latter area, but in their educational role models assist the thinking of students, farmers and, in an in-service training context, advisers. (The help which linear programming models provide in a logical approach to budgeting was quoted as an example.) Because of the inherent complexity of agricultural systems models, their application in practical decision situations at the individual farm level ultimately has to be through the medium of the expert adviser, which may tend to limit their usefulness. In this sense, the model builder faces a conflict between the generality and representativeness of his system and its relevance to a particular farm situation. Models can, however, help in the evaluation of farm-scale development by establishing what type of farms and natural conditions should be included in such studies. They could also indicate by forecasting what could be expected on individual farms which are being farmed below the level of the best so that variance in performance between actual and estimated could be investigated at a research level in technical, economic, sociological and psychological terms. A further possible practical contribution by models lies in indicating by synthesis the likely results of integrating the best of all systems and by testing and monitoring this under controlled conditions, either on the farms of universities and colleges or under commercial farming conditions.

There is more to be gained from investigating and evaluating 'apparently efficient' farm-scale systems in both technical and economic terms than in drawing heavily on experimental data. There could be grave danger in systems analysis, particularly in whole farm development work, but even in enterprise or skeletal systems, through the erroneous assumption that experimental results necessarily have a direct relevance to real-world situations on the farm. In particular, farmers' tactics and strategy in handling risk and uncertainty can result in major adjustments to input–output balances which invalidate research findings. One parameter in model building which so far appears to have been neglected, perhaps due to difficulties of both measurement and interpretation is the interaction of the farmer's management ability, inherent and acquired, with his environment.

While the foregoing probably represents a fair claim for potential advantages of model building, in practice these have not been realised. Perhaps part of the reason for this is that workers with a particular interest in systems modelling and synthesis have tended to build this activity into almost a separate and distinct discipline which 'outsiders' have regarded as esoteric and of no relevance to their needs. In reality, workers from a wide range of disciplines, both in the technical sciences and in social sciences, have for many years been active in systems modelling and synthesis, though they may not have recognised their activities as a growing specialism. Systems analysis can span a wide spectrum of activities, from consideration of the technical and economic relationships of part of a crop or livestock system to modelling the economy of a whole country or studies by environmentalists and ecologists of continents or oceans. Many speakers commented that they now viewed their work as being within the context of the systems approach whereas originally it had been conceived of in a different framework.

Perhaps the most important contribution from systems work it that it is an aid to thinking and is a method of approach in tackling problems in their proper perspective, that is, in a practical setting. The systems philosophy should benefit here in that workers ought to consider techniques as part of an integral approach rather than concentrating on techniques in isolation as an end in themselves. Furthermore, the systems approach, to be effective, must aid the decision-making process within systems.

The symposium had succeeded in bringing together representatives of a wide range of disciplines and, though a recurring theme had been the difficulties of obtaining true interdisciplinary teamwork in system analysis, it had brought scientists and economists closer together, however briefly. Two potential problem areas had emerged. One of the major ones had been that of communication. This was not merely a difficulty over semantics or differences of opinion as to whether or not there was a case for the development of a taxonomy of terms, or of models or of decisions. There was a need for an understanding of each other's philosophy as well as a need to use the same language, or at least to know that when A says 'aeroplane' that is synonymous with B's 'bus'. To those mainly concerned at the national or regional level with education, advisory and development/research work, then, all senses of communication are of paramount importance if the subject is to gain general credence.

It was felt that an undue emphasis on the importance of a practical approach could lead to an undesirable exclusion of more fundamental work which was so essential for future development of systems study. Against this it was argued that a vast fund of research and development work already existed, with insufficient effort being made to carry this through to the practical decision-making environment, there now being a need to exploit the available approaches before any more emphasis was directed towards further pioneering study. A plea was made for an appropriate balance between new knowledge and practical utilisation; too often in academic life the pendulum is located at the extremes, though no clear idea emerged as to where this optimum balance lies. However, it was important to recognise that decision-making is a costly process, in terms of the time, information, understanding and expertise that is required. At any one time a point is inevitably reached where it is not worthwhile to strive for greater perfection in systems modelling, for as models become more and more complex their marginal returns in a practical setting no longer cover the marginal costs of their application. For example, the claimed cost-benefit ratios and internal rates of return to the farmer for management techniques purveyed by the CanFarm organisation were regarded by some as being vastly overestimated.

Bearing in mind that farmers will find the provision of a large amount of detailed information a difficult (if not impossible) task, it is important to build models which explicitly recognise and adjust to the constraints of available data. While research workers can validly attempt to gather all the information they need, models for practical use should not be developed as though the data they require unquestionably deserve to be collected. This point highlights the problem of the level of complexity and detail to adopt in model development if the final product is to be understandable by, and acceptable and useful to, farmers or advisers. The obvious answer to this question is that it depends on the objective that the model is designed to serve, bringing us full circle back to the role of systems study in a practical setting.

There is much appeal in keeping models as simple as possible, concentrating on the key factors which directly influence output and/or the level of variable costs. But if models become too simple, then one can hardly describe them as part of a systems approach. An alternative might be to construct subsystem models. The problem here is how does one separate subsystem models from the whole system without interfering with important relationships or, conversely, how can one aggregate subsystem models without overlooking important interactions?

From a practical farm management point of view it was considered worthwhile to emphasise *that the objective of all systems modelling exercises should be their practicality and usefulness to research workers and farmers.*

The uptake of new knowledge by farmers in many countries has been disappointingly slow. Merely by applying existing knowledge, agriculture could still make major strides forward in productivity. The most effective methods of ensuring widespread adoption of new practices is through concentration on those factors which have a major impingement on production. For example, in dairying in the United Kingdom, it seems most sensible to concentrate on the largest item of variable costs, the feeding of the dairy herd. By focusing attention on the efficiency of concentrate usage and on grassland management factors such as stocking rate, grazing management and conservation (which affects production from forage), more rapid progress is made. It may be necessary to adopt fairly crude measurements and to use a 'blunt instrument' approach to put the message over to the majority of farmers.

It is thus initially at the enterprise level and in a somewhat unsophisticated approach that most impact can be made on the farmer's perception and thinking. There is therefore no need to apologise for the use of simple 'calculator' models. Indeed, they have strength in the above context in that they should give an immediate 'payoff' to the farmer. Meanwhile, researchers can tackle refinements to the basic calculator model for the benefit of the more sophisticated manager at some leisure. There is a great deal to be claimed for co-operation between institutions in the validating, testing and use of the more universally applicable enterprise models.

There was some evidence of a dichotomy of views as to whether farmers think in a whole-farm context or adopt a more 'bitty' approach. There is a need to identify which of these approaches individual farmers are most concerned with. Farmers have, in the past, been frequently more concerned with tactical rather than strategic problems. But as economic conditions in agriculture have become more and more volatile, so informed farmers are showing a closer interest in strategic matters and are more interested in future prospects than in the consideration of past history.

The recent feed grain and energy crises illustrate potential problem areas in using forecasting models such as the Nottingham University Dairy Enterprise Simulator and enterprise growth models. The uncertainty over harvest yields and, perhaps more importantly, over trends in international trade for feed grains could perhaps partly be reduced by choosing a forecasting period which runs from one harvest period to another, when more will be known about the availability and likely price of feed. In budgeting growth models under the increasing volatility and uncertainty of national economic situations, the length of planning horizon becomes critical. Under such conditions, few large commercial companies would care to risk forecasting cash flows for more than 18 months in advance.

A further problem in such economic circumstances also attaches to the use of standard data for farm planning. Such data is, by definition, historical and may well be obsolete at the planning stage, let alone at the time of implementation. Caution is therefore required in avoiding uncritical use of data banks and standard production functions which may become outdated by large shifts in resource balance and input–output ratios in response to changing economic conditions.

Discussions of macro-economic models ranged widely. There was no evidence that governments produce computer packages for farm plans in implementing agricultural policy, and doubts were expressed over the probability of farmers responding in a predictable manner to programmes written for them by policy-makers. Indeed, aggregate models can represent a massive workload in return for benefits of questionable magnitude, if only because no way has been found to incorporate consideration of the human element.

Aggregate linear programming models for detailed sector planning are particularly susceptible in this respect, given the demands they make on data and computational facilities while still yielding results of little more than indicative value. As such they are expensive devices for exploratory work in decision situations. Indeed, it was questioned whether such models were really consistent with the conception of a 'systems approach', since the tightly defined structure of the model in terms of fixed constraints seems to deny the possibility of interactions with and feedbacks from other parts of the wider economic system. Furthermore, linear programming models can be constraining if a unique optimum is looked for as an operational goal and aggregate models may incorporate many optima, none of them unique. In the national planning context, such optima need to be robust.

One basic problem in aggregate, large-scale linear programming models used for future planning is that, though a solution may be derived for some point in the future, no detail is given of the ways this solution can be achieved in practice; it is like a vicar telling his flock that they must be 'saved' without giving detailed directions how to reach this exalted state. An alternative might be to use a multiperiod model, somewhat simpler in concept.

It was generally agreed that there are fewer problems in conceiving models than in implementing them. This can be a time-consuming and expensive task. The measurement of the *effectiveness* of models in farm advisory work is a laudable aim. But there are difficulties in evaluating one of the main 'spin-offs'; notably the educational function of these in training the farmer in perception and thinking and, in addition, the value to attach to passing advice from adviser to recipient farmer who then disseminates this to fellow farmers. Benefits to the farming community cannot be measured in terms of the number of farm plans which are implemented because very few are implemented to the letter. Farm development has often to be 'played by ear' and flexibility in the use of resources must be maintained. Benefits might be measured by the extent to which farmers' own decision-making processes are aided by consideration of computer-produced plans. But how can this be properly evaluated and, indeed, by using computer print-outs in an undiscerning fashion, are we not hiding some of the structure of a decision tree which the farmer should be aware of?

Despite the fact that many models are calculators which make no economic decisions in the sense of allocating scarce resources between competing ends, and despite the failure as yet to incorporate psychological and social aspects of farm operation into the optimising function, systems study would seem to have more to offer than developing and tackling algorithms in isolation. Programming methods, simulation and network analysis have all been used alone and in combinations, but the use of these will not give the same perspective as the systems approach. The results of systems study have nevertheless repeatedly emphasised the fundamental point that there is rarely a single and unique programme of action that will achieve a particular target. In identifying the relevant alternative courses of action, the approach can only lead to more rational and considered decision-making behaviour.

PART 3

Applications of a Systems Approach to Research

11

The Use of a Systems Approach in Biological Research

G. S. INNIS

Faculty of Wild Life Science, Utah State University, Logan, Utah, USA

11.1. INTRODUCTION

To initiate a discussion of the uses of systems approaches to biological research, it is convenient to define a few terms, put the subject in a larger context, bound the area of interest and talk about some classification schemes for models and for utility.

11.1.1 The Systems Approach

A system, according to *Webster's Collegiate Dictionary*, is 'an assemblage of objects united by some form of regular interaction or interdependence'. This is a good systems definition for our purposes. It focuses on two major entities, the assemblage and the regular interaction or interdependence. In the 'state space' description of systems derived from engineering, the condition of the assemblage of objects at any point in time is the state of the system. This condition may be the amount of a thing (such as water in a pool), a logical condition (such as fertile or sterile), a more qualitative condition (such as quality of life), or a combination of such things.

The regular interaction or interdependence guarantees that the change of condition (change of state) with time of any object in the assemblage depends on the state of the other objects in some degree. Simulation of system dynamics focuses attention on the interactions of the mechanisms whereby the states change. To build a simulation model of a biological system, we need *only* identify the objects and interactions of interest, get some measure of the initial state of the system (condition of the assemblage) and describe analytically the interactions and interdependencies. These steps are reviewed again later. For now it suffices to emphasise the point that dynamics are determined by initial conditions and the interactions. The first of these is a single measurement or other determination of the condition of the objects in the assemblage at some point in time. The interactions, on the other hand, are derived only after a series of experimental studies have been conducted to elucidate the processes which make up the interactions and the variables which control these processes. Stated mathematically:

$$x_{t+\Delta t} = A_t x_t + z_t \qquad (1)$$

and

$$x_0 = c \qquad (2)$$

where c is the vector of initial conditions, A_t is a state transition matrix which depends on time t and z_t is a vector of driving variables. Equation (1) is written in difference form simply to be specific.

The relationship between the systems approach and the scientific method makes clear that the above specificity is *not* necessary to systems studies. Systems studies require that the investigator define his objects of interest and design experiments to expose the interactions that control the unfolding dynamics of the system. This investigator accepts Charles Darwin's statement that there can be no observation without hypothesis. The experiments that he conducts are designed to test these hypotheses. Described in this way, the systems approach is just an embodiment of the scientific method for empirical science. The tools of system analysis (often, for biological systems, including computers) put another dimension on the method. As we shall see below, new potential exists for hypothesis testing, new emphasis is placed on precision, etc. Systems approaches of the type to be emphasised in this paper enforce the scientific method.

The tools of systems analysis go far beyond the structuring of computer-based models. Some questions relating to the analysis of models will be discussed below. For the most part, however, systems analysis will be limited in this paper to the building, analysis and use of simulation models.

11.1.2. Biological Research in Context

Biological studies may be pursued out of scientific curiosity, to achieve definite managerial or social goals or for many other reasons. In any event it is encumbent on the investigator to recognise the environment in which his biological system resides. This is a part of the definition of the objects and interactions of interest, for if the investigator treats a variable as external to his system which affects and is affected by his system during his experiments, then his results can be completely spurious. The papers for this symposium indicate acceptance that biological research is one area of the agricultural system. The agricultural system is part of an economic system. Additionally, there are regional planning systems and social systems of interest as inferred from the titles of the papers.

In this context, biological research is viewed and discussed differently (though with many points in common) than in the context of teaching pure research at the university. This context, important in delimiting the portion of the field that will be discussed, implies macro-level concerns with micro-level studies to be included only as needed. Further, the context implies a rather immediate applicability of results as opposed to the more speculative studies and focuses our attention on certain spatial areas and certain species of plants and animals, and thereby begins to limit the experimental methods of concern.

There has recently been a nice paper on this context for biological research in the United States by Richard Levins.[1] He points out that the separation of biological research from agricultural research, even to the extent of putting them in separate institutions, is detrimental to both. Some biological research has gone awry by becoming too theoretical and by not maintaining adequate contact with a major user of their research, the agriculturalist. The agriculturalist has become anti-intellectual to his own detriment. He treats the research results of many biologists as irrelevant without adequately considering them.

At this point I would interject what may be the most important single utility for systems approaches in biological research, namely that from the systems viewpoint, the applied/pure or agriculturalist/biologist dichotomy is false. If one takes a systems approach to any interesting agricultural problem, one will find quickly that basic biological research (a) has allowed the analysis to go this far and (b) is needed to progress further. With the guidance of the systems approach, the need of the agriculturalist and the research results of the biologist are part of one unified attack on social need and scientific ignorance.

11.1.3. Boundaries for Discussion

To reiterate, we shall be concerned with the simulation modelling component of systems analysis and with that biological research which is appropriate to the context of the analysis of agricultural systems contained in economic, social and planning systems. This still leaves a great deal of material to be covered.

While many of the uses described in the second section are sufficiently broad to encompass additional studies, I must admit to certain prejudices that shape the classification of utility and the choice of examples. First, the time frames of interest are fairly short (up to ten years), and any genetic shifts in the objects under study will be assumed to come from outside the system (exogenous variables). By this limitation, succession as a shift in relative abundances of the species present is treated, but evolutionary changes are not. Introduced objects (species) would fall within the prejudices described. There is a rich literature on the simulation of genetic change to which the recent book by Bartlett & Hiorns[2] provides an introduction.

Secondly, weather vagaries from day to day and year to year may be treated, but climatic changes resulting in, for example, shifts in the long-term mean of climatic variables are not treated. Such shifts may be treated in the short term as weather vagaries but if allowed to continue over long times would probably result in ecosystem changes beyond our experience and biological understanding.

We are rather sharply limited at the present to simulation of systems with which we have much experience. Our models generally require considerable data describing the system of interest. This is contrasted with the desirable, if remote, situation in which an adequate biological theory might buffer us from the large data demands of most current simulations.

11.1.4. Classification Schemes

In the sequel we shall be discussing models and their utility. It is useful here to review some of the model classification schemes that have been presented and to mention a classification of utility that has been suggested.

Forrester's[3] classic on the simulation of industrial systems contains a model classification scheme that is paraphrased in Fig. 1. This scheme allows one to choose a collection of adjectives for the word model ('a miniaturisation of a thing') so as to make an appropriately descriptive phrase. This scheme omits certain additional adjectives that are useful, such as stochastic, deterministic, mechanistic, holistic, and statistical, to name a few. As one would expect, these dichotomies do not survive really careful scrutiny, and large models are made up of components with many of the indicated characteristics.

Holling[4] identified strategic versus tactical models. The strategic models were more theoretically based and concerned with the explication and display of hypotheses and their dynamic interaction. Tactical models were more data-based and designed to meet relatively immediate management objectives. Holling saw each modelling effort as lying somewhere on the

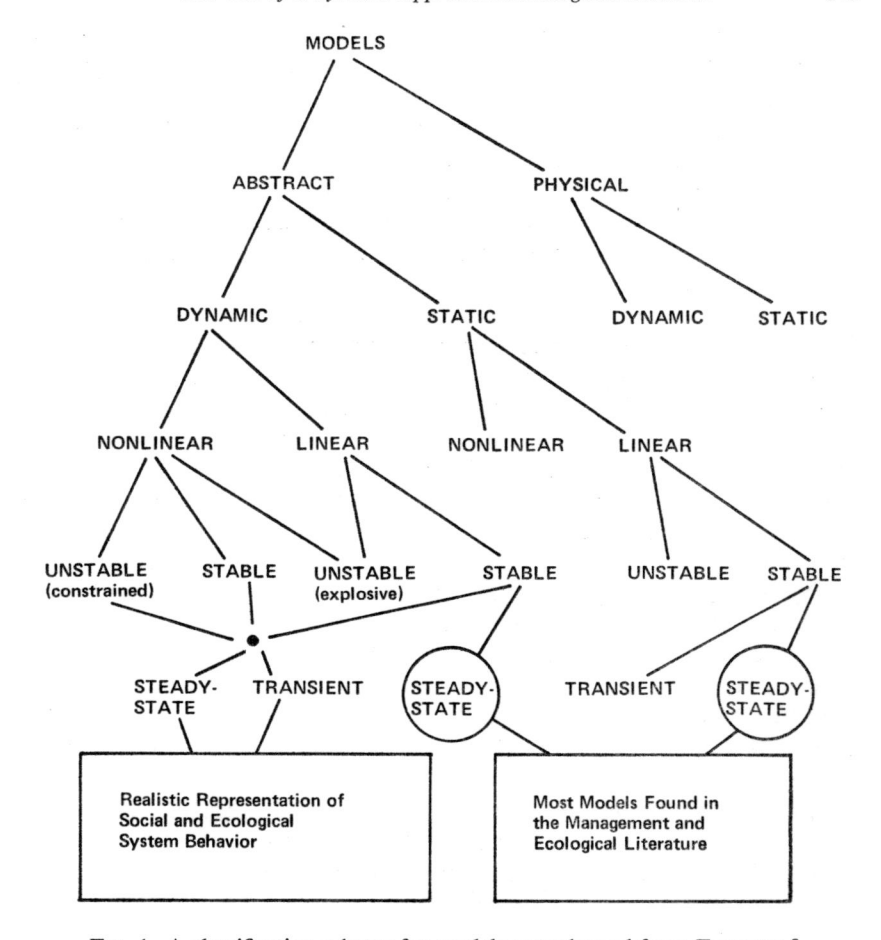

FIG. 1. A classification scheme for models paraphrased from Forrester.[3]

tactical–strategic spectrum. No models were totally one or the other. The choice of trade-off between the two is part of the modelling art and is further discussed in Goodall.[5]

Levins[6] categorises models along dimensions of precision, realism and generality. He points out that it is difficult to make a precise, realistic model very general and that general models are usually either imprecise or unrealistic or both. A good bit of later work has borne out these observations of Levins.

Pielou[7] offers four dichotomies of the Forrester sort for characterising models. These are (a) time continuous or time discrete, (b) analytic or simulation, (c) deterministic or stochastic and (d) empirical or theoretical. The time continuous/time discrete classification is one which I consider of fundamental interest.[8] Pielou's discussion of analytic versus simulation is too brief for me to be sure of her intention. It seems different from the analysis versus synthesis classification sometimes used, but I am not sure. The empirical versus theoretical distinction is Holling's tactical versus strategic separation.

The analysis versus synthesis distinction deserves some emphasis because of its effect on the tools we use in systems analysis. Systems analytic tools have developed largely in physical science studies where the objective has often been the synthesis of a system that achieves certain goals. (Churchman[9] approaches the subject this way.) The properties of interest were those developed from the interacting properties of the components incorporated into the design. This is the mechanistic philosophy of systems in action. In biological studies, however, the system is basically given even though many manipulations are possible, and the problem is one of analysis. System objectives give way to study objectives and the modeller knows that his system is going to contain numerous operants that are not modelled. Thus there will be properties of his system that are pertinent to the dynamics and that are not included in the model. These holistic features are difficult to incorporate using the modelling tools of engineering and physics. These tools were developed to study systems that were mechanistic in the large and holistic in the small (basic components are treated holistically). Consideration of holistic properties in the large in addition to mechanistic properties in the large is difficult using these tools.

While I have no doubt that there are other model classification schemes, these at least focus attention on the concerns that persons classifying models have had. Let us now consider utility for a moment.

In a paper on the simulation of ill-defined systems,[10] I defined conceptual, developmental and output utility. The conceptual utility derived from thought organisation and classification as well as the focus on systems. The developmental utility derived from the design of experiments, hypothesis clarification and testing and interpolation and output utility derived from extrapolations of sufficiently high quality that management decisions could be based on them. All models have conceptual utility, some have developmental utility, and a few have output utility. The major headings of the next section expand on these forms of utility and present examples of biological simulation models that have served in those capacities.

11.2. USES OF THE SYSTEMS APPROACH IN BIOLOGICAL RESEARCH

Before entering into the various forms of utility that are identified for the systems approach to biological research, it is useful to identify clearly the structure of a systems approach (particularly model building). Later in the discussion, many potentially difficult questions will be resolved or avoided by reference to these steps.

11.2.1. Model Building

Forrester[3] discusses the question of things to include in a model but does not provide a step-by-step enumeration of the procedure. Kowal[11] identifies five steps in the formulation of a mathematical model. These are as follows:

(i) Specification of variables of interest.
(ii) Construction of control diagram.
(iii) Classification, operational definition and specification of units for variables.
(iv) Specification of forms of equations.
(v) Evaluation of constants.

Other writers, such as Nolan[12] and Watt,[13] have also made such lists of steps, although in Nolan's case the steps are more general and the discussion is concerned with validation and verification more than model construction. An analysis of Hollings[4] component analysis approach or Van Dyne's[14] paper leads to a series of steps to be taken in model building.

The steps indicated in Fig. 2 constitute my version of the model-building effort. The effort, like any scientific research project, begins with a statement of *objectives*. This process is sometimes rather excruciating because it forces a more careful consideration of the problem at an early stage than many people feel appropriate. It (setting objectives) is referred to as 'specification of the variables of interest' or 'defining system boundaries' or in other terms by other writers. These other terms are not necessarily equivalent and often indicate the modeller's emphasis on certain facets of the objective setting (or even modelling) exercise.

Objectives are sometimes stated as questions (i.e. the objective is to build a model which addresses the question). They are often too general to perform their basic task which is to provide a foundation for decisions.

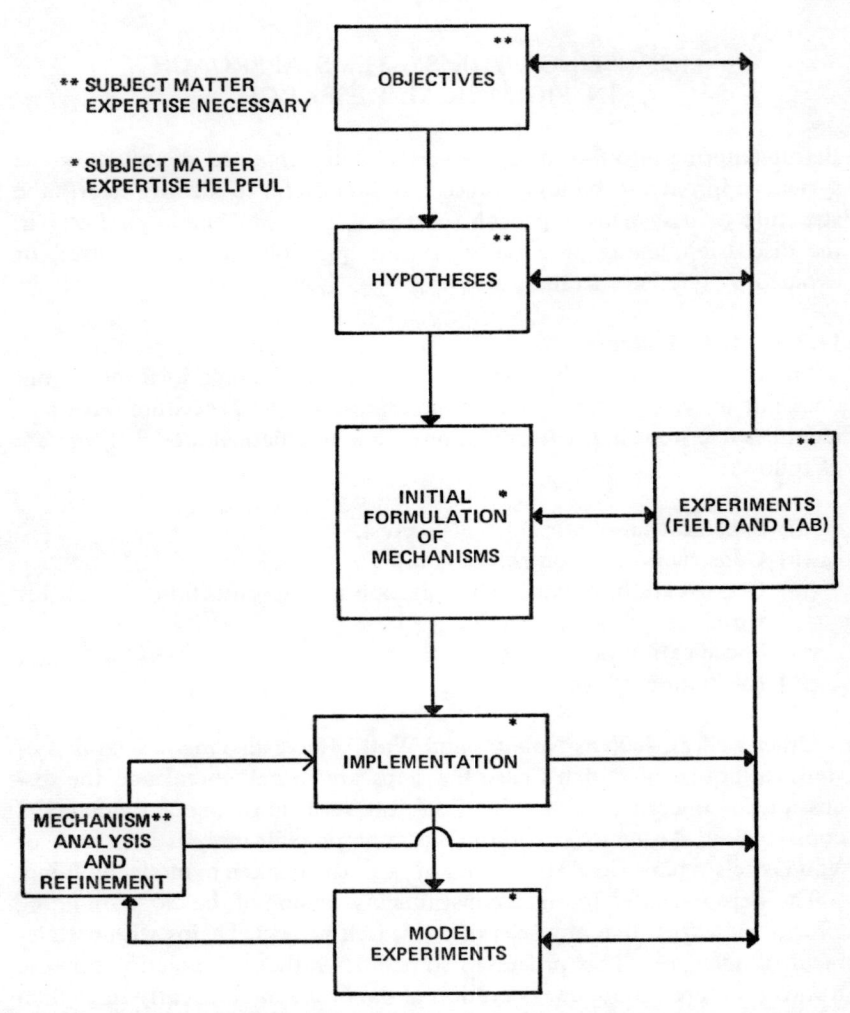

```
** SUBJECT MATTER
   EXPERTISE NECESSARY

 * SUBJECT MATTER
   EXPERTISE HELPFUL
```

FIG. 2. The steps in building a simulation model. The study follows the path down the centre of the page and around the loop at the bottom before feedback through the loop to the right.

The model components, system boundaries, hypotheses, levels of temporal, spatial and biological resolution all depend on what the model is to accomplish.

Specification of objectives is more clearly imperative for group efforts than for individual ones. If more than technician level contributions are to be expected from participants, they must be able to use the objectives in their decision-making processes. An acid test of the objectives is to see if others in the project (other than the one or ones who specified the objectives) can use them to make the necessary modelling decisions.

Next, a set of *hypotheses*, regarding the biological processes that must be incorporated in the model to achieve the objectives, is needed. These hypotheses may be theoretical or empirical, deterministic or stochastic, mechanistic or holistic. They may be posed verbally, graphically, in tabular form or mathematically. The few limitations on their form are as follows:

(i) Flows and levels (Forrester[3]) must be dimensionally compatible in a given submodel. The units on those flows into and out of a given compartment must be the units of the compartment *per unit* time.
(ii) Flows do not depend on flows either biologically or physically. If one's model has a flow dependent on a flow, one must realise that this is a modelling convenience, not a true representation of the biology or physics.
(iii) Flows that depend on time, explicitly, are also compromises for modelling convenience. Flows generally depend on state variables of the system (levels) or on exogenous variables.

Actually, one can make quite a list of such constraints; however, these few facilitate the evaluation and the critique of many published biological models. For example, many diagrams show water and nitrogen flowing into carbon in contradiction to point (i); rates of flow of water and dissolved nutrients are often treated as identical (point (ii) above); finally, flows and flow structures change with Julian date violating point (iii).

The *initial formulation of mechanisms* involves converting the variety of forms that the hypotheses were posed in into a common (usually mathematical) language. A point to call attention to here is that there is little to be gained in expending large amounts of energy in elaborate formulation of mechanisms if the hypotheses are known to be rough. Indeed, one gains a false sense of security and accuracy by using advanced mathematical techniques to formulate mechanisms that are poorly known. Some biological modelling efforts may get lost at this point in studies of applied mathematics, cybernetics, numerical analysis and computer science.

Implementation is usually computer coding. There are decisions to be made here regarding the language to use and these decisions are quite important. A simulation language can reduce by an order of magnitude the time required to implement a model as compared to a FORTRAN implementation. For some problems an additional order of magnitude reduction is possible by the use of time-shared, interactive systems. There are few general rules in this area except to recommend that local computer centre personnel be involved in making some of the implementation decisions.

Model experiments are debugging exercises early and simulation validation efforts later. Skipping the debugging exercise, model experiments are another form of hypothesis testing that is offered by the simulation modelling exercise. The model is just a hypothesis. Exercising the model is a way of investigating the consequences of this hypothesis. If the model is not in accord with our biological knowledge or intuition, then the hypothesis can be rejected. Rejection is not a simple matter, however, because of the complexity of the hypothesis. Rejection of a complex hypothesis says that at least one of the many subhypotheses of which it is made is rejected—but which one? It could be an overt one that fails to describe a biological phenomenon appropriately or a covert one that has cropped up in the formulation of mechanisms or implementation. Time and skill both may be needed to unravel the consequences of the rejection of such a hypothesis.

Mechanism analysis and refinement is this unravelling process followed by a modification of the model to correct the flaws.

The implementation/model experiment/mechanism analysis and refinement loop may bring the model to as high a level of development as possible without resort to additional experiments. This is another of the points at which the systems approach is currently having a significant impact. Both field and laboratory experiments, designed to fill the knowledge gaps preventing a model from meeting its objectives, are different from those that are traditional in biological research. The flow orientation results in more care being given to controls, more emphasis on non-destructive techniques, and more attention to the similarities among organisms rather than a focus on their differences.

Experimental results can require the initial formulation of new mechanisms, can result in new hypotheses and even cause objectives to change. Models are 'unstable with respect to objectives' and, therefore, a change of objectives should be interpreted as a new modelling effort, not a variation of an old one.

11.2.2. Model Uses

11.2.2.1. Thought Organisation

A design such as Fig. 2 is helpful in getting one's (or several people's) wits together at the beginning of an investigation. The constraints imposed are consistent with scientific method and, while any structure imposes limitations, limitations are often needed to focus attention and effort on problem areas.

The classical tutorial models in diagrammatic form are ideal examples of the thought organisation use of models. Such models as growth models, growth with limitations, logistic and Lotka–Volterra models help identify major components and important flows in a system relative to certain objectives. This thought organisation utility is further emphasised by the extent to which such tutorial models are used as the starting point for more advanced studies.

11.2.2.2. Thought Clarification

The diagrammatic tools of flow diagrams such as those of Forrester[3] or Odum[15] help to visualise both the system and its component parts. Keeping both the whole and the parts constantly in view promotes the subdivision of the problem into component parts appropriate to the objectives and avoids reductionism (the willy-nilly subdivision of a problem guided by personal interest).

Additionally, the constraints imposed on the flows as discussed above promote clearer thinking about the mechanisms operating within the system. Physicists learned long ago that dimensional analyses promoted clearer thinking and better theoretical structures. Attention to the relationship between flows helps avoid model structures with implicit hypotheses that may affect dynamic characteristics in an unknown or unrecognised way. Thirdly, a system responds to the environment with time as the independent variable. The system responds over time but not to time itself. The main effect of operating within such constraints is to avoid logical errors and investigate blind alleys.

The 'splitters' emphasise the differences among species, organisms, organs, etc., whereas the 'lumpers' emphasise the similarities among these entities. Under certain conditions and treatments and at some levels of observation or abstraction, distinct biological entities may perform 'alike'. Under other conditions, treatments, or at other levels of observation and abstraction, these distinct entities perform in clearly distinct ways. Part of the challenge of constructing a model of a biological system is lumping the entities together so that, with respect to the conditions and treatments of concern and the levels of observation and abstraction of interest, the grouped items are homogeneous enough to permit their uniform treatment in the model. The sense of 'permit' returns to the objectives of the analysis.

Ernst Mayr[16] points out that 'causality in biology is a far cry from causality in classical mechanics'. However, our current systems approaches are adopted from classical mechanics (and engineering) and remain a far cry from being ideally suited to the analysis of biological systems. Building models with inappropriate tools is like doing surgery with wood-working equipment. The quality of the tool does not mediate its lack of utility for a task for which is it not suited.

The debate about the utility of physical science tools and theories in biological studies continues. In the book *Global Systems Dynamics*[17] various authors and discussants defend several of the classical positions. The issue is not resolved, however, except in the obvious generality that some of the tools and theories from physical systems will be and are useful in the biological arena while others are not. I have argued[8] that even the differential equation representation of system dynamics can mislead as well as lead if the user is not careful.

One of the effects of the thought clarification type, that is often commented on, is associated with mechanism formulation; that is, conversion from verbal or graphical form to mathematical form. This step imposes the precision of mathematics on the imprecise and heuristic discussions of the hypotheses that may have preceded it. Units are assigned to the variables and the relations among them written as equations. The dimensional units of some of the conversion parameters lead to questions of the realism (biological) of a certain relationship. Interesting basic equations arise of the form, 'Do plants respond to soil-water deficit or to volumetric water content?' or 'what controls the rates of conversion of various organic compounds of nitrogen into forms of nitrogen useable by plants?' It is often learned at this stage that additional expertise is needed to carry the study forward or that more reading and better understanding on the part of the participants is the next logical step.

One can imagine that many modelling studies do not progress beyond this point. After clarifying thoughts, the investigators return to the objectives, choose a less ambitious study and start again.

11.2.2.3. Systems View

A systems view, as described and used above to put biological research into context, is a useful consequence of the systems approach. One result of this view is an appreciation of what may be called essentially open and essentially closed systems. First, any system of biological concern will probably be solar powered and, hence, also probably treated as open; the sun is not modelled. Open systems are no particular problem to analyse (relative to closed systems) *unless* the system affects the environment in which the system operates in such a way as to be important to system dynamics in the time frame of interest. Let us rephrase that by defining an essentially closed system as one whose environment is not affected by system dynamics in any way that is significant to the system during the time frame of interest. Thus, in time frames of years, the biosphere is essentially closed because sunlight and particulate matter from space are not significantly affected by biospheric dynamics. On the other hand, the water flow system in the soil of a grassland in a several-year period is markedly affected by plant growth. An essentially open system is defined as one which affects its environment in a way which is significant as far as the system is concerned in the time frame of interest.

The importance of the time frame in this discussion needs to be emphasised. Over the span of a few minutes, plant growth might not significantly affect soil-water movement and thus could be ignored. On a one-year (or growing season) basis, however, plant growth might be expected to affect soil-water dynamics markedly.

Essentially open systems are no more difficult, conceptually, than closed systems. The environment may be treated as independent of the system and simply as a 'driver' of the system. Essentially open systems are analytical impossibilities. The dynamics of the system modify the environment in ways that are important to the system; yet these modifications are considered as external to the system. Concern for this point is exemplified by emphasis on fixing model boundaries as part of the objective setting exercise.

Other pertinent values of the systems view include the enforcement of the scientific method and guided subdivision of the problem, both mentioned above. A systems view also enforces a concern for all the components of the problem needed to achieve objectives. This point is elaborated upon in the next section. Systems views work to guarantee that all the components will be compatible in the final stages of analysis since a concern for all parts is one element of that view. Although only mentioned briefly, each of these points is critical to the extent that failure to achieve any one of them can spell disaster for the modelling effort.

11.2.2.4. Reticular View

The analysis of systems almost always results in attention being paid to interlinkages among components that had been ignored, or in other words to holes in the system. For biological systems, with their large dimensionality, the number of holes is great and some guidance is needed to invest limited resources well in a given biological research effort. The reticular view focuses on these holes, evaluates them in the systems context and guides the choice of research projects needed to achieve objectives.

The systems analyst and the biologist must constantly be on their guard against letting interesting facets of the reticulum divert them from their objectives. Much of the criticism that is heaped on model's and modeller's heads is based in the reticular concerns of the critic. He sees some one or few ideas as all-important in his view of the system. If those ideas are absent, or poorly represented, his reaction may be negative. I am reminded of the discussion between the decomposer expert (DE) and the animal demography expert (ADE) on reviewing a large ecosystem model. The DE remarked that the demography model was looking good but the decomposition still had a long way to go. The ADE replied that he felt the decomposer model was far advanced but that animal demography was still a crude approximation.

11.2.2.5. Design of Experiments

The first step in empirical science is the design of appropriate experiments. Lindeman's[18] trophic dynamic view of ecological and biological systems has shifted the focus of experiments from the states to the rates. This shift has left the field with something of a void in terms of experimental design procedures which fit well into dynamic simulation studies. The systems view and the reticular view combine to provide, in the systems context, a tool for constraining and shaping experiments. Concern for flow rates and the confounding of flow rates in state variable measurements (the only measurements that can be made directly) leads to more careful attention to alternate flow paths, non-destructive sampling, high-rate, short-term sampling, etc. These studies are not new to the field, but the emphasis that they are receiving is increasing.

One specific example is that maximum uptake rate of phosphorus by plant roots is a parameter in a Michaelis–Menten equation. The rate has often been estimated experimentally using a several-day interval. The rate is then converted to a relative rate (rate per unit root weight) to make it more 'useable'. Some recent simulation model experiments have indicated that the rate so determined is perhaps three orders of magnitude low because only a small part of the root weight participates actively in uptake and because rates over shorter intervals can clearly be higher. As a result of these model analysis results, new experiments have been designed to determine the active fraction of roots for phosphorus uptake and to estimate uptake over shorter intervals (such as an hour). Furthermore, the design of the experiment will take into consideration a select number of other variables that affect uptake and are of primary concern to achieve the given objectives.

11.2.2.6. Hypothesis Clarification and Storage

The role of the mathematical formulation of mechanisms in clarifying hypotheses was discussed above. It needs no further elaboration here.

Hypothesis storage in the form of model components is changing the approach of a number of biological investigators to their system. The reason for this change seems to be that the complexity of the system is great and its understanding involves the simultaneous treatment of several, if not many, hypotheses. Contradictions among these hypotheses and the consequences of these hypotheses acting together are analysable mentally, only for a few time steps and few coacting mechanisms. A simulation model wherein a collection of these mechanisms have evolved into a form that is compatible with observations serves as a platform from which new insights may be achieved. It seems that the investigator has been freed of some ties and, having resolved at least temporarily one part of the puzzle, he can now turn his talents to other parts.

This bootstrapping process (lifting oneself by one's own bootstraps) is the way science proceeds. The store of knowledge that resides in the model is a collection of consistent (internally and with some observations) hypotheses. These stored and tested hypotheses are part of the environment in which other hypotheses are tested and at the same time are continually changing in response to new information.

11.2.2.7. Hypothesis Testing

Simulation models provide a new arena for hypothesis testing. The model is a collection of hypotheses that act together in a simulation run. If that run or model experiment produces results that are not in accord with known system behaviour, the hypothesis (model) is rejected. The point was made earlier that the rejection of such complex hypotheses requires careful analysis to be useful but, if we have built up a good store of tested hypotheses, we can look to recent additions and weak hypotheses for the source of the error.

The phosphorus modelling experience mentioned above is an example of this hypothesis-testing mode of operation. A series of model experiments showed inconsistencies with field and laboratory observations. The weak hypothesis (uptake rate) was discovered by eliminating the other hypotheses as error sources. This process of elimination was largely one of seeking corroborative evidence in support of the hypotheses. Finally, the phosphorus expert was convinced enough to allow a model experiment with a much higher uptake rate. When the results of this experiment indicated consistency with field and laboratory experiments, the initially hypothesised rate was rejected and experiments initiated to obtain more accurate estimates.

The use of simulation models for hypothesis testing was anticipated by Slobodkin[19] and is beginning to play an important role in research on biological systems. I have proposed the development of a 'self-organising system' for biological modelling, a digital computer-based simulation model-building system, that uses this form of hypothesis testing to develop an ecological system theory, *viz.* a consistent set of ecological system hypotheses.

11.2.2.8. Hierarchies of Hypotheses

L. B. Slobodkin's[20] suggestion that meta-models might be more useful in the development of ecological theory is interesting. He defines a meta-model as a classification of models of ecosystems. He uses, as examples, MacArthur's[21] paper on community stability, his (Slobodkin's[22]) work on the types of stability in single species populations and Rickers'[23] population equilibrium models. Slobodkin points out that each of these meta-models (a) use simple criteria for problem area subdivision and organisation, (b) require little information and (c) are each applicable to any given ecological event. He states that 'analysing a phenomenon on the basis of several independent meta-models may provide almost as good a basis for prediction as would the construction of a precise predictive model'.

Since Slobodkin's[20] paper appeared, there have been several additional model subdivision/classification schemes produced. However, the promise of predictive capability from such meta-models has yet to be extensively realised.

In a 1960 paper, Slobodkin,[19] suggests a role for models and meta-models that is beginning to be realised. He states that 'when a sufficient number of these generalisations have been stated and tested, a comprehensive predictive general theory of community ecology will appear, if only by the elimination of all conceivable theories whose predictions do not conform to the generalisations'.

As Kowal[11] states so well, hypotheses about the dynamics of biological systems are each complexes of hypotheses. In many cases there are several alternative hypotheses that could be used. We can often eliminate one or several of these hypotheses using the technique that Slobodkin suggests, namely casting out those that do not conform to the generalisations. Going a bit further, we cast out those that do not conform with more tested hypotheses. The arena for such experiments is a computer model. Forrester[24] states that complex systems are counter-intuitive. By this he means that the consequences of the simultaneous interaction of a group of hypotheses is difficult to follow mentally. Well-built computer models of biological systems allow the expert to focus his attention where it is best applied—on small components of the system. The logical consequences of the resulting complex of hypotheses then follows from the exercise of the model. Ideally, this is Norbert Wiener's[25] human use of human beings at its best: minds applied to creative efforts while computers take care of the drudgery.

This hierarchy of hypotheses takes another form of value in developing ecological systems theory. The form is that of determining which hypotheses are 'fundamental' and which are 'derived'. Mathematics is concerned, among other things, with the question of minimal collections of fundamental entities and the rules whereby the derived entities follow from the fundamental ones. An excellent example of the use of models in biology to discover such hypotheses occurred in an attempt to simulate root dynamics under grazing stress. The investigator wanted to determine which of several hypotheses about plant response to grazing would be needed in addition to the coded hypotheses for plant growth to achieve the experimentally observed results. Somewhat surprisingly, the simulated root dynamics under grazing followed the observed values quite well without the addition of new hypotheses. It obtained that one of the hypotheses proposed for addition was found to be the consequence of the hypotheses in the model, a result not at all expected by the modellers or the biologists!

11.2.2.9. Interpolation

Interpolation is concerned with filling in between given points whereas extrapolation (the next major heading) is concerned with estimation of values beyond the range spanned by the given points. This difference (known values at either end of the interval versus known values at only one end of the interval of interest) is fundamental both biologically and mathematically.

Both interpolation and extrapolation have two main mathematical roots, properties of orthogonal polynomials and properties of differential equations. The orthogonal polynomial results (Szego[26] and Jackson[27]) are the basis for a number of the statistical routines that are part of the biologist's stock-in-trade. A weakness relates to the formal, purely mathematical nature of the analysis which takes no cognisance of the biological mechanisms involved.

The differential equation properties are similarly formal but, if the equation represents the biology, the results have biological meaning. Because of a prejudice for this ability to interpret the results in terms of the application, I shall limit attention to this latter form of interpolation (see also Forrester[3]).

Differential equations and other models structured from them have proven to be quite useful in interpolation. Abiotic phenomena, such as soil temperature, can be quite accurately estimated at points between the soil surface and a deep (isothermal or slowly changing) layer using the Fourier Heat Equation (Churchill[28]). Agronomic plant models represent growing season dynamics of certain species of plants with such accuracy that additional production studies are unneeded and indeed have been terminated. (This agronomic example is termed interpolation rather than extrapolation because of the dependence on both initial and terminal (winter kill) conditions. Models of perennials will be mentioned under the heading of extrapolation.)

The obvious advantage to good interpolation models is that they reduce the expense, time and effort needed to gain sufficient information about the modelled variables for both scientific and management purposes. Resources freed from these efforts can be redirected into more fruitful studies, in the sense of analysing less known system components.

In that segment of the biological community where modelling scepticism is the rule rather than the exception, interpolation is often more palatable than extrapolation. Interpolation can, therefore, be used as an introduction to the modelling arts.

11.2.2.10 Extrapolation

Extrapolation with polynomials is troubled not only by the difficulty of interpreting the coefficients in biological terms but also by their inherently unbounded nature. Statistical methods of extrapolation using time series (Fourier) techniques produce bounded, periodic estimates of the future course of events (Wiener[29]). Since coeffecnt interpretation in biological terms is still difficult, only the analyses based on differential or difference equations will be discussed.

Extrapolation implies prediction and the paper by Mayr[16] leads one to be sceptical of prediction in biological systems. Forrester,[3] speaking of social systems, distinguishes between trend prediction (a reasonable goal) and event prediction (an unreasonable goal in complex systems). Certain components of the biological sciences are sufficiently advanced that the direction, magnitude and period of response to given events can be predicted with some certainty. On the other hand, there is little hope and perhaps no reason for predicting events such as the numbers of young a given mammal will have, the date on which the litter will appear or the ratio of males to females in the litter. Care must be taken to see that extrapolation and prediction are interpreted in the trend sense rather than in the event sense.

Generally speaking, extrapolation models are not as accurate as interpolation models. Interpolation is bounded to data at both extremes of the interval of interest whereas the further extrapolation proceeds, the further it gets from a known point. In spite of these limitations, certain crop models and models of natural systems are capable of relatively sound extrapolations. Typically these models are used for such purposes as the evaluation of certain management strategies (such as irrigation and fertilisation plans or livestock rotation plans). In the areas of irrigation and fertilisation, the models are well enough advanced to be used to determine the best time (to within a few days to a week) for irrigation and/or fertilisation. For parts of Idaho, USA, in 1973 such plans were estimated to save agriculturalists $200 per hectare in excess of the cost of the service!

Extrapolation and prediction are areas where great improvements in our current capability are desired. From purely biological questions of stability and diversity of communities to the practical questions of the response of communities to perturbations, the need is felt for better predictive capability. The lack of such capability is not the inadequacy of the analytical tools but the inadequacy of the biological information base. Models and model buildings are an integral part of the effort needed to establish this base. The use of models as hypothesis storage devices that can be exercised together will allow us to recognise that certain areas are well enough developed to be de-emphasised for a while and that other areas need immediate attention. Modelling efforts will guide the experimental design. The ever higher level of hypotheses will be tested, in part, in the models and incorporated into the theory (if acceptable).

11.3. CONCLUSIONS

With all the uses of the preceding section available to the researchers who use the systems approach in biology, one wonders why there is such controversy over its use. Several reasons can be given. First, there is some history of mathematical models being presented as more than they are. The furore over the world model as presented by Forrester[30] and Meadows et al.[31] stems largely from perceived abuses of the systems approach. Few people would try to describe world dynamics with five state variables, and the objectives for the model are so poorly stated that each reader can imagine what he will of the designer's objectives. Thus, the models seem hopelessly simple and unrealistic. Similar, if less spectacular, abuses have occurred in many areas. In biology the logistic and Lotka–Voltera models are such oversimplifications as to be quite limited for all but tutorial purposes; yet they continue to appear in research papers. Many scientists interpret this to mean that systems efforts are still at this preliminary stage.

Secondly, mathematics and mathematical sciences are impositions on and threats to some of the practising biologists. In their own defence they must defame the systems approach insofar as it is identified with maths and computers.

Thirdly, many of our attempts to go beyond the simple models have taken the form of trying to force the biology into the extant mathematical tools. Foremost among these trials has been the insistence on linear representations because of their analytical simplicity. I do not think anyone argues any longer that biological systems are generally linear; yet the use of linear representations continues.

Fourthly, by forcing the biology into the extant mathematical structure, we also inherit all the current mathematical analysis jargon and sophistication. Biological simulations often get completely lost in numerical analysis, stability theory, parameter estimation procedures, etc., which may or may not be appropriate to the problem in question. When the biological principles used to formulate the model have errors of estimates that approximate the mean, does it matter whether an Adams–Moulton or a fourth-order Runge–Kutta scheme is used? It may, but most likely it does not. To have the principles so subject to error and to put emphasis on the numerical details seems most inappropriate.

It is clear, now, that new mathematical tools are called for to further the field. Forrester[32] claims that, in feedback systems, the overall system sensitivity to errors in parameter or initial condition values is reduced as complexity increases. No proof or conditions are given, however, and this needs to be investigated. One has the sense that the argument might be similar to MacArthur's[21] and perhaps be subject to similar limitations.

Fifthly, the problems of pursuing an interdisciplinary career, particularly in academic institutions, is great. The disciplines, via departmental structures, hold sway. To survive, the interdisciplinarian must be good in a discipline *and* become involved in the efforts of other disciplines. Except for a few, these demands are too great. It appears that the only viable solution is to establish one's credentials in a discipline (a several-year task) before pursuing interdisciplinary interests. The careers of the successful interdisciplinarians (or multidisciplinarians) illustrates, I think, that they have established themselves in one field before tackling another. Such an approach precludes our training students for interdisciplinary careers. My present advice to students is to put their energies into really learning one field and ignore the rest. They should get their degrees and work for several years in that field. Then, if they are still interested in other fields, they can safely put effort into those studies. By seeking students with mathematics degrees at the undergraduate or master's level, but with a desire to study biology, several biology and natural resource colleges are succeeding in increasing the systems efforts in biological curricula. (This seems to work better than taking biology undergraduates and trying to give them graduate mathematics training. This latter seems to fail because of the more structured and abstract nature of mathematics. Elderly people (past 20) have difficulty learning maths.)

In conclusion, the systems approach will probably revolutionise biological research in the next few decades. The revolution will be based on the realisation of the utilities described in Section 11.2. It will take decades because of the human, institutional and knowledge limitations that we face and must conquer. The most that we at this symposium can hope to do is to lead the revolution and encourage our students to do the same. They and their students will see the fruits of these labours.

REFERENCES

1. Levins, R. (1973). Fundamental and applied research in agriculture, *Science*, **181**, 523–4.
2. Bartlett, M. S. & Hiorns, R. W. (eds.) (1973). *The Mathematical Theory of the Dynamics of Biological Populations*, Academic Press, London, 347 pp.
3. Forrester, J. W. (1961). *Industrial Dynamics*, MIT Press, Cambridge, Mass., 464 pp.
4. Holling, C. S. (1966). The strategy of building models of complex ecological systems, in: *Systems Analysis in Ecology* (ed. K. E. F. Watt), Academic Press, New York. pp. 195–214.
5. Goodall, D. W. (1972). Building and testing ecosystem models, in: *Mathematical Models in Ecology* (ed. J. N. R. Jeffers), Blackwell, Oxford, pp. 173–94.

6. Levins, R. (1968). *Evolution in Changing Environments*, Princeton Univ. Press, Princeton, 120 pp.
7. Pielou, E. C. (1972). On kinds of models, *Science*, **177**, 981–2.
8. Innis, G. S. (in press). Dynamic analysis in 'soft science' studies: in defence of difference equations', in: *Lecture Notes in Biomathematics* (ed. P. Van den Driessche), Springer-Verlag Publishing Co., New York.
9. Churchman, C. W. (1968). *The Systems Approach*, Dell, New York, 243 pp.
10. Innis, G. (1972). Simulation of ill-defined systems: some problems and progress, *Simulation*, **19**, (6), 33–6.
11. Kowal, N. E. (1971). A rationale for modelling dynamic ecological systems, in: *Systems Analysis and Simulation in Ecology*, vol. I (ed. B. C. Patten), Academic Press, New York, pp. 123–94.
12. Nolan, R. L. (1972). *Verification/Validation of Computer Simulation Models*. Summer Computer Simulation Conference, pp. 1254–65.
13. Watt, K. E. F. (ed.) (1966). The nature of systems analysis, in: *Systems Analysis in Ecology* (ed. K. E. F. Watt), Academic Press, New York, pp. 1–14.
14. Van Dyne, G. M. (1972). Organisation and management of an integrated ecological research program—with special emphasis on systems analysis, universities, and scientific cooperation, in: *Mathematical Models in Ecology* (ed. J. N. R. Jeffers), Blackwell, Oxford, pp. 111–72.
15. Odum, H. T. (1973). An energy circuit language for ecological and social systems: its physical basis, in: *Systems Analysis and Simulation in Ecology*, vol. II (ed. B. C. Patten), Academic Press, New York, pp. 140–211.
16. Mayr, E. (1961). Cause and effect in biology, *Science*, **134**, 1501.
17. Attinger, E. O. (ed.), (1970). *Global Systems Dynamics*, Wiley-Interscience, New York, 353 pp.
18. Lindeman, R. L. (1942). The trophic–dynamic aspect of ecology, *Ecology*, **23**, 399–418.
19. Slobodkin, L. B. (1960). Ecological energy relationships at the population level, *American Naturalist*, **94**, 213–60.
20. Slobodkin, L. B. (1958). Meta-models in the theoretical ecology, *Ecology*, **39**, 550–1.
21. MacArthur, R. (1955). Fluctuations of animal populations and a measure of community stability, *Ecology*, **36**, 533–6.
22. Slobodkin, L. B. (1955). Conditions for population equilibrium, *Ecology*, **36**, 530–3.
23. Ricker, W. E. (1954). Stock and recruitment, *J. Fish. Res. Bd. Canada*, **11**, 559–623.
24. Forrester, J. W. (1971), Counterintuitive behaviour of social systems, *Technology Review*, **73**, 1–16.
25. Wiener, N. (1956). *The Human Use of Human Beings: Cybernetics and Society*, Doubleday, Garden City, N.Y., 199 pp.
26. Szego, G. (1939). *Orthogonal Polynomials*, American Mathematical Society Colloquium Publication Vol. 23, New York, 401 pp.
27. Jackson, D. (1941). *Fourier Series and Orthogonal Polynomials*, Mathematical Association of America, 234 pp.

28. Churchill, R. V. (1963). *Fourier Series and Boundary Value Problems*, 2nd ed., McGraw-Hill, New York, 248 pp.
29. Wiener, N. (1964). *Time Series*, MIT Press, Cambridge, Mass., 163 pp.
30. Forrester, J. W. (1971). *World Dynamics*, Wright-Allen Press, Cambridge, Mass., 142 pp.
31. Meadows, Donella, Meadows, Dennis, Randers, J. and Behrens, W., III (1972). *The Limits to Growth*, Universe Books, New York, 205 pp.
32. Forrester, J. W. (1968). Industrial dynamics—after the first decade, *Management Science*, 14, 398–415.

Discussion Report

N. R. BROCKINGTON

Systems Synthesis Department, Grassland Research Institute, Hurley, Berkshire, England

Professor Innis introduced his paper by suggesting that many of the important problems in studying biological and agricultural systems which had been referred to earlier in the symposium might be solved by developing a 'self-organising modelling system'.[1] He believed it would be possible to automate the model-building process to a large extent, reducing the extreme data-dependence of current biological models by substituting biological/ecological theory. A library of submodels would generate an appropriate overall model from minimal data inputs specifying the long-term physical environment, recent weather and other perturbations and the objectives of the user. The objectives would have a major influence on the structure of the model produced. He thought that this approach would be helpful in resolving such difficulties as finding a suitable compromise between realism and precision on the one hand and generality on the other. It would aid the process of matching model structure and performance to the objectives and speed up the model construction process.

Reactions to this proposal varied from a horrified conclusion that it was designed to mislead decision-makers with false evidence, unsupported by sound data, to a suggestion that the attempt to construct the necessary submodels would be extremely valuable in itself, even if the ultimate objective could not be attained in a reasonable period of time.

Undoubtedly there is a dilemma in the application of systems analysis to biological and agricultural systems at this time. Immediate practical applications demand 'hard' models, based on 'hard' data. But these models normally have only a limited scope because of the paucity of available data; in extreme cases they take the form of computerised budgeting exercises. Such 'calculator' models apparently cannot be faulted on their validity, although there can be a greater dependence on unstated, and hence often untested, assumptions about the biology of the systems they represent than is appreciated by some of their protagonists. They tend to be very dependent on specific data inputs for individual applications and use the data in a simple arithmetic structure which engenders considerable confidence by its very simplicity. The major snag is that the problems which can be tackled in this way are strictly limited and by no means cover the total range with which the farmer has to cope.

Attempts to construct models which take more account of the real complexity of the biological processes in agriculture rapidly run out of the area in which they can be supported by well-established statistical theory, normally because the central assumption of linearity in much of the theory is violated. From then on, even if the trap of attempting to construct virtually isomorphic models, which are so massive and complicated as to be unusable, is avoided, there is great difficulty in establishing an adequate framework for model-building. A simple arithmetic structure tends to be inadequate because it requires a large quantity of specific data which it is uneconomic to collect, and the dangers of attempting to substitute data collected for other purposes in inappropriate ways have been rightly emphasised by Jeffers.[2]

The need for a suitable framework has been recognised by Professor Innis and a serious effort to devise it deserves support. Just as in the case of managing ecosystems in general,[3] the customer for agricultural research needs answers to the difficult problems as well as the simple ones and will be increasingly dissatisfied if he does not get them. Part of the solution may lie in helping the customer to state his objectives and ask his questions in a way which makes it feasible to answer them more adequately. In the discussion of the paper a suggestion was made that one 'good' form of question is that which can be directly related to sensitivity analysis: undoubtedly this is more meaningful than such woolly questions as 'How do I make more money?' More effort is certainly justified to explore the area of objective-setting and asking pertinent questions because it is inextricably linked to the formulation of an appropriate overall structure for modelling.

REFERENCES

1. Innis, G. S. (1975). One direction for improving ecosystem modelling, *Behavioural Science* (in press).
2. Jeffers, J. N. R. (1975). Constraints and limitations of data sources for systems models, in: *Study of Agricultural Systems* (ed. G. E. Dalton), Applied Science Publishers, London.
3. Jeffers, J. N. R. (1974). Future prospects of systems analysis in ecology, *Proc. of 1st International Congress of Ecology*, pp. 255–61.

12

Systems Research in Hill Sheep Farming

J. EADIE and T. J. MAXWELL

Hill Farming Research Organisation, Edinburgh, Scotland

12.1. INTRODUCTION

One of the major objectives of the Hill Farming Research Organisation is improvement in the economic efficiency of hill sheep production in the United Kingdom.

Hill sheep farming, like many other kinds of ruminant animal production, is basically a biological system in which saleable animal products are obtained from the integration of an animal population and a given set of land resources together with the co-operant factors of labour and capital. In such systems the farmer attempts to integrate their various components to maximum economic advantage consistent with maintaining the capacity of the basic land resources and other private objectives. He does this using his skill, knowledge and experience, and within the various branches of livestock production there is a considerable store of traditional knowledge and lore available to him. One important reason for the very wide range of profitability in farming lies in the widely differing abilities of farmers to integrate the various components of their production systems.

In response to the need implied by this fact a range of operational research techniques have been developed. However, techniques based on studies of existing systems are in the main incapable of taking adequate account of the kind of system changes discussed in this paper, in relation to hill sheep farming.

Operations research methods have been most widely applied in the field of agricultural economics. They attempt to deal with problems of investment, farm organisation and the operation of individual farm enterprises. Most studies in the latter area have attempted to deal with intensive enterprises. In more extensive forms of production, and perhaps particularly in those involving animal production from pasture, the lack of quantitative information on some of the more crucial biological relationships is a major limitation.

Much agricultural research is necessarily carried out on a limited number of specific processes and functional mechanisms. Components are isolated for this purpose from the systems of which they are part. But the ultimate purpose of agricultural research is to improve the operation of whole production systems. The component parts do not exist in isolation. Each component influences and is influenced by the others. It follows that the evaluation of the consequences of change in one component cannot be limited to a consideration of that component alone. An understanding of its relationships with others and a measure of its impact on the system as a whole is required. Very often the limiting deficiency is not a more detailed knowledge of a particular component, but a knowledge of how changes in that component interact with others in the system.

The repeated plea for more relevant and more practical research is often an expression of this need on the part of those who use the results of agricultural research.

It has generally been left to the progressive farmer and his adviser to deal with the problem of incorporating new knowledge and technical information into farming systems. The process is essentially one of trial and error. The evaluation is usually crude; it is often inefficient and may sometimes be misleading.

There is a clear need to recognise that the synthesis of the results of research is a pressing and important area of research in itself.

This view has been accepted in the Hill Farming Research Organisation for some years, and what follows is a brief account of what has been done to date and a discussion of some current issues and problems.

12.2. THE PROBLEM

The basic resources of hill sheep farming are, in the main, the grazings classified in the agricultural statistics as 'rough grazings'. These are uncultivated pastures, often at elevations of upwards of 230 m above sea level. They encompass a wide range of soils and climatic conditions and consequently include a wide variety of vegetation types from the various hill grassland pastures of the drier east, to the *Calluna*, *Eriophorum* and *Tricophorum* plant communities found on blanket peat in the west and north of Scotland.

Traditionally, hill sheep are set-stocked in a free range grazing system. They are expected to obtain their nutrient needs from grazed pasture the whole year round.

TABLE 1
Stocking rate distribution of hill sheep in Scotland (1967)[a]

| | \multicolumn{6}{c}{Acres of rough pasture per sheep} | |
	0–2	2–4	4–6	6–8	8–10	>10	Total
Highland area	41 850	323 939	248 236	103 257	137 138	214 515	1 068 935
North-east	1 772	17 841	32 661	19 987	30 040	19 700	112 001
East-central	37 954	166 685	65 015	33 414	5 555	14 578	323 201
South-east	50 086	221 431	3 309	880	723	500	276 929
South-west	77 193	552 899	32 684	6 357	50	1 105	640 287
Total	208 855	1 252 794	381 905	63 895	173 506	250 398	2 431 353
Percentage	8·59	51·56	15·71	6·74	7·13	10·30	100·0

[a] Taken from J. M. M. Cunningham et al., HFRO 5th Report.

The average size of a hill sheep flock in Scotland is in the region of one thousand ewes of one or other of the hill breeds. Such a flock would normally comprise two units, each accommodated on its own grazings or 'hirsel'. The flocks are self-replenishing and output is in the form of weaned lambs, wool, and ewes culled for age or other reasons. Output per acre, although varying greatly from one hill farming region to another, is low. This is a consequence of low stocking rates (Table 1) and low levels of individual sheep performance. Average weaning percentages are in the region of 80 and range from below 60 to over 100; lamb liveweights at weaning range from below 20 kg to over 30 kg.

The economic status of hill sheep farming has changed considerably over the last two decades, but the industry has a long history of relative economic depression. The impact of technology, mechanisation and so on has been slight, and one of the basic long-term problems arises from the comparatively small size of the hill farming business.[1]

The Hill Farming Research Organisation was set up in 1954 to carry out biological research in hill sheep farming. In its early years the staff was small and the work mainly descriptive in character. By the early 1960s the research programme had been expanded and studies were being undertaken in such topics as the nutritive value and productivity of some hill pasture types, the nutritional status of the sheep on hill grazings, the potential of the hill sheep, and its responses to nutrition in terms of the various components of performance.

12.3. SYSTEMS DEVELOPMENT

By 1967 a good deal of information of a variety of kinds had been accumulated and interest in examining its implication with respect to the development of improved systems of production was increasing.

Improved systems conceived by means other than the systematic, detailed and explicit approach, characteristic of mathematical modelling, are often described as intuitive. The use of this word implies a process less consciously deliberate and orderly than is in fact the case.

The first step was to examine the body of knowledge available about the current system and its component parts, with a view to preparing an analysis of the problem in biological terms.

12.3.1. Analysis

In systems involving animal production from pastoral resources the major component parts are pasture production and animal production.

Each of these may be considered in greater detail under a number of headings.

Under each heading the analysis requires (a) an evaluation of the performance, or output of the component, and (b) an assessment of the major factors influencing it in the system with particular reference to those over which the operator of the system exercises some control. The first of these requires a reference base, which may variously be an output under more favourable environmental circumstances, a genetic potential, and so on. In animal/pasture systems the reasons for unsatisfactory performance of a number of the components will have to do with the grazing system or grazing management. These various references to the grazing system and grazing management must be satisfactorily accounted for in terms of an overall hypothesis.

By way of example, a very brief biological analysis of the hill sheep farming problem follows:

12.3.1.1. Pasture Production

Total annual pasture production which averages around 1800 kg of dry matter per hectare (DM/ha) is poor. The reasons are partly climatic, partly soil physical conditions, but also occur because of low available plant nutrient content of soils and soil acidity.[2,3]

Seasonal distribution of pasture production which is highly seasonal. The growing season is less than six months; 75% of the grass produced grows in less than three months due mainly to climatic reasons, while existing grazing system makes no deliberate provision for pasture saving for future use.

Pasture quality of feed on offer is poor. This is partly a function of the intrinsic nature of many hill species and partly because the available fund of material usually contains a large proportion of previously ungrazed material which in turn is a consequence of the grazing system.[4]

12.3.1.2. Animal production

Total output of animal product/unit area is low. The average is around 14–15 kg weaned lamb liveweight/ha/year. Animal output per unit area is a function of stocking rate and individual animal output.[5]

Stocking rates, averaging one sheep to 1·2 ha (range 1:0·7–1:4 ha) are low, even in relation to poor levels of pasture production. Overall levels of pasture utilisation are poor; they seldom exceed 20% and are usually very much less due mainly to the grazing system.[6]

Individual sheep performance averaging some 18 kg weaned lamb/ewe/-year is poor. This is contributed to by poor performance in the various components of performance.[7]

(i) *Fertility* is poor, the average weaning percentage being around 80; ranging from 60 to over 100 despite the fact that the genetic potential of breeds of interest is in excess of 170%. Ovulation rate is a function mainly of body condition at conception which in turn is a consequence of nutrition over the 'recovery period' from end of lactation to mating[8] (see 'pasture quality' in Section 12.3.1.1. above).

(ii) *Lamb deaths* at or around birth are high at a level of 15–25%, because of the poor climate but also due in part to poor body condition of the ewe in the latter stages of pregnancy and in part to poor nutrition of the ewe over that period.[9]

(iii) *Lamb growth* rates at around 200 g per head per day from birth to weaning are moderate in relation to the potential of hill lambs to grow in good nutritional conditions. Slow growth rates are due to inadequate nutrition during lactation in the ewe and to moderate quality of the pasture feed ingested by the lamb.[10] (See 'pasture quality', Section 12.3.1.1 above).

12.3.1.3. The Grazing System and Grazing Management

Grazing management and the grazing system have been implicated in explanation of many of the inadequacies outlined above.

The grazing system is one in which the sheep are set-stocked on free range over the whole year. The sheep are stocked at rates which will provide the minimum tolerable level of winter nutrition. But pasture growth is highly seasonal and stocking rates so set lead to substantial underutilisation of the summer pasture growth. The unutilised material accumulates during the growing season and its quality deteriorates. The available fund of feed at the beginning of winter therefore contains a large proportion of poor-quality herbage. This, in turn, means that a high degree of diet selection opportunity must be allowed. Hence the low stocking rate, and so on.

The available feed, even in early summer, always contains a high proportion of poor-quality material so that even on those pasture types which contain material of intrinsically high nutritive value the sheep are unable to select a good quality diet.[11]

12.3.2. Synthesis

The first step is to use the analysis, together with knowledge of the responses to the various possible manipulations, to outline the biological possibilities. In any synthesis which is to be of practical value, the biological possibilities are inevitably constrained by various non-biological factors, e.g. physical, operational, economic. The second step is therefore to outline the constraints and how they limit the biological possibilities. The main framework of the synthesis can then be elaborated.

12.3.2.1. Pasture Production

Total pasture production can be improved by means of soil nutrient status improvement (mainly Ca and P), and the replacement of indigenous species by sown grasses and clovers. Yield increases of the order of 2–3 fold can be obtained by this means. A range of hill land improvement techniques, applicable to a wide range of hill soils/pastures, is available.

The seasonality of pasture production. By grazing management some part of the mid-to-late summer growth may be conserved *in situ* for later use. Improvements in pasture production will tend to exacerbate the seasonality of pasture production.

Pasture quality will be improved by replacement of indigenous vegetation with sown grasses and clovers. Higher overall levels of pasture utilisation would reduce the amount of utilised poor-quality herbage present in the available feed.[12]

12.3.2.2. Animal Production

Animal output/unit area increases can come from increases in stocking rate and/or improvements in individual sheep performance.

Stocking rate increases require improvements in pasture production and/or improvement in the efficiency with which the existing pasture production is utilised. They could also be improved by removing sheep from their grazings in winter, and stocking at rates more consonant with levels of summer pasture production.

Individual animal performance improvement requires improvement in the various components of performance:

(i) *Fertility* can be improved by better body condition at conception which requires nutritional improvement in all, or some part, of the recovery period.[8] It can also be increased by better pasture quality and/or feed introduced from without the system.
(ii) *Lamb survivability* at or around birth can be improved (to below 10%) by better nutrition in late pregnancy by means of feed introduced from outside the system.[13]
(iii) *Lamb growth rates* can be increased by better ewe nutrition in lactation and better-quality solid feed for the lamb. This may be achieved by better quality pasture and/or feed introduced from outside the system.[10]

12.3.2.3. The Grazing System and Grazing Management

In the above examination of the biological possibilities, changes in the grazing system and in grazing management are repeatedly referred to in the context of improving pasture quality and pasture utilisation. There are, however, important biological constraints on both utilisation and pasture quality.

On a limited range of hill pasture types, notably the *Agrostis–Festuca* pastures which constitute some 15–40% of the land area of many of the grassy hill grazings, higher levels of pasture utilisation have been shown to eliminate most of the backlog of underutilised herbage, and to give rise to regrowths of quite high-quality pasture.[12]

On other pasture types (e.g. *Nardus* and *Molinia* dominant grass heath) better utilisation gives rise to a substantial short-term nutritional penalty to the stock,[14] and to a much less productive and nutritionally poorer pasture than that achieved with *Agrostis–Festuca*.[15]

On yet others (e.g. the vegetation types found on blanket peat) the more valuable species are likely to be grazed out at grazing intensities well below those consistent with high levels of utilisation of the primary production. However, the evidence also suggests that much higher levels of pasture utilisation than those currently achieved could be sustained, although improvement in pasture quality would be slight.[16]

Most hill grazings are vegetationally heterogeneous. In the light of these observations and the hypothesis advanced in the analysis to account for existing levels of pasture utilisation and quality, it can be argued that a form of grazing control which takes account of the characteristics of these various classes of vegetation is a necessary prerequisite to higher animal output from existing indigenous pastures.

Existing indigenous pastures can of course be replaced by sown pastures. But there seems little point in increasing pasture production on a part of a grazing unless steps are taken to ensure its efficient utilisation. This again implies grazing control.

A distinction can be made between those pasture types (sown or indigenous), which can be managed to provide feed of much improved quality, and those from which only better utilisation (and that often only in limited degree) can be expected. Animal performance improvement requires better nutrition, especially at certain times. The synthesis therefore requires a two-pasture conception. The objectives are better utilisation of the primary production of both pasture components, and the utilisation of that component capable of offering high-quality feed at such times as to maximise its impact on individual animal performance.

12.3.2.4. Non-biological Constraints

The physical constraints referred to earlier mainly concern pasture improvement. This, in its various forms, requires access for vehicles and materials, and also slopes and terrain safe enough for tractors and equipment. The topography of much hill country often inhibits access to quite large areas of hill land, and renders reseeding impossible on substantial parts of many grazings.

Operation constraints arise for a variety of reasons. Many of them are concerned with various aspects of labour use. Apart from the fact that labour costs are a significantly greater proportion of total costs in hill sheep farming than in other types of livestock production, labour is often in limited supply. Improved systems must therefore provide for more efficient and more effective labour use. Such fencing as is done must not only relate to biological objectives; it must also provide a framework within which labour can be more efficiently used.

In any system which has as its objective improved economic efficiency, the economic constraints are as real and important as the biological constraints. The more important economic constraints arise because money is often scarce and always expensive to borrow. System change, especially if it involves hill pasture improvement, is costly. Stock purchases usually have to be made. There is a point in investing capital in a hill sheep enterprise beyond which the need to service the capital expenditures would give rise to short-term cash flow problems from which ultimate financial recovery would be difficult.[17]

Economic arguments also lie behind the decision not to improve nutrition during the summer, e.g. with respect to lactation, lamb growth, and body condition recovery by means of concentrate feed inputs.

One biological possibility which was mentioned but not pursued further was that of offwintering. Substantial increases in variable costs due to winter feeding require large increases in animal performance in the short term. This fact, together with the need to investigate the land utilisation consequences of removing sheep in winter, has led to a series of investigations independent of the current synthesis.

When the need for ways of handling the economic problems became urgent, a break-even budget technique appropriate to the purpose was being developed.[18] This has since been considerably elaborated and computerised,[19] and has proved invaluable in handling the economic aspects of the Organisation's systems programme.

12.4. IMPLEMENTATION

For biological, practical and economic reasons the proposed system is based on two pasture components, one of which, depending upon the nature of the pastoral resources and other factors, is enclosed indigenous pasture or reseeded pasture. The other comprises the remaining unimproved hill.

The second part of the synthesis concerns the integration of the two components. Individual animal performance improvement requires improved nutrition ideally on a year-round basis, but essentially in the latter part of pregnancy, during lactation and early lamb growth, and again in the period prior to mating.

Late pregnancy nutrition requires the use of feed from outside the system. The other periods occur during the pasture growing season. The impact of the two pasture components on performance will be maximised by utilising the improved pasture as much as possible during lactation and lamb growth; by accumulating *in situ* the regrowth produced during the late summer and utilising it in the several weeks before and during mating which takes place in late November. The available evidence suggests that such a management scheme is also compatible with the long-term productivity of the improved pasture component, and also, within the limits visualised at present, with maintaining a positive long-term dynamic in the unimproved hill component.

The synthesis was general and fairly crude, but there were several arguments which seemed to justify an early attempt to examine it in practice. The industry's need for some indication of future possibilities was urgent. It would be several years before the research programme could lead to a more precise and detailed synthesis. Such calculations of a technical and economic nature as could be done with the data which then existed suggested a real possibility of substantially increased output, obtainable in an economically sound fashion. The synthesis suggested a production system markedly different in a number of respects from the current traditional system. It was likely to throw up and highlight problems for further research and lend another and perhaps more valid perspective against which to evaluate the ongoing research programme. There seemed good reason to attempt to test the synthesis in terms of its practicability and profitability.

With respect to implementation, one of the first decisions to be made concerned scale. In the traditional system there is an important behavioural component. Hill sheep display a diurnal movement between the lower and higher ground of their grazing. In addition, they exhibit a well-defined territorial behaviour which is known to be nutritionally significant.[20] The proposed system changes would grossly interfere with current behaviour. The scale of the resources used in a system is positively correlated with vegetational diversity, the size of the open unimproved hill component, distances between fence lines, etc., all of which could interact with behaviour.

The positive correlation of size of unit and vegetational diversity is likely to be especially important where the improved pasture component is created by management alone or by that together with soils upgrading, i.e. where no replacement of indigenous vegetation by sown species is contemplated.

Furthermore, as part of the synthesis a number of husbandry changes, including control of lambing, raddling rams at mating, the establishment of semi-permanent feeding points, and so on, were to be initiated. For all these reasons it seemed more valid, safer and more relevant to establish the systems studies on units of a size approaching those to be found in practice.

A second major decision arose out of the very nature of hill country. No two hill grazings are alike. They differ greatly, even on the same farm, in terms of the proportions of the various pasture types they contain, and in the distribution of these pasture types throughout a grazing. They differ in slope aspect and altitude and the possibility of finding two units within the boundary of one property sufficiently alike to warrant serious consideration as replicates is very remote.

This rules out the classical experimental approach of treatment control and replication. The only possibility in practice is to initiate the synthesis on a given grazing and to measure the changes which take place. The experimental weaknesses of this approach are obvious.

A major redeeming feature was that the proposed synthesis appeared to offer the possibility of very substantial increases in output. Whilst some doubt could attach to the significance of increases in output of 20% or even 30% it would be reasonable to assume that increases of upwards of 50% had something to do with the system changes. Good records of ouput and sheep performance and its components going back over a ten-year period existed on many of the Organisation's stations' grazing units. It was concluded that, provided an adequate monitoring of events was instituted, the proposed procedure would be acceptable and its limitations less than crippling.

In the first instance two studies were initiated, one on a grassy hill on the Organisation's Sourhope research station in the Eastern Cheviots (rainfall 880 mm), and one on its Lephinmore station in Argyll on blanket peat (rainfall 1900 mm).

The application of the broad synthesis to a given piece of land requires a rigorous and detailed description of resources, components and of the system itself. This is inevitably lengthy, and space does not allow a specimen presentation here.

Important initial decisions include those relating to the area of land to be improved, sites for improvement and improvement procedures. In the Sourhope study it was decided to limit the objectives to better utilisation of the existing indigenous pastures. On a unit of some 280 hectares a total of 77 ha of *Agrostis–Festuca* pasture was enclosed over a three-year period. The enclosed *Agrostis–Festuca* pastures, four in all, function in the system as the improved pasture component.

A summary of the output and financial data to date is given in Tables 2 and 3. A more detailed outline of the study has been published.[21]

The restriction of the objectives of this study to the consequences of better utilisation had the advantages of limiting the size of the initial capital investment and of providing a very stringent test of the ideas embodied in the synthesis. However, pasture production can be improved in this environment as in others, and a more recent study has been established which includes such improvement as a significant part of the synthesis.

TABLE 2
Sourhope: Production data

	1969	1970	1971	1972	1973
Stock numbers	398	451	518	529	573
Weaning percentage	84·7	86·5	103·3	104·7	99·5
Total weight lamb weaned (kg)	7 359	8 893	14 700	13 953	14 202
Total weight of wool (kg)	787	1 005	1 273	1 369	1 560

TABLE 3
Sourhope: Gross margin/ewe (costs and prices as of 1973)

	1969	1970	1971	1972	1973	Base
Income (£)						
Lamb @ 42p/kg	2153	2724	4194	4113	4458	2092
Wool @ 52p/kg	442	529	652	712	812	348
Cast ewes @ £10	520	420	820	840	910	600
Subsidy @ £1·65	657	744	855	873	945	561
	3772	4417	6521	6538	7125	3601
Expenditure (£)						
Feed[a]	1007	1567	890	1036	1446	286
Grazing	32	32	92	122	119	—
Other costs[b]	494	559	642	656	711	292
	1533	2158	1624	1814	2276	578
Gross margin	2239	2259	4897	4724	4849	3023
Gross margin/ewe	5·62	5·01	9·45	8·93	8·46	8·89

[a] Feed costs: Hay £34/t, sugar beet pulp £49/t, concentrates £70/t, grass cubes £57/t.
[b] Other costs: Vet 55p/ewe, ram feed 7p/ewe, ram replacement 50p/ewe, haulage 12p/ewe.

On the blanket peat at Lephinmore the synthesis has of necessity to include reseeded pasture. On a unit of 444 ha, 50 ha have been improved; 36 ha of it is totally enclosed and the remainder distributed in a mosaic of improved patches amounting to some 20% of the total enclosed area.
A summary of the results of this study in physical and financial terms is

given in Tables 4 and 5. A more detailed preliminary account of the project has been published elsewhere.[22]

In the years since this programme began in 1968–69 other studies have been started. Some are concerned with examining essentially the same synthesis in the two environments already referred to, but involving larger proportionate inputs of improved pasture and hence greater capital expenditures. The impact of removing sheep from their hill grazings in

TABLE 4
Lephinmore: Production data

	1969	1970	1971	1972	1973	1974
Stock numbers	339	361	373	384	422	433
Weaning percentage	85·0	92·5	103·5	103·6	103·3	
Total weight lamb weaned (kg)	7 207	8 500	10 268	9 924	10 218	
Total weight of wool (kg)	652	772	772	814	815	

TABLE 5
Lephinmore: Gross margin/ewe (costs and prices 1973–74)

	1969	1970	1971	1972	1973	Base
Income (£)						
Lamb @ 30p/kg	1459	1780	2251	2164	2244	569
Wool @ 51p/kg	333	393	393	415	416	282
Cast ewes @ £9	360	414	558	540	747	342
Subsidy @ £1·65	549	595	615	634	696	338
	2701	3182	3817	3753	4103	1531
Expenditure (£)						
Feed[a]	512	446	420	420	584	—
Hogg wintering @ £2·50	250	245	250	250	313	68
Other costs[b]	421	448	462	476	523	254
	1183	1139	1132	1146	1420	322
Gross margin	1518	2044	2685	2621	2683	1209
Gross margin/ewe	4·48	5·66	7·20	6·83	6·36	5·90

[a] Feed costs: Sugar beet pulp £49/t, concentrates £70/t.
[b] Other costs: Vet 65p/ewe, ram replacement 40p/ewe, ram feed 7p/ewe, haulage 12p/ewe.

winter, in systems in which improved pasture has, and has not, been provided, is being examined. More recently preliminary work has started on heather (*Calluna*) dominant hill land, which presents special systems problems.[23]

In an attempt to widen the range of hill environments from which information is being obtained, an association of HFRO and the Scottish Agricultural Colleges is in the early stages of setting up a development programme on private farms based on the synthesis discussed here.

In respect of the objectives set for the programme to date it can reasonably be argued that good progress has been made. The general synthesis has been shown to be successful so far as it has gone. Much information has been collected from the monitoring of various aspects of the studies.

If systems studies in the future are to be based entirely, or even largely, on physical studies in the field, progress will be slow.

Within the terms of the objectives of the current field programme, for example, what is ultimately required is information on responses to an adequate range of inputs in a variety of hill environments. Even if the scale of operation currently thought to be necessary were to be reduced several-fold, there would still be considerable physical limitations to what could be undertaken over any given period of time. In relation to the need, the resources of the Organisation, although considerable, are limited.

Whilst it is vital to the current programme to collect the information necessary to characterise output adequately, and desirable to acquire as much important subsidiary information as possible, many desirable measurements are too difficult, or expensive, or time-consuming to make in the field and some could only be made at the price of an unacceptable degree of interference with the systems under study.

In addition, the opportunities for within-system experimentation in the current programme are small in relation to need. So far little within-system experimentation has been attempted, partly because of this and the desire to ensure that it is devoted to really crucial issues which cannot be satisfactorily dealt with in any other way. Current thinking, arising out of the results of some of the ongoing systems studies and the research programme which is revealing potentially important differences between the important breeds of hill sheep, is to devote some of this capacity to quantifying these differences.

It has become increasingly clear, partly because of its nature, partly because of its major objectives and partly due to the various limitations briefly referred to above, that physical systems experimentation does not, of itself, provide an adequate means of dealing with the problems which arise at the systems level. Its contribution to providing insight into functionally meaningful properties of systems is limited. Many important practical questions which arise within the present systems conception concerning, for example, the prediction of animal output responses to pasture improvement, stocking rate/animal performance relationships, the significance of pasture quality (and particularly of white clover content) in the improved pasture component to animal output, and many others, depend upon quantifying functionally important relationships. The sheer volume of practical experimentation at the systems level implied by these needs is impossible to contemplate. There is too the problem of examining the continuing flow of new information not only in the context of existing systems ideas but also with respect to new systems possibilities. The mathematical approach to systems analysis and simulation would appear to present a powerful means of helping to resolve these problems.

If the tenor of this observation suggests a conclusion rather reluctantly arrived at, and an attitude somewhat reminiscent of the scepticism of many practising biological researchers towards mathematical modelling, the reasons are not the more usual ones of an unwillingness to recognise systems questions as real questions, or of fear, whether it be of mathematics or systems supermen. Nor do we require any persuasion that a systematic approach to complex problems and a rigorous methodology have a great deal to offer. Rather it is the implications which pose formidable problems, some of which go well beyond systems studies to the organisation and management of conventional compartmentalised research.

For the last year or so a small team in the Organisation has been making an exploratory attempt at mathematical systems simulation. It has been a salutory experience, which has fully borne out the findings of others who have engaged in similar exercises in a range of problems. The need to express a whole process in terms of a rigorous conception and a precise formulation of relationships between sub-systems sharply exposes vagueness and imprecision where one had previously assumed certainty.

Gaps in knowledge are revealed and the new perspective gained leads to a re-evaluation of their significance. Many of these recognised in relation to hill sheep farming are common to grazing systems generally. One example is the central importance of an adequate modelling of the grazed pasture/grazing animal interface. Our experience to date would suggest that this presents an even more important, and substantially more difficult, problem in the hill pasture context than in the more intensive lowland pasture environment. Lowland ruminant grazing systems are summer grazing systems, in which pasture utilisation is maintained throughout the grazing period at a relatively high level. Because of this and the limited species content of lowland pastures, within-sward digestibility ranges and diet selection opportunities are comparatively restricted. This contrasts greatly with the situation in the hills, with its species complexity, large funds of available herbage, considerable within-sward digestibility ranges and high levels of diet selection opportunity. The rather simple models currently used in some lowland grazing systems studies are unlikely to be conceptually adequate. Their functional relationships are unlikely to bear extrapolation over the ranges encountered in hill grazings to the point at which they would be useful. This interface is of central importance in systems studies in the hill sheep context and research in this area has been recently strengthened by the creation of a new unit of grazing studies in the Organisation.

In addition to the usual problems inherent in simulating animal/pasture interactions in grazing systems, hill sheep systems present an additional and difficult one which arises out of vegetational heterogeneity and sheep grazing behaviour. Hill sheep have considerable opportunities for selective grazing. Even in more intensive systems this will continue to be the case. The sheep select not only from within the fund of material available to them on individual pasture types; they also exhibit preferences among the various vegetation types which together make up a grazing. Their preferences in this respect vary with season of the year. It may be that conceptually a 'compartment' approach, in principle similar to that described by Goodall,[24] can be developed as an acceptable and convenient simplification, but much work remains to be done on this question.

The mathematical modelling work done to date has not only highlighted gaps in knowledge, it has also revealed important deficiencies in the information available in areas which have been the subject of a great deal of study. For example, quantitative information on relations between nutrition and lactation performance is important in modelling any sheep production system. A considerable amount of research has been carried out in recent years on lactation performance of sheep. Much of it has involved studies in which concentrate diets have been used, in which sheep have been fed rationed quantities of food, and in which animals have been penfed indoors. But sheep are almost invariably at pasture during lactation, where the diet is fresh grass, and where intake is *ad libitum*. During lactation the intake achieved is a function, in part, of the quality of the ingested pasture, and in part a function of the physiological state of the ewe. The intake may also be mediated by characteristics of the grazing situation, such as the available amount/unit area, pasture structure, and so on. Extrapolation from the available body of information is therefore hazardous and necessitates many assumptions, some weakly based, including those concerned with the partition of intake. The point is not that the available information is valueless. There can be many objectives in studying lactation, but they must often include a desire to provide information to be used in systems of production. In respect of this objective, and in relation to the current state of knowledge, the value of much current information is very limited indeed. In the fullness of time it will be possible to incorporate it into a framework which includes a quantification of these factors affecting intake and those determining the partition of nutrient intake. But this is some way off, and in the meantime some quantitative information from grazing sheep, in which the responses to changes in pasture quality in terms of intake, milk production, bodyweight and body compositional change are measured, is urgently required.

This example could be repeated with respect to other components of sheep performance. The problem it highlights is common to other aspects of the production process. The acceptance of mathematical modelling as an important part of a systems research programme requires a very close integration of systems synthesis and component research. This must go well beyond the mere identification of gaps in knowledge to include a consideration of how the gaps should be filled in relation to the conception of models and in relation to the current state of knowledge in related fields. This in turn goes right to the heart of the question of research organisation. Component researchers themselves must be encouraged to take part in the modelling process. There can be no question of an organisational separation of systems synthesis and analytical research.

ACKNOWLEDGMENT

The authors have pleasure in acknowledging the valuable contributions of the Research Stations and scientific staffs to the ideas and work reported under systems development.

REFERENCES

1. Allen, G. R. (1972). The economic outlook for hill farming in the 1970's, *3rd Coll. Potassium Institute Ltd.*, Edinburgh, p. 55.
2. Hill sheep farming to-day and to-morrow: a Workshop Report, Bull. No. 13, Agricultural Adjustment Unit, University of Newcastle upon Tyne.
3. Floate, M. J. S. (1970). Plant nutrient cycling, *5th Report*, Hill Farming Research Organisation.

4. Eadie, J. (1967). The nutrition of grazing hill sheep: improved utilisation of hill pastures, *4th Report*, Hill Farming Research Organisation.
5. Eadie, J. & Cunningham, J. M. M. (1971). Efficiency of hill sheep production systems, in: *Potential Crop Production* (ed. P. F. Wareing and J. P. Cooper), Heinemann, London.
6. Eadie, J. (1971). Hill pastoral resources and sheep production, *Proceedings of Nutrition Society*, 30, 204–10.
7. Russel, A. J. F. (1971). Relationships between energy intake and productivity in hill sheep, *Proceedings of Nutrition Society*, 30, 197.
8. Gunn, R. G. & Doney, J. (1970). Some factors affecting reproductive rate in hill sheep, *5th Report*, Hill Farming Research Organisation,
9. Russel, A. J. F. (1967). Nutrition of the pregnant ewe, *4th Report*, Hill Farming Research Organisation.
10. Peart, J. (1970). Factors influencing lactation of hill ewes, *5th Report*, Hill Farming Research Organisation.
11. Eadie, J. (1970). Sheep production and pastoral resources, in: *Animal Populations in Relation to their Food Resources*, British Ecological Society Symposium No. 10, p. 7.
12. Eadie, J. & Black, J. S. (1968). Herbage utilisation on hill pastures, *Occasional Symposium British Grassland Society*, 4, 191.
13. Russel, A. J. F., Maxwell, T. J. & Foot, J. Z. (1974). Nutrition of the hill ewe in late pregnancy, *6th Report*, Hill Farming Research Organisation.
14. Floate, M. J. S., Eadie, J., Black, J. S. & Nicholson, I. A. (1972). The improvement of *Nardus* dominant hill pasture by grazing control and fertiliser treatment and its economic assessment, *3rd Coll. Potassium Inst. Edinburgh*, p. 33.
15. Eadie, J. (1967). Unpublished data.
16. Grant, S. A. (1974). Unpublished data.
17. Maxwell, T. J., Eadie, J. & Sibbald, A. R. (1973). Methods of economic appraisal of hill sheep production systems, *3rd Coll. Potassium Institute, Edinburgh*, pp. 103–13.
18. Harkins, J. M. (1968). Assessing new capital investment on hill sheep farms, *Scottish Journal of Agriculture*, 47, 196.
19. Maxwell, T. J., Eadie, J. & Sibbald, A. R. (1974). Economic appraisal of investments in hill sheep production, *6th Report*, Hill Farming Research Organisation.
20. Hunter, R. F. (1962). Hill sheep and their pasture: a study of sheep grazing in S.E. Scotland, *Journal of Ecology*, 50, 651.
21. Eadie, J., Armstrong, R. H. & Maxwell, T. J. (1972). Hill sheep production systems—development in the Cheviots, *3rd Coll. Potassium Institute, Edinburgh*, p. 139.
22. Eadie, J., Maxwell, T. J., Kerr, C. D. & Currie, D. C. (1972). Development on blanket peat, *3rd Coll. Potassium Institute, Edinburgh*, p. 145.
23. Grant, S. A. & Milne, J. A. (1972). Factors affecting the role of heather in grazing systems, *3rd Coll. Potassium Institute, Edinburgh*, p. 41.
24. Goodall, D. W. (1970). Use of computers in grazing management of semi-arid lands, *Proc. XIth International Grassland Congress*, p. 917.

13

Systems Analysis in Relation to Agricultural Policy and Marketing

G. R. ALLEN

Department of Agriculture, University of Aberdeen, Scotland

A recent introduction to the use of systems analysis in public decision-making states:

'Systems have essentially three characteristics: they are made up of *separate elements* that *interact* with each other for the purpose of *obtaining some goal* or objective. Those three characteristics are all essential. There are individual elements, each separate, each identifiable, but none of them in themselves are the system about which we are talking. The elements must interact with each other to create that system. They must interact in such a way as to resemble goal-directed behaviour. One more condition is needed to define a system: it must be bounded. Determining the boundary of the system is a task, performed by an analyst or a decision-maker, that is essential to performing any analysis, since we must have some notion of the creature being studied. It also illustrates the artificiality of any particular system.'[1]

Systems analysis can provide a formal procedure for planning, giving a framework to ensure a definition and reconciliation of objectives, consideration of all relevant exogenous and endogenous influences (and so of the disciplines and subject areas involved), a feedback mechanism for revision of objectives and plans in the light of actual or simulated experience, recognition of the externalities associated with particular systems relationships and, above all, specification of the plan and its means of implementation. Formal specification of sub-systems is usually essential.

We have through systems analysis, then, the means of providing a framework of analysis and review of results which ensure that our thinking will be as complete and flexible as possible in defining objectives, means and ends.

It follows that one important element of systems analysis, which will be barely touched upon in the remainder of this paper concerns the structuring of channels of information flow and decision-making so that planning can be more effective in determining goals, problems and opportunities, and encouraging consensus for achieving goals (or, as the Americans say, getting everybody on board) than it would be otherwise. Suffice to describe the pre-systems thinking situation in an American hospital:

> 'In the past, each department more or less went its own way. A kind of feudal system existed in which the hospital represented a series of principalities, loosely organised under the aegis of the hospital administrator. Each department was possessive about its own jurisdictional rights and privileges. Occasionally two or more would informally reach some accommodation over certain types of changes. This represented a treaty of sorts between respective dukedoms.
>
> An implicit struggle between departments continued over budget allocations, with each attempting to maximise its own departmental development. Hence, each exercised continual pressure upward towards the hospital administrator in relation to requests for additional equipment, space, for personnel, for programmes that each honestly felt would improve overall departmental performance.'[2]

It sounds a bit like the organisation of agricultural research in Britain pre- (and maybe post-) Rothschild[3] or, for that matter, the organisation of higher education. But enough of such controversial diversion.

It may be helpful to expand a recent schematic presentation of Dent[4] to highlight the systems considerations that are likely to arise in the analysis of agricultural policy and marketing situations. In a real-world situation the interrelationships are feedbacks between the various components of the suggested system and are very much more complicated than shown in Fig. 1, but hopefully it will serve in its present form as a precursor of what is to come.

Systems analysis, whether or not it is so-called, contributes the key conceptual elements of a planning framework:

(i) *The determination of objectives*, to be achieved after the specification and reconciliation of conflicts between those concerned and, which is exceptionally important although often overlooked, assessment of the feasibility of objectives in the light of the organisation's special problems and opportunities which respectively may block or further them.

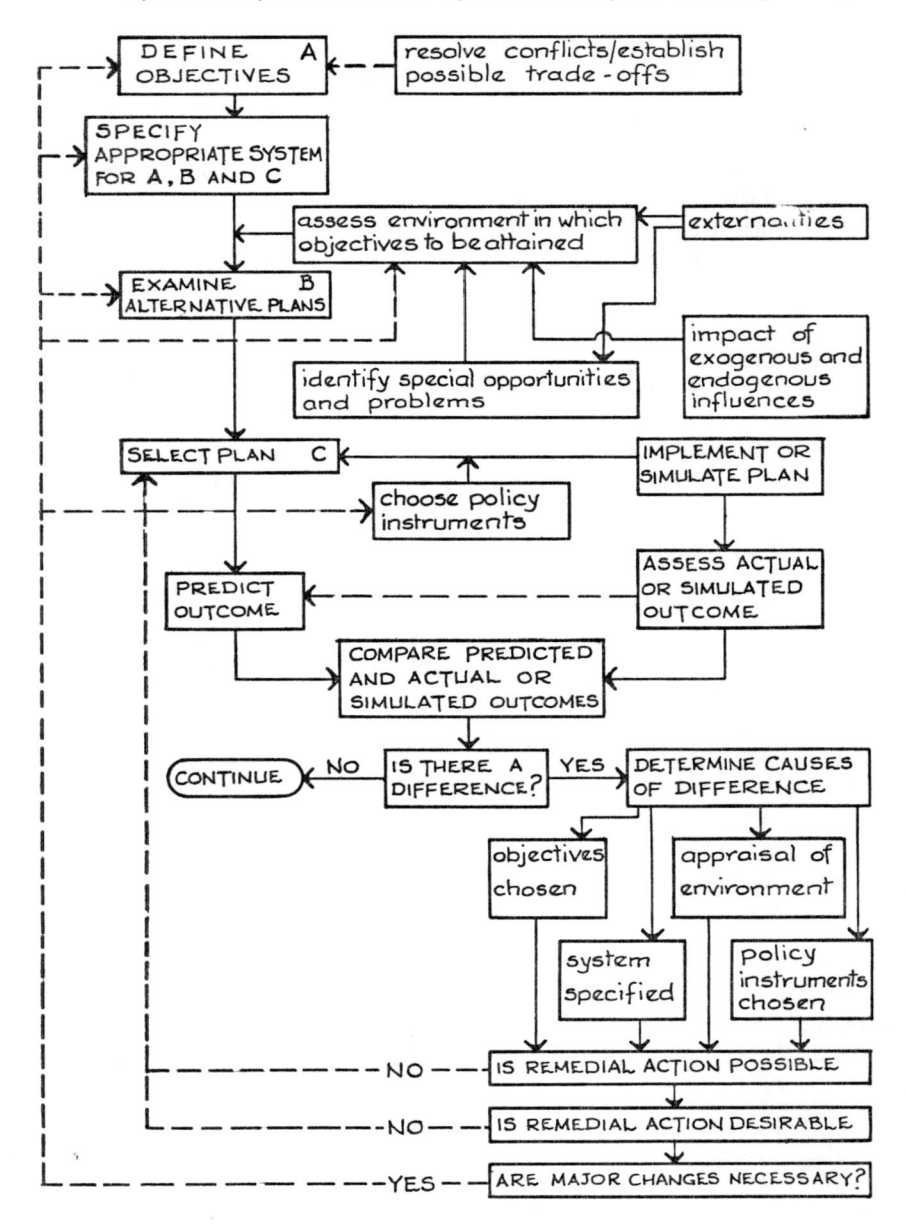

Fig. 1

(ii) *Determining and creating the institutional framework to ensure a systems view*, which must occur simultaneously with (i), to minimise the difficulties in specifying objectives and to ensure the most favourable setting for their attainment.

(iii) *Specify and describe the working of appropriate systems and sub-systems.* This stage of the analysis will involve the definition and measurement of (a) linkages which permit the identification of appropriate sub-systems and measure the extent of their interconnection, (b) environmental influences (exogenous and endogenous), (c) appropriate functional relationships to describe the working of systems and sub-systems in the light of environmental influences, and (d) external effects of the individual systems or sub-systems when in operation (i.e. externalities).

(iv) *Formulate a plan.* This means specifying goals (that is, *attainable* objectives), priorities, means (policy instruments), and results expected from the plan for subsequent comparison with achievement.

(v) *Establish information flow and analysis for maximising benefits from the learning process*, which is essentially the formal appraisal of performance (that is, continuing performance reports) in the light of experience and of simulation of experience; or in other words, feedback procedures.

(vi) *Recasting the plan in the light of experience.* Insofar as differences between planned and actual or simulated performances show that the original plan was erroneous or inadequate due to inappropriate goals or priorities or incorrect policy instruments; in other words, the influence of (v) on (i), (iii) and (iv).

(vii) *Recasting of the systems institutional framework in the light of experience* may be necessary.

A very simple alternative definition of systems analysis might be 'an intellectual approach which enables us to consider everything which is relevant and put each component in true perspective'. To this end the distinctive components of systems thinking in the total approach are:

(i) Identification of an appropriate institutional framework.
(ii) Sorting out objectives.
(iii) Identifying linkages, including externalities.
(iv) Specifying and describing the working of appropriate systems and sub-systems.
(v) Specifying policy instruments in relation to (iii) and (iv).
(vi) Optimising the learning process (feedback) through (a) specifying

appropriate relationships in systems and sub-systems and in linkages, and (b) observation and analysis of plan results or their simulation.

The pure planning function, stripped of the systems component, is concerned with specifying (a) realism in relation to problems, opportunities and means, (b) goals, (c) necessary tasks, and (d) revision of any one or more of the three foregoing components in the light of experience.

Economists often claim to be planners and, as is about to be argued, the thought processes of economics are conceptually very close to those of systems analysis. But how far do economists adequately handle the systems and pure planning components of agricultural policy or marketing issues?

The term 'systems analysis' has no place, or certainly not an established one, in the vocabulary of economics. Yet it is not hard to identify the similarities. If one reads, for example, L. von Bertalanffy's 'The Theory of Open Systems in Physics and Biology',[5] one is struck by the similarity of concepts, which is not surprising seeing that the theories of statics and dynamics in physics have been the building blocks of economic techniques of analysis over the last hundred years.

It has often been pointed out, I believe, that the *General Theory of Employment, Interest and Money* by Lord Keynes[6] represents a systems framework for analysing the working of the economy at the *macro economic* level and that the book and its derivatives provided a more successful basis for managing the general level of economic activity of the economy than had previously existed—although Keynes would never have countenanced the increasing monetary and physical mismanagement of the last decade. In terms of the *micro-economic* subject matter of the present paper a systems framework was, in effect, set out in considerable detail in one of the other recent classics of British economics, A. C. Pigou's *Economics of Welfare*,[7] which had its first edition in 1924 and a final and more substantial edition in 1932 and which, it is worth remembering, contained the essential elements of a modern benefit–cost analysis. My own book[8] on agricultural marketing, which is now very much dated in its empirical content, illustrates quite explicitly an attempt to apply the Pigovian approach in agricultural marketing.

The intellectual tradition within which Keynes, Pigou and so many British economists have been trained and work, namely that of partial equilibrium analysis, might seem to be a weakness from the point of view of systems analysis. Partial equilibrium analysis deliberately sets out to limit the number of independent variables which need to be considered in any particular situation by treating as constants any others which are judged to be of minor importance. However, the approach has the strength of training one to identify appropriate sub-systems according to the strength of the relationships between the variables involved or rejected. In any case, the general equilibrium economic framework of analysis in which everything is determinate (although sometimes very little can be determined!) is well established and, of course, reflected in agricultural economics by the use of linear programming and other tools.

The quantitative developments of the last 25 years, whether econometrics or optimisation models and whether or not used for simulation purposes, complete the affinity between economics and systems analysis in the context of agricultural policy and marketing, but much more of this anon.

From the general economic framework so briefly outlined above come several major contributions to the development of systems analysis even though, as already stated, the term hardly exists in the economist's vocabulary.

Identifying, Reconciling and Specifying Objectives

Economics claims to be the science, or at least the art, of making choices to employ scarce resources between competing ends. The economic analysis of multiple objectives in decision-making for the individual firm, such as maximising versus satisficing, is well known.[9,10] But the vast economic literature on more general questions of economic choices and the usually associated problems of the distribution of real income between individuals may be less well known to many systems analysts. Economics has failed to find unequivocal answers to the critical issues of balancing or choosing between alternative economic objectives, but the failure does not lie within the framework of analysis or its concepts. How can one scientifically evaluate one man's meat against another man's meat, let alone his poison? But there is an extremely thorough, and often extremely complex, literature to point up the issues. (For a general introduction see Phelps,[11] especially his own introduction.) The literature on agricultural policy is, of course, replete with studies which can serve as illustrations; for example, that of Josling.[12,13]

In passing, one may note that agricultural economists, together with others in the agricultural teaching and research establishments, have been accused quite often of being insensitive to the inegalitarian effects of much new agricultural technology and of many aspects of policy (or rather they have in the USA). There is an element of truth in this accusation, not least because it is usually easier to pursue the application of one's knowledge among the better-educated sectors of farming and, in any case, many economists, such as the writer, have been and are 'growth orientated' which may, or may not, reflect a proper sense of social values. However, the criticism can easily be exaggerated as pointed out by Heady.[14]

Wright[1] is probably going too far when he states:

'That sometimes venturesome area of economics known as cost–benefit analysis is, when applied to real and interesting problems, essentially the same as systems analysis. We will use systems analysis as synonymous with cost–benefit analysis, cost–effectiveness analysis, or cost–utility analysis.'

But cost–benefit appraisal certainly forms an important component of systems analysis and its formal approach to the evaluation of gains and losses associated with a particular decision, albeit often necessarily imperfect, represents a valuable tool. (In the field of agricultural policy see, for example, Power & Harris[15] and in marketing the papers by LeFevre & Pickering[16] and by Kirk[17] on the economics of moving Covent Garden Market.)

Linkages

The notion of discrepancies between the costs of an input or output to the individual user or group of users and the total costs to society, or of similar discrepancies with respect to benefit, has been, of course, an integral part of economics since the *Economics of Welfare* and is one of the prime elements in cost–benefit analysis. These concepts of externalities are an important contribution in helping to identify systems and sub-systems and the links between them—although the economist has to make the leap to see that in a complete systems framework the extra analysis may be between economic and non-economic variables.

Specification of Systems and Sub-systems

Whether or not the economist works within a framework of partial equilibrium or one of general equilibrium his analytical techniques lead naturally to the identification of systems and sub-systems, even though the terms are not used, and suggest appropriate hierarchical structures. For example, for some purposes agriculture might be regarded as a single industry since, except for certain industrial uses (mainly of fibres) where there is direct competition from manufactured goods, farm products have a low cross-elasticity of demand with those of other industries. (Another complication, but still minor, would be fishing.) But the agricultural industry can be broken into a hierarchy of sub-groups according to the cross-elasticities of demand for the individual foods, with the values of the particular elasticities indicating significant distinctions between separate sub-groups and the strength of the residual links between them. (Examples are legion but, to select only two, see Colman & Miah[18] and MacLaren[19].)

Econometric or optimisation models to describe the working of systems or sub-systems are equally numerous as will appear later. Either category can be used, and often is, for purposes of simulation. Models frequently fail to achieve the success expected of them, but this difficulty is inherent in the statistical problems of measurement and could not be improved by adding a systems element not already implicitly included.

Policy Instruments

This is a terminology introduced by Dutch economic planners in the 1950s. Economists spend much time in discussing the relative merits of policy instruments in agricultural policy. An example is the discussion by Josling[12] of the relative merits of deficiency payments, variable levies, minimum import prices or combinations of these with, or without, so-called compensation policies as alternatives in British agricultural policy.

The Learning Process Through Feedback Mechanisms

Feedback is an integral part of economic analysis. One of the simpler examples is the Cobweb Theorem. (It flourishes on the feedback of misinformation!)

While one links systems analysis with planning (whether it be for a hospital or a road network, or policies for redistributing income in some specified way within agriculture or between agriculture and the rest of the economy, or arrangements for reducing fluctuations in supplies and prices of a particular farm product, or whatever), there are certain integral elements in a plan which do not necessarily involve systems thinking (either explicitly or implicitly), as already noted. Here the most important consideration is that a plan, however reached, must have some sense of realism (by definition, the realism will not be fully achieved without systems analysis), particularly by specifying the important problems and opportunities which would need to be considered before setting tasks and goals. Economics *per se* contributes no greater sense of the need for realism than any other scientific discipline, and perhaps less than, say, history.

The success or failure of the economic discipline in this context will usually depend upon the extent of the practitioner's experience in some goal-orientated practical situation, which is perhaps why agricultural economics has been quite often praised, at least in the United States, for its sense of relevance as compared with many other branches of applied economics.

Wherever one turns, therefore, in the literature of agricultural policy or agricultural marketing one should be able to find a rich quarry of material which can be conveniently and naturally slotted into a systems analysis, whether the techniques are rigorously quantitative or appropriately qualitative where quantification is impracticable and even undesirable. Examples include macro-models ranging from the FAO provisional world plan for agricultural development[20] to models of government farm programmes[21,22,23] and supply response.[24] More specific models examine such problems as price support for individual products,[26] the location of cattle feeding,[27] agricultural stabilisation policies[28] and product marketing studies.[29,30]

No doubt other reviewers would make a different selection of material intended to represent those economic studies in the fields of agricultural policy and marketing which come closest to a formal systems analysis. Some of the foregoing justify their choice more by the spirit of their approach than from the formal presentation and structuring of the argument.

1. The formal language and schematic presentation of systems analysis appear hardly at all in the foregoing examples. Ray & Heady[22] use a diagrammatic summary of their econometric formulations which would bring joy to the heart of any systems analyst, but one must turn to the Judge Symposium[31] to find an explicit connection between economics and systems thinking:

'There appears to be a definite need to combine the approaches of the econometrician and the systems analyst in formulating models of complex economic systems. To the systems analyst, an economic model consists of a set of mathematical inequalities which reflect the various conditional statements, logical branchings, and complex feedback mechanisms that depicts the economy as a dynamic, self-regulating system. Although economists have made considerable progress in building econometric models and developing techniques to estimate their parameters, little or no attention has been given to alternative model structures such as those used by systems analysis.'

This provoked the protest by Holland[32] at the same Symposium:

'Several...economists in this session are entitled to feel somewhat slighted by statements that economists have given little or no attention to alternative model structures such as those used by systems analysts'.

It will be clear, by now, that one intention of this paper has been to support Holland's contention and to go further, namely to suggest that economics *is* systems analysis in regard to the subjects under review.

2. Some of the quantitative techniques used in the sample of studies are normative, but others are positive. However, it is doubtful whether this is a meaningful distinction in a systems context. Linear programming can be predictive, as in the Newcastle programme, and econometric analysis can become normative through the use of simulation (e.g. Ray & Heady[22]).

3. The degree of formal quantification varies considerably between the various studies listed above, such as Evans,[26] Agarwala[28] and Ray & Heady.[22] Others remain thoroughly numerate although not entirely quantitative in the strict sense. This is particularly the case of the supply response systems based on linear programming by Sharples & Schaller[33] in the Carter Symposium. The Newcastle workers (Davey & Weightman[25]) impose production constraints exogenously, quite pragmatically and correctly in my opinion, according to changes in the overall farm structure through the review period to 1980. (To the linear programmer this last comment may seem a statement of the obvious. So it would be were not the objective to point up the fact that a well-developed systems analysis of an economic situation must explicitly point up all relevant issues for review.)

4. Others of the chosen studies are a mixture of quantitative, numerate and qualitative analyses. This combination is best illustrated by the Indicative World Plan[20] where only the projected demand for agricultural products has a quantitative base (see Appendix I).

5. Many of the selected studies contain no assessment of objectives, although they are often implicit in the terms of reference and orientation of the analysis, and the absence of a formal feedback system for re-appraising objectives or plans is common. For example, one eminent analyst, J. M. Chacel[34] (Director of Research of the Brazilian Institute of Economics), has criticised the Indicative World Plan[20] for its failure to employ simulation techniques and operational games. One can see the theoretical case for regarding a formal feedback mechanism as an integral part of systems analysis. But such a result would often be difficult to attain in view of the complexity of the problem to be analysed. Put another way, the lack of

formal feedback is due to the complexity of the problem to be reviewed and not to the inadequacy of economic analysis to appreciate the need.

6. The selected studies vary greatly in their scope when set against the systems outline shown in Fig. 1. Conceptually the Indicative World Plan[20] is the most ambitious, although there are some important gaps and possibly errors of emphasis in its analysis which will be mentioned below. Next one might rank the book by Blakeslee et al.[21]

The supply response models based on linear programming, aimed at measuring total national supply response, would come next, namely the Newcastle study for the United Kingdom[25] and the work by Sharples & Schaller[33] for the United States.

They would be closely followed by Halter et al.[23] and by Ray & Heady.[22] The articles by Evans[26] and by Agarwala[28] illustrate the effective analysis of economic sub-systems which can be safely isolated from the much larger agricultural system of which they are elements.

By reading the critical reviews that have accompanied or followed the selected studies one can soon round up a good selection of the criticisms which economists can make as a result of the sheer complexity of developing complex quantitative models successfully. A very enlightening review in this respect is the discussion which follows the paper by Davey & Weightman;[25] for example, linear programming will not sufficiently take account of the multiple objectives of farmers in their decision-making, or farm structure ought to be determined endogenously as a consequence of the adjustment process to changing prices and costs. However, this sort of criticism is not my concern. It reflects the state of the arts and is not attributable to a lack of formal systems analysis.

My concerns relate to the dangers arising from excessive zeal for quantification and model building as pointed out by Halter et al.:[23]

'A broader issue involves interpreting and implementing the results from simulation. Even with the best model that can be formulated from existing data, it must be pointed out that knowledge about an economic system is not at the same level as that known about most physical systems. The fact that results come from a computer certainly does not make them divine revelations or make them any more valid than the basic information inputs and perceived interrelationships determining the results.'

First of all, I would like to know more about the management problems of bringing an economic systems analysis based on large models to a successful conclusion. Thomson,[35] who is opening the discussion on my paper, addressed the Agricultural Economics Society on this subject a few months ago and might be prepared to cover the issue now. There has, after all, been an enormous investment in macro-economic agricultural policy models in the United States in the last 15 years and there seems very little to show for it in terms of a more successful analysis and understanding than has been obtained by less ambitious quantitative efforts.

But my greatest concern is that priorities in terms of determining and improving the most critical gaps in knowledge and of bringing forward speedy results give way before the tantalising challenges of technical quantitative problems. The best becomes the enemy of the good.

One might feel a small regret here that we have not seen more of the interim Newcastle studies of the kind produced by Davey & Weightman[25] and wonder where the balance lies between more information of this sort and the attainment of a complete and internally consistent model for the whole of English or British agriculture. However, the foregoing is a comparatively minor worry. My real doubts are best illustrated by the two studies at the top of the list of selected publications given earlier, namely those relating to an appraisal of the overall world food situation.

In both of these—the Indicative World Plan[20] and the Iowa publication[21]—econometric demand analysis seems to be worked to death and out of all proportion in relation to the possible payoff compared to the returns obtainable by a more thorough understanding of factors on the side of supply. The outstanding gap in both these studies is a sufficient analysis of prospective changes in technology and in climatic conditions. The latter is not considered at all and the former seems to rest on a rather superficial analysis of historical trends; certainly superficial in relation to the analysis which has gone into the study of demand.*

The Indicative World Plan[20] was concerned with 1975 and 1985, but with no consideration of how we should get from the mid-1960s (when the study was being made) to 1975. We now know that that is proving to be a difficult journey. In other words, the preparers of the Plan failed to define objectives accurately.

The Iowa analysis[21] has a detailed transportation model to show expected trade patterns for grains and fertilisers in 1985 and in the year 2000. One wonders whether such projections have much planning significance and whether the time could not have been spent in getting more

* There are very good reasons for believing that past trends in agricultural technology could be extremely misleading as regards the prospects for the next 10 or 15 years, but that is another story.

information, for example on the fertiliser response relationships in the less developed countries which is so critical in forming a medium-term view of food supplies at different levels of prices.

FAO is strong in discussing the qualitative issues in developing its Indicative World Plan. Iowa is not.

I wonder, therefore, whether there are not two clearly established dangers in the present emphasis on attaining quantitative perfection in what amounts to economic systems analysis, namely (a) tunnel vision, and (b) a disinclination to consider those critical considerations which cannot be readily quantified and fitted into the model.

I conclude, therefore, with three questions relating to systems analysis in the context of agricultural policy and marketing:

(i) Will systems analysis be anything other than the use of well-established economic techniques as long as there is an appropriate balance between the quantitative, the numerate, and the qualitative issues involved in appraising the means to achieve any particular objective?
(ii) How far are formal feedback mechanisms an essential element of systems analysis when applied to agricultural policy or marketing? How critical is the need for simulation?
(iii) How great is the danger that the intellectual challenge of quantitative methods will lead to 'tunnel vision' in the analysis of economic systems and to an imbalance between the collection and refinement of data on the one hand and its analysis of models on the other?

REFERENCES

1. Wright, C. (1973). *Economics and Systems Analysis: Introduction for Public Managers*, Addison-Wesley.
2. Young, S. (1969). Designing the management system, in: *Decision Theory and Information Systems* (ed. W. T. Greenwood), South Western Publishing Company.
3. Government White Paper (1972). *A Framework for Government Research and Development*, Cmnd 5046, HMSO.
4. Dent, J. B. (1974). *Application of Systems Concepts and Simulation in Agriculture*, Department of Agriculture, Aberdeen University, Miscellaneous Publication.
5. Von Bertalanffy, L. (1960). The theory of open systems in physics and biology, in: *Systems Thinking* (ed. F. E. Emery), Penguin Modern Management Readings.

6. Keynes, J. M. (1936). *General Theory of Employment, Interest and Money*, Macmillan (2nd ed., 1951).
7. Pigou, A. C. (1932). *Economics of Welfare*, 2nd ed., Macmillan.
8. Allen, G. R. (1959). *Agricultural Marketing Policies*, Basil Blackwell, Oxford.
9. Baumol, W. J. (1972). *Economic Theory and Operations Analysis*, 3rd ed., Prentice-Hall International.
10. Cyert, R. M. & March, J. G. (1963). *A Behavioural Theory of the Firm*, Prentice-Hall International.
11. Phelps, E. S. (ed.) (1973). *Economic Justice*, Penguin Education.
12. Josling, T. (1969). A formal approach to agricultural policy, *Journal of Agricultural Economics*, 20, (2).
13. Josling, T. (1974). Agricultural policies in developed countries: a review, *Journal of Agricultural Economics*, 25, (3).
14. Heady, E. O. (1972). Allocations of colleges and economists, *American Journal of Agricultural Economics*, 54, (5).
15. Power, A. P. & Harris, S. A. (1973). A cost–benefit evaluation of alternative control policies for foot-and-mouth disease in Great Britain, *Journal of Agricultural Economics*, 24 (3).
16. Lefevre, A. J. & Pickering, J. F. (1972). The economics of moving Covent Garden market, *Journal of Agricultural Economics*, 23 (1), 35.
17. Kirk, J. H. (1972). The economics of moving Covent Garden: a reply, *Journal of Agricultural Economics*, 23(2), 161.
18. Colman, D. R. & Miah, M. (1973). On source estimates of price flexibilities for meat and their interpretation, *Journal of Agricultural Economics*, 24(2).
19. MacLaren, D. (1974). Forecasts of wholesale prices for five categories of meat, in: *The Outlook for Beef in the United Kingdom, 1974 and 1975* (ed. G. R. Allen), School of Agriculture, Aberdeen University.
20. FAO (1970). *Provisional Indicative World Plan for Agricultural Development: A Synthesis and Analysis of Factors Relevant to World Agricultural Development*, Rome, UN/FAO.
21. Blakeslee, L. L., Heady, E. O. & Framingham, C. F. (1973). *World Food Supply, Demand and Trade*, Iowa State University Press.
22. Ray, D. E. & Heady, E. O. (1972). Government farm programmes and commodity interaction: a simulation analysis, *American Journal of Agricultural Economics*, 53(4), 1.
23. Judge, G. C. (1970). (Chairman). Symposium on macrosimulation models, *American Journal of Agricultural Economics*, 52(2), with special reference to: Halter, A. N., Hayenga, M. L. & Manetsch, T. J. Simulating a developing agricultural economy: methodology and planning capability.
24. Carter, H. O. (1968). (Chairman) Symposium on Macromodels of US agriculture, *American Journal of Agricultural Economics*, 50(5).
25. Davey, E. H. & Weightman, P. W. H. (1971). A micro-economic approach to the analysis of supply response in British agriculture, *Journal of Agricultural Economics*, 22(3).
26. Evans, M. (1974). Guaranteed price adjustment and market stability in the United Kingdom: the case of beef and milk, *American Journal of Agricultural Economics*, 56(1).

27. Langmeier, L. N. & Finlay, R. M. (1971). Effects of split demand and slaughter capacity assumptions on optimal locations of cattle feeding, *American Journal of Agricultural Economics*, **53**(2).
28. Agarwala, R. (1971). A simulation approach to the analysis of stabilisation policies in agricultural markets: a case study, *Journal of Agricultural Economics*, **22**(1).
29. Baron, P. J., Brayshaw, G. H. & Hinks, C. E. (1968). *The Commercial Prospects for Frozen August Lamb*, Department of Agricultural Marketing, University of Newcastle-upon-Tyne, Report No. 12.
30. Cumberland, J. H. (1973). *The Market for Yellow-skinned Chickens*, Department of Agricultural Marketing, University of Newcastle-upon-Tyne, Report No. 17.
31. Naylor, T. H. (1970). Policy simulation experiments with macroeconometric models: the state of the art, Winter Meeting of the American Agricultural Economics Society with the Econometrics Society, *American Journal of Agricultural Economics*, **52**(2).
32. Holland, E. P. (1970). Discussion at Winter Meeting of the American Agricultural Economics Society with the Econometrics Society, *American Journal of Agricultural Economics*, **52**(2).
33. Sharples, J. A. & Schaller, W. N. (1968). Macromodels of US agriculture: predicting short-run aggregate adjustment to policy alternatives, *American Journal of Agricultural Economics*, **49**.
34. Chacel, J. M. (1971). On the indicative world plan: a proposal for using simulation techniques and operational games for economic forecasting, *Ceres*, **4**(1).
35. Thomsom, K. J. (1974). Are large-scale models worth it?, discussion paper at Annual Conference of the Agricultural Economics Society, Cambridge, unpublished.

APPENDIX I

A Comment on the Indicative World Model by Chacel[34]

'With 17 endogenous* variables and 11 behaviour parameters or exogenous† variables, the model is developed through 17 equations and identities. It is based upon an initial rate of growth of the Harrod–Domar‡ type in an attempt to determine the gross and net value of the agricultural output of each country. Translating the analytical relationships between the aggregates into words, the mechanics of the model may be described as follows:

The rate of growth of the national economies corresponding to the rate

---

* Endogenous variable—whose value is determined inside the model.

† Exogenous variable—whose value is determined outside the model.

‡ Economists Harrod and Domar created a growth model where growth is essentially explained by the accumulation of capital.

of growth of the Gross Domestic Product is given by the interaction of the rate of investments and the incremental capital–output ratio. The Gross Domestic Product (at factor cost) for the current period may be determined once we know its value at two different periods of time, that is, its rate of growth between time 0 and time 1, and its level at time 0.

Since the rate of investment is a conjectural variable, the level of investment for the current period may be obtained. It therefore follows that the question is to define the level of savings which were induced domestically and the savings inflow from abroad assuming the equality *a posteriori* between savings and investment. In the next step, the net income remitted abroad is determined as a function of the external indebtedness, taking into account the growing lag between domestic savings and investment in the model. Exports are calculated from the Gross Domestic Product since their income elasticity is an exogenous variable, that is to say, is a given value not to be explained by the model. Imports are obtained as residuals since the inflow of savings, the exports and the net income remitted abroad are already known. In other words, this is the equation which corresponds to the surplus or deficit in current account of the balance of payments. A coefficient corresponding to the net balance between indirect taxes and subsidies makes it possible to estimate the GDP at market prices.

Once this aggregate is estimated and exports and investments are deducted from, and imports added to it, aggregate consumption is obtained as a residual. By breaking down consumption into private and government aggregates, the domestic consumption of agricultural goods may then be estimated. This consumption is calculated in physical units by means of demand functions (for type of goods) taking private consumption as an explanatory variable. By adding the domestic consumption of agricultural goods to their exports, exogenously determined, we obtain the national aggregate demand for agricultural commodities. The next step is to estimate the value of the gross agricultural output at the equilibrium point between supply and said demand. Finally, based on some given assumption, the value of inputs is calculated; the value added of the agricultural sector is then determined by the difference between the value of the gross agricultural output and the value of inputs.'

Discussion Report

A. N. DUCKHAM, C.B.E.
Department of Agriculture and Horticulture, University of Reading, England

K. J. THOMSON
Department of Agricultural Economics, University of Newcastle upon Tyne, England

A comparison was made as to the role of the objectives in these two papers. Eadie and Maxwell merely state that they wish to maximise economic advantage and immediately proceed to more technical matters. Such an uncertain livelihood as hill farming would suggest that rather more attention be paid to the kind of objectives such farmers have, and that the response of any system to particularly adverse conditions be analysed as well as its behaviour in normal conditions. In contrast, Professor Allen spends a considerable amount of time in considering objectives for policy models, where one might have thought that in view of the complexity of society and the non-dictatorial role of the economist it would be best to confine ourselves to modelling the economic system under study without unnecessary attention as to whether its behaviour conforms to our own ideas of optimality. It is noticeable that Professor Allen uses the word planning rather than management, which suggests that he recognises that the builder of policy models is more distant from the makers of policy than the adviser from his farmers, and/or that policy decisions are not in fact very important, at least in the short run, in affecting the progress of large economic systems.

Several points were raised about the value of the systems approach even though there was debate about what it actually is. It does allow a greater diversity of variables to be brought into models which helps to ensure some degree of consistency between them and also makes possible the simultaneous analysis of a relatively widely dispersed system. A danger to avoid as mentioned by Professor Allen is that of 'tunnel vision'. This was felt to be a timely warning not only to those involved in the day-to-day details of such work but also to senior consultants who must make it their task to ensure that large models remain useful to their intended clientele by sufficient concentration on input/output matters, data handling and clear organisation of both personnel and model components. The particular policies to be tested on the model and the relevant sensitivity analyses must be considered as carefully as the choice of econometric technique. The systems approach is also useful for explanation, for diagnosis and for suggesting treatments which should improve the biological and/or economic efficiency of agricultural production. But to date it does not appear to have been used for assessing the biological efficiency of food consumption, although Joy[1] and others from Orr[2] onwards have systematically examined the economic and some of the social aspects of this key counterpart of production efficiency.

Duckham & Masefield[3] and Ruthenberg[4] have shown the value of the systems approach in *explaining* the location, nature and intensity of agricultural systems.

Eadie and Maxwell's paper showed how their earlier *diagnostic* systems analysis of Scottish hill sheep farming had great value in locating, identifying and quantifying the biological and sometimes the economic interactions and weaknesses in a given system. Allen agreed that such analysis could be usefully applied to the physical side of marketing. Current work at Reading, in collaboration with other centres, on 'Human Food Chains and Nutrient Cycles', a much larger and more complex system than hill sheep farming or milk marketing, confirms the great diagnostic value of systems analysis for pinpointing and then measuring the inefficiencies of weak links. Such weak link data is therefore useful both in formulating research on isolatable scientific or technical specifics (e.g. lamb mortality) involving only a few disciplines, and in drawing up scientific and socio-economic research or investigation programmes for wider multi-disciplinary fields.

Moving on from diagnosis to *remedial* treatments, Eadie and Maxwell synthesised new and potentially better forms of hill sheep farming and are testing them *as systems*. They compared five or six years' results from the new systems they had synthesised with the results obtained from old traditional systems in the five or six years preceding the change. Their data showed the apparent success of their new systems. But in classical field plot experiments the season is not regarded as a variable because all the treatments are applied simultaneously in the same season or series of seasons. Time is not regarded as a variable in the experimental design. Eadie and Maxwell had compared their critical measurements *before* with the same measurements *after* the introduction of the experimental treatments, i.e. the new systems. Time and the season had become uncontrollable variables. This inevitable introduction of between-season variations did not, of course, nullify their experimental results. But it did emphasise the need, which the authors fully recognised, to keep careful weather, disease and other diaries and records and perhaps to introduce regular metabolic monitoring on the lines evolved at the Animal Diseases Research Institute, Compton, UK. This need to keep what might, in the nutritionist's sense, be termed 'epidemiological' records would be emphasised when the major climatic parameters, e.g. the lengths of the growing season or the annual rainfall, had large coefficients of variation.

If diagnostic systems analysis leads to further experiments with newly designed or redesigned farming systems, i.e. with radical rather than palliative treatments, then it may be necessary to rethink traditional 'one controllable variable at a time' experimental methods and statistical analyses. From the *farmer's* point of view the one variable in the experiment may be a complete new farming system. Are we satisfied that we have the methodology to quantify and verify the value of whole new systems?

If such an experimental methodology is or could be available, then the Hill Farming Research Organisation's approach to analysing and redesigning a farming system, albeit in their case on the surface a simple one, ought to be extended to other whole systems. For instance, in the UK the more or less continuous growing of cereals has been questioned, not only on ecological environmental grounds, but also as to its bio-economic long-term viability. Systems analysis might well be used to answer these questions. Again, and perhaps more important, with increasing population growth, shifting cultivation is coming under pressure in Africa and elsewhere. This and other tropical crop-producing systems should be analysed diagnostically *as systems* and, hopefully, new remedial treatments would emerge. It was good to hear that some such work had started in Kenya.

Macro-economic models tend to assume for the most part normative behaviour or at least confine themselves to those aspects of collective human behaviour that has been measured historically. The danger of using these models for predictive purposes is that human behaviour is often irrational in economic terms.

There was considerable discussion about the virtues and specific defects or inaccuracies of the various world or national macro-models for agricultural or world resource policy that had been published. Academically, it is obviously right to refine such models; if accuracy was not at an academic premium, how would argon have been discovered? But if it is accepted that macro-economic treatments are no more than rough-and-ready prognoses and may have relatively unpredicted results, then in the real world such models are perhaps more politically and administratively, than statistically, useful. They can provide propaganda targets, guidelines or upper and lower negotiating limits for international agencies, national governments and corporation policy-makers. A generation ago, agricultural policy was often framed (over a convivial drink with one's opposite number) on the back of an envelope. Has the 'systems approach' improved on this very pragmatic method? Yes it has. For current international and national macro-models have substantially reduced the content of 'normative' guesswork in the final, politically accepted policy. They *should* also have increased the predictability of the actions of governments and of human response by farmers, merchants or consumers to the final policy. But whether they have in fact must remain to be demonstrated.

REFERENCES

1. Joy, L. (1973). Food and nutrition planning, *Journal of Agricultural Economics*, **24**(1), 165–93.
2. Orr, J. B. (1936). *Food, Health and Income*, Macmillan, London.
3. Duckham, A. N. & Masefield, G. B. (1970). *Farming Systems of the World*, Chatto & Windus, London.
4. Ruthenberg, H. (1971). *Farming Systems in the Tropics*, Oxford University Press.

Index

Advisory services, 310–12, 314
Agricultural system
  analysis perspectives, 61–74
  classification, 6–8
  complexity, 4
  content of, 17
  definition, 24
  description, 8–10
  hierarchical arrangement, 25
  identification, 5
  levels of, 24–5
  models in—*see* Models
  nature of, 129–31
  organisational implications, 17–18
  study of, 3–19
  total range of, 25
  see also System(s)
Autonomous modules, 113

Bio-economic models, 13, 17, 235–65
Bio-economic systems, 262
Biological components, 23, 25
Biological constraints, 131
Biological research
  systems approach, 369–91
  uses 375–87
Biological tolerances, 271, 283

Capacity in farm operations, 268
Cash flow, 307
Cereals
  crop yields, 272, 285
  harvesting models
    components, 269–77
    costs, 288–93

Cereals
  harvesting models—*contd.*
    interpretation of output, 288–93
    results of experiments, 289–93
    specifications, 277–81
  seeding models
    components, 282–7
    costs, 293–9
    interpretation of outputs, 293–9
    results of experiments, 293–9
    specifications, 287–8
Chemosterilants, 225
Circular diagrams, 11, 14
Cluster analysis techniques, 114–15
Cobweb Theorem, 422
Cold storage facilities, 343
Commodity production cycles, 237
Computers, 13, 23, 58, 60, 71, 72, 114, 118, 120, 149, 168, 184, 185, 235, 244–7, 309, 377
Constant-coefficient approach, 77
Consumer's surplus, 320
Control function, 194, 198
Control theory techniques, 87
Cost-benefit analysis, 313, 419, 421
Cost-benefit ratio for long-term investments, 359–60
Cost optimisation, 164–5
Crop yield, cereal, 272, 285
Cybernetics, 237

Dairy forecasting model, 250–62
  alternative systems designs, 257
  background and objectives, 250–3
  data specification, 258
  flow diagram, 254
  iterative facilities, 259

Dairy forecasting model—*contd.*
 processing problems, 260
 simulation section, 256
 structure and function, 253–62
 sub-routines, 254–6
 validation, 261–2
Damage function, 194
Data banks, 175, 312
 compilation constraints, 180–3
 concept of, 179–80
Data collection, 303
 constraints, 180–3, 303
 philosophies of, 178–80
 population limitations, 181
 structure imposed on, 180–3
Data sources
 models, 175–86
 types of modelling activities related to, 176–9
Data specification, 258
Data storage, 183
 languages, 185
DECIDE fertiliser model, 165
 implementation, 166–8
Decision-making, 13, 134–5, 188–9, 231, 268, 307, 415, 416, 425
Decision rules, 155–9
Demand curve, 320, 321
Demand functions, 326, 329, 360
Density dependence, 207–15
Development analysis, 327
Development programs, regional planning, 327–32
Differential equations, 386
Driving variables, 27
Dynamic programming, 88
Dynamic system theory, 26

Economic analysis, 307–16
Economic components, 25
Economic system in four dimensions, 317
Economic theory and optimisation, 135–7
Economics, 17, 420, 422, 423
 models, 78
Engineering models, 78

Events, 34
Experimentation, 4, 5
Exports, Portugal, 353
Extensive services, 311, 312
Extrapolation, 385–7

Farm operations
 mechanisation, 268
 simulation, 269–88
 systems approach, 267–304
 tropics, 7
Feedback, 422–7
Fertiliser model, DECIDE, 165
 implementation, 166–8
Fertilisers, 338
Firm growth
 computer model, 244–7
 conceptual model, 241–4
 simulation model, 240–50
  research and extension use, 247–50
 theoretical aspects, 240–1
Fishery models, 54–5
Flow diagrams, 28, 85, 254, 379
Flow equations, 77
Flow rate, 77
Forecasting, 313
Forestry investment, 330
Forestry products, 335
Future-events chain, 34

Gauss-Seidel method, 35
Grade losses, 273, 280
Grain
 drying, 271, 280, 292
 moisture content, 274, 280
 quality effects, 273
Grazing
 management, 400, 402
 models, 42–3, 121
 system, 400, 402
Gross Domestic Product, 430
Group extension, 311

Harvest timeliness penalty, 283

Index

Harvesting of cereals, models
  components, 269–77
  costs, 288–93
  interpretation of output, 288–93
  results of experiments, 289–93
  specifications, 277–81
Hill sheep farming
  animal production, 399, 401–2
  basic resources, 396
  biological analysis, 399
  economic status, 398
  grazing management, 400, 402
  grazing system, 400, 402
  non-biological constraints, 403
  pasture production, 399, 401
  stocking rates, 400, 401
  synthesis, 401
  systems development, 398–404
  systems research, 395–413
    implementation, 404–12
    problem, 396–8
Hypotheses, 12, 377
  clarification and storage, 383
  hierarchies, 384–5
  testing, 179, 383

Imports, Portugal, 352–3
Indicative World Plan, 424–9
Information collection, 312
Information storage, 183
Information system, 110–12, 115, 120, 123
  pig herd, 118, 122
Insecticide spraying, 215–23
  cost-effectiveness, 232
  resistance development, 225
  simulation, 203
Insecticides, 194
Insects, sterilisation of, 225
Interpolation, 385
Investments, 330–44
  long-term, benefit-cost ratio for, 359–60
Irrigation models, 42–3
Irrigation schemes, 330, 335
Iterative facilities, 259

Leslie Matrix model, 200
Limiting factor approach, 77
Linear programming, 13, 61, 62, 72, 84, 145, 146, 149, 190, 240, 311, 319, 323, 324, 365, 366, 424, 425
Lotka–Volterra type models, 66
Lupin crop, optimisation, 148–52

Machine performance, 276, 285
Machine systems, 270, 282
Macro-economic models, 365
Macro-economics, 419, 426, 433–4
Macro-models, 423
Magnetic drums, 184
Magnetic tape, 184
Maize, physiology and growth, 12
Management aids, 309
Management games, 80
Management modelling, 69–71
Management strategy, optimal, 152
Mantis, hunger in, 12
Market equilibrium, 320
Markov chain models, 34
Maximum-reduction approach, 77
Measurement error, 78
Meat
  processing facilities, 335
  storage, 342
Mechanisation of farm operations, 268
Meteorological Office, 184
Michaelis–Menten equation, 382
Micro-economics, 419
Milk
  cow herd improvement, 334
  production systems, 8, 14
Mitscherlich representation, 166
Model builders, 66–7, 133
Model codes, 83
Models, 23–106, 109–17
  bio-economic, 13, 17, 235–65
  calculator, 364–393
  characterisation, 37–61, 374
  classification, 372
  comparison with engineering and economic approaches, 78

Models—*contd.*
  compartment, 28
  complexity, 65
  computer, 244–7
  conceptual, 176, 239, 241–4, 252, 374
  construction, 114
  costs, 71–2
  coupled, 116, 117, 121
  dairy forecasting—*see* Dairy forecasting model
  dangers arising from, 425
  data
    relationship, 67
    sources, 175–86
  data-based, post analysis, 176
  DECIDE, 165–8
  description, 82
  design, 177–9
  design and reporting, 74–86
  development, 68, 79, 114–16
  developmental utility, 374
  diagrammatic, 74, 238–9
  discussion, 361
  domain, 62
  econometric, 34, 423
  economics, 78
  ecosystem, 65, 67, 382
  effectiveness, 366
  empirical dynamic prediction, 87
  engineering, 78
  events in, 34
  evolution, 68, 71
  experiments, 378, 382
  exploration, 179
  farm planning, 188
  firm growth, 241–7
  fishery, 54–5
  formulation procedure, 375
  function, 64
  future of, 67, 186
  general impressions, 58–61
  general problems, 187, 239
  generalised, 62–3, 188
  generality, 61
  grazing, 42–3, 121
  harvesting of cereals
    components, 269–77

Models
  harvesting of cereals—*contd.*
    costs, 288–93
    interpretation of output, 288–93
    results of experiments, 289–93
    specifications, 277–81
  hierarchies, 82
  implementation, 377
  initial formulation of mechanisms, 377
  irrigation, 42–3
  journal specifications, 84
  Leslie Matrix, 200
  limitations, 86, 355–6
  Lotka–Volterra type, 66
  macro-economic, 365
  managerial problems, 69–71
  Markov, 34
  mathematical, 10, 23, 80, 83, 195, 243
  meta, 384
  Monte Carlo type, 267
  multiperiod, 366
  multiple-flow, 60
  normative, 116
  notation, 32–4, 85
  objective, 364, 375–7
  opportunities in agricultural systems, 86–9
  optimisation, 32, 58, 59, 60, 72, 84, 86, 132, 140
    see also Optimisation
  output precision, 60
  overall, 146
  pest control, 40–1
  plant growth, 48–9
  precision, 61
  realism, 61
  regional planning, 319–27, 355–6
  relationship of objectives, structure and output, 80–2
  relative importance, 12
  resolution, 61, 62
  review, 35–61
  seeding of cereals
    components, 282–7
    costs, 293–9
    interpretation of outputs, 293–9

Models
  seeding of cereals—*contd.*
    results of experiments, 293–9
    specifications, 287–8
  simplicity, 364
  simulation, 27, 58–61, 87, 112, 120, 145, 146, 152, 155, 165, 253, 267, 369, 383–4
  simulation/optimisation, 58, 60, 72, 86
  size, 187
  skeleton, 123–4, 189, 253
    application of new technology, 123
    assessment of new technology, 122
    construction of, 116–17
    enterprise system, 118–20
    pig herd, 118–20, 122
    research and development, 122–3
    use of, 117–23
    validation of, 120–2
  standard outputs and diagrams, 85
  standard symbolism, 83–4
  statistical, 32
  stochastic, 145
  strategic, 63, 372
  structure, 65
  subdivision/classification schemes, 384
  systems approach, 238–40
  tactical, 63, 372
  techniques, 87
  time requirements, 67
  time trends, 73–4
  top-down versus bottom-up, 68–70
  utility of, 10–17, 64, 65, 88, 361
  validation, 120–2, 179, 299–303
  wildlife, 57
  see also Sub-models
Modules, autonomous, 113
Monte Carlo type models, 267
Moth borers, 225, 232

Non-linear program, 149
Non-linear simplex method, 164

Objective function, 321, 323

Observation, problems of, 4
Open systems, 380–1
Operational-conceptual approaches, 27
Operations research, 395
  systems approach, and 187
Optimal management strategy, 152
Optimisation, 32, 58, 188–9
  costs, 164–5
  dynamic agricultural systems, 146–64
  dynamic pasture system, 152–62
  economic theory, and 135–7
  is it necessary? 135
  linear and non-linear, 150
  lupin crop, 148–52
  management, 129–31
  models—*see* Models
  pasture system, 152–62
  reason for, 132–5
  techniques, 137–40
  use, 140–5
Optimum solution, 129–73
  definition, 129
  implementation, 165–8
Organisation, levels of, 65
Orthogonal polynomials, 385

Paraffin lamps, 225
Parameters, 28
Pasture system optimisation, 152–62
Pest control, 40–1, 193–5, 232
Physical components, 25
Pig herd
  information system, 118, 122
  skeleton model, 118–20, 122
Planning
  agriculture, of, 310
  regional—*see* Regional planning
  sub-system, 112
Plant growth models, 48–9
Policy in agriculture, 309–10
Polynomials, 386
Population growth, 327
Portugal
  conventional and new techniques, 338

Portugal—*contd.*
  foreign trade, 352–5
  improved and new production activities, 331–2
Precipitation level, 274, 279
Precompilers, 83
Prediction, 386–7
Prediction error, 78
Principal component analysis, 114–15
Producer's surplus, 320
Production systems, 8
Programming techniques in resource management, 144

Rainfall, 274, 279
Rate of work, 268, 276, 277, 285
Rate process, 28
Regional consumption patterns, 326
Regional demands, 326
Regional planning, 317–60
  changes in production consumption and price levels, 349–52
  development programs, 327–32
  models, 319–27
    limitations, 355–6
  programming results, 332–44
Relational diagram, 74
Relative humidity, 274
Research, 17–18
  modelling, 78–80
  programming techniques, 144
Ripening time, 271

Seasonal conditions, 159–62
Seeding of cereals, models
  components, 282–7
  costs, 293–9
  interpretation of outputs, 293–9
  results of experiments, 293–9
  specifications, 287–8
Self-organising system, 384
Sensitivity analysis, 206
Sex pheromone, 225
Sheep production, 9–10

Shelling losses, 272, 280
Short-cut techniques, 166
Simplex method, 149, 164
Simulation techniques, 13, 27, 58, 269–88
Social constraints, 131
Soil types, 283
Spatial problems, 61
Statistical models, 32
Sterilisation of insects, 225
Stocking rates, 152–5, 159–62, 400, 401
Sub-groups, hierarchy of, 422
Sub-hypotheses, 179
Sub-models, 12, 146, 164–5
  autonomous, 113
  canonical, 72
  coupling between, 69
  inputs, 75
  outputs, 75
  specifications, 74–6
Sub-optimal programming, 166
Sub-systems, 14–17, 28, 111–12, 318, 415, 420
  planning, 112
  specifications, 421
Sugar cane froghopper control, 193–229, 231
  age distribution of population, 200
  alternative forms, 225
  density dependence, 207–15
  fecundity as function of female age, 201
  four broods, 221
  hatching as function of rainfall, 201
  male population in simulation, 203
  management scheme, 223
  outline of problem, 197
  population dynamics, 198–207
  single brood, 217–21
  spraying, by, 215–23
  sterilisation, 225
  two broods, 221
Supply curves, 320, 321
Syrphid predators, 232

System-dependent approach, 77
System state variable, 27, 28
Systematic bias, 78
Systems analysis, 186, 195, 231, 362, 370, 415–30
  agricultural policy and marketing, in, 427
  application of, 393
  contribution of, 416
  definition, 415–20
  diagnostic, 432, 433
  economic framework, 420
  economics, and, 420–1
  identifying, reconciling and specifying objectives, 420
  planning function, 419
  remedial, 432
Systems approach, 26, 193–229, 231, 236–8, 308, 365
  biological research, 369–91
    uses, 375–87
  farm operations, 267–304
  operational research, and, 187
  problems of, 188
  scientific method, and, 370
Systems behaviour, 27
Systems characteristics, 415
Systems concepts, 26–35, 109–10
Systems description, 27
Systems designs, testing alternative, 257
Systems diagram, 323
Systems methodology, 187, 237
Systems research, 238
  hill sheep farming—*see* Hill sheep farming
  levels of, 108
Systems specification, 42,

Systems theory, 26–7, 107–27
  approaches, 27
  research, 107–9
Systems view, 380
  see also Agricultural System; Sub-systems

Thought clarification, 379–80
Thought organisation, 378
Time available, 269
Time-varying approach, 77
Transfer function, 28–9
  diagrams, 28–9
Transition probability matrix, 87
Tropics, farming systems, 7
Tunnel vision, 431

UNESCO, 79
Utility function, 323, 357

Volterra predator-prey systems, 63

Weather
  conditions, 274
  constraints, 273, 285
  data, 270
  effects, 283
Wildlife models, 57

Yield penalty for untimely seeding, 283